Molecular Genetics of
Plant Development

The purpose of this book is to present classical plant development in modern, molecular–genetic terms. The study of plant development is rapidly changing as plant genome projects uncover a multitude of new genes. This book provides a framework for integrating gene discovery and genome analysis into the context of plant development. Concepts in plant development are compared with those in animal development, and complex processes, such as flowering and photomorphogenesis, are presented as pathways of gene action regulated by positional and environmental cues. Emphasis is placed on organ formation, such as the development of roots, shoots, and leaves, and life-cycle events, for example, embryogenesis, seedling development, and the transition to flowering. Examples are drawn primarily from model plants with well-studied genetic systems, particularly Arabidopsis and maize.

Molecular Genetics of Plant Development is designed to be used as a textbook for upper-division or graduate courses in plant development. The book will also serve as a reference book for scientists in the field of plant molecular biology or plant molecular genetics. The book is also useful for general development courses in which both animal and plant development are presented.

Stephen Howell is currently Boyce Schulze Downey Scientist at the Boyce Thompson Institute for Plant Research and adjunct professor of Plant Biology at Cornell University.

Molecular Genetics of
Plant Development

STEPHEN H. HOWELL

CAMBRIDGE
UNIVERSITY PRESS

PUBLISHED BY THE PRESS SYNDICATE OF THE UNIVERSITY OF CAMBRIDGE
The Pitt Building, Trumpington Street, Cambridge CB2 1RP, United Kingdom

CAMBRIDGE UNIVERSITY PRESS
The Edinburgh Building, Cambridge CB2 2RU, UK   http://www.cup.cam.ac.uk
40 West 20th Street, New York, NY 10011-4211, USA   http://www.cup.org
10 Stamford Road, Oakleigh, Melbourne 3166, Australia

First published 1998

Printed in the United States of America

Typeset in Meridien, in QuarkXPress™ [GH]

*A catalog record for this book is available from
the British Library*

Library of Congress Cataloging-in-Publication Data

Howell, Stephen H. (Stephen Herbert), 1941–
    Molecular genetics of plant development / Stephen H. Howell.
        p.   cm.
    Includes bibliographical references.
    ISBN 0-521-58255-5 (hb). — ISBN 0-521-58784-0 (pb)
    1. Plants—Development.   2. Plant molecular genetics.   I. Title.
QK731.HG8 1998
572.8'2—dc21   97-35238
                                                  CIP

ISBN 0 521 58255 5 hardback
ISBN 0 521 58784 0 paperback

To my family

Contents

Preface

This book was written as a text for upper-division undergraduates and first-year graduate students and as a resource for others who are interested in the field. It is aimed at students who have already taken molecular biology and genetics and who want to apply the concepts in these areas to plant development. The book is not an elementary presentation of molecular biology for plant scientists but an advanced treatment of a science that had made phenomenal progress during the past few years.

The book provides an up-to-date account of a fascinating and rapidly evolving field. It was written not because older approaches to plant development have become obsolete but because many new observations and concepts have enriched the science. Mastery of plant development in today's world requires bridging classical plant anatomy and development to modern molecular genetics.

This text represents a systems approach to plant development, organized on the basis of life-cycle events (flowering, pollination, embryogenesis, and seedling development) and organ systems (roots, leaves, and vascular systems). Chapter 1 reviews the principles of pattern formation, the major paradigm for early development in animal systems. Also in Chapter 1, plant and animal development are compared, and the features unique to plant development are emphasized, such as the ability of plants to regenerate from vegetative cells. In addition, the virtues of studying development in model plants with well-studied genetic systems are presented.

Chapter 2 focuses on the role of cell lineages and positional information in plant development. Steeves and Sussex (1989) dwelled on cell lineage analysis in their celebrated text *Patterns in Plant Development*. The reasons for doing so then are no less compelling than they are now, even though it is now recognized that the role of cell lineage in determining cell fate in plants had been highly overvalued. Instead, cell lineage information allows us to trace the history of cells in development so that we can understand the forces that determine cell fates. The importance of position information in determining cell fate is stressed in Chapter 2, and examples of positional information are discussed.

The remaining chapters of the book deal with various life-cycle stages and organ systems. Special emphasis is given to the stages of plant development that involve organ formation, such as embryogenesis in Chapter 3 and shoot development in Chapter 5. Chapter 8 on flower development may be most fascinating to readers because of our attraction to flowers. Remarkably, our understanding of flower development is the most advanced of any organ system, and models of flower development are absolutely elegant. However, don't overlook the later chapters on root and vascular system development (Chapters 12 and 13). These systems are the new frontiers, and many of the exciting questions in plant development have been raised by studies in these systems.

Several years ago I was involved in planning a new graduate course in plant molecular biology at Cornell, and I became aware of the need for a text in the area of plant development. The course in planning was made up of separate modules offered in Chinese-menu style to upper-division undergraduates and incoming graduate students. At the time, I had the rare opportunity to choose any one of several modules to teach. We had planned for two modules on plant development. A colleague of mine, June Nasrallah, an accomplished developmental biologist, volunteered to teach one module, and I agreed to do the other. For me it was a brash move because at the time I was not involved in research in this area. However, the topic fascinated me, and I wanted to devote my intellectual energy to the subject.

In my first two years of teaching, I was generally satisfied by picking current papers and discussing them in class. However, in the next few years, it became increasingly difficult to teach from current papers because the concepts in the field were becoming more weighty, and individual papers required a broader understanding. In hindsight, I was witnessing the development of a field in which ideas were building one on another.

My frustrations about not having a suitable text in plant development were shared by instructors in many other colleges and universities. As an aid to teaching in this area, Tom Jacobs at the University of Illinois collected, informally published, and distributed course synopses and lecture notes from plant development courses taught at various institutions. His effort was noble; however, I was finally moved to action by an e-mail letter in the Arabidopsis news group from a new instructor bemoaning the fact that she could not find an up-to-date text that offered a genetic approach to plant development.

I was determined to write a single-author text even though my colleagues warned me that it would be an impossible job. The field is too broad and too fast paced. Nonetheless, I wanted to develop common themes that could be carried through the book and to provide a coherent treatment of a body of work that had become a legitimate discipline for students to study. I doubt if the first edition of this book will satisfy all the interests in the audience I hope to address. However, I welcome comments from both students and instructors to improve and modify subsequent editions, which I hope to write.

When I was an undergraduate at Grinnell College many years ago, I studied embryology under a professor named Guillermo Mendoza. Mendoza was much loved, always teasing students with ideas and never averse to mixing a bit of philosophy with facts during his lecture. One day while musing about the limits of science, he told us that although biochemistry and physiology could be explained mechanistically, development could not. Development was too complex — beyond human comprehension. Development could be described in morphological and phenomenological terms, but never really understood. I don't remember whether I recorded his statements in my notebook, but his words stuck with me as a challenge for the future.

In the days since Mendoza, the world has changed. Marvelous technological advances have extended knowledge beyond conceivable limits. Space probes have landed on Mars, computer networks link the world, and we have cracked some of the codes of life. Tools have been found for probing into the mysteries of development — and asking some of the questions that even a generation ago Mendoza thought were unaskable.

My thanks to the many scientists who shared information with me and gave me permission to use illustrations, and to my students who read and commented on the text as it evolved during the past two years. Thanks also to my wife, Liz, for all the help in proofing and review and for all her patience and encouragement.

Ithaca, New York   S. H. H.
December 1997

A Word on Genetic Nomenclature

Genetic notations used in this book conform to the standards for each organism and are not uniform throughout. For historical reasons, different genetic notations have been developed for different organisms. Although there are differences in notation, some general features about nomenclature are in common. For example, the names of genes, loci, and mutants are generally italicized, while the names of proteins encoded by genes are not.

In the genetic nomenclature for Arabidopsis, the name of a gene (or locus) is italicized and capitalized, such as *SCARECROW* (*SCA*) or *ABSCISIC ACID INSENSITIVE3* (*ABI3*). The name of a mutation in that gene or at that locus is italicized but not capitalized, such as *scarecrow* (*sca*) or *abscisic acid insensitive3* (*abi3*). In Arabidopsis, genes and their corresponding loci are usually given three-letter names. If more than one locus bears the same gene name, the loci are numbered sequentially, such as *ABI1, ABI2,* and *ABI3*. Different mutations can occur in a single gene or at a single locus, and the different mutant alleles are numbered sequentially after a hyphen, such as *abi1*-1, *abi1*-2, and so forth. If there is only one allele at a locus, then that allele is not given a number and is referred to only as *abi3*. Dominance relationships in Arabidopsis are not indicated in the name because these relationships can be complex and are not easily represented by upper and lower case letters. The protein product encoded by a gene is capitalized, but not italicized, such as ABI3.

In Arabidopsis, wild type is designated wt or +. Plants with a mutation in homozygous condition can be represented as *abi1*-1/*abi1*-1, or simply as *abi1*-1. Plants with mutations in heterozygous condition can be represented, for example, as *abi1*/+ or *abi1*-1/*abi1*-2. Double mutants, in which both mutations are in a homozygous state, are written with a space between the mutant names, such as the double *abi1*-1 *abi2*-1 mutant.

In maize, the name of a gene or locus is italicized, but is not capitalized, such as *shrunken1* (*sh1*) or *alcohol dehydrogenase1* (*adh1*); however, the first letter is capitalized if the name is used as a noun, such as a mutation in *Shrunken* or alleles of *Sh.* The names of mutants or mutations at that locus are italicized, but they are not capitalized if they are recessive mutations, such as *shrunken2* or *abi3.* Dominant alleles are indicated by capitalizing the first letter of the name, as in the *Knotted-1* mutant.

Other aspects of maize nomenclature can be found in "A Standard for Maize Nomenclature" from the *Maize Newsletter* (available at http://www.agron.missouri.edu/maize_nomenclature.html). Mutants in pea (*Pisum*), petunia, snapdragon (*Antirrhinum*), tobacco (*Nicotiana*), and so on, are also discussed, and they will be explained in context if the notation differs significantly from Arabidopsis or maize.

The terms *gene* and *locus* are often used interchangeably; however, the words are not synonymous. A gene is a DNA sequence that includes the region transcribed (with both exons and introns) plus the *cis* control elements that may lie 5' and 3' to the transcribed region. Alleles are different forms of the same gene in that one allele has a slightly different DNA sequence (or bears a different mutation) than another allele. A locus is the site in the genome where the gene is located. Likewise, the terms *mutation* and *mutant* are not synonymous. A mutation is an alteration in a gene, such as a base-substitution, deletion, insertion, and so on. When *mutant* is used as a noun, it generally refers to the organism in which a mutation has occurred, such as "the mutant was an albino."

A *transgenic plant* is a plant into which a transgene (usually a gene from another organism) has been introduced. A *transgene* is often a chimeric gene in which various parts of the gene have been swapped with another gene. Many examples in this book involve chimeric gene constructs in which the promoter from one gene has been exchanged with another. One of the more common promoters is the cauliflower mosaic virus (CaMV) 35S RNA promoter (35S). A construct in which CaMV 35S RNA promoter has been linked to the *ALCOHOL DEHYDRO-GENASE1* gene (*ADH1*) is designated 35S:*ADH1*.

1

Approaches to the Study of Plant Development

Molecular genetics has revolutionized our understanding of plant development. For many years, plant developmental biology was the domain of plant anatomists, who provided us with a wealth of valuable information and careful descriptions of plant development. Experimental plant biologists and physiologists later contributed extensively to our understanding of plant processes. Today, modern plant genetics has renewed our interest in the systems studied by taxonomists, anatomists, and physiologists. Many modern studies have revisited classical problems with the new tools of molecular genetics. Molecular genetics has equipped us with the means to explore developmental processes that are too complex to approach with older scientific methods. We now recognize that processes as complicated and exotic as flower development can be described in operational terms as an ordered set of events directed by a few master regulatory genes.

At present, almost every developmental process in the life cycle of plants is being scrutinized with molecular genetic tools. In this book we will examine various events in the life cycles of plants, such as embryogenesis, vegetative development, gametophyte formation, and pollination (Fig. 1.1). In doing so, we will focus on the development of various organs or organ systems, such as seedlings, roots, and leaves. We will emphasize how these developmental processes, which were only described in phenomenological terms in the past, have yielded to molecular genetic analysis.

It is important to understand why genetics played the leading role in untangling the complexities of development. Genetics is a powerful tool that can be used to guide the efforts of an investigator in the search for genes that control development. A geneticist operates by using a toolbox of mutants that affect a developmental process. A mutation can be recognized by its phenotype, and for our purposes, the phenotypes of interest

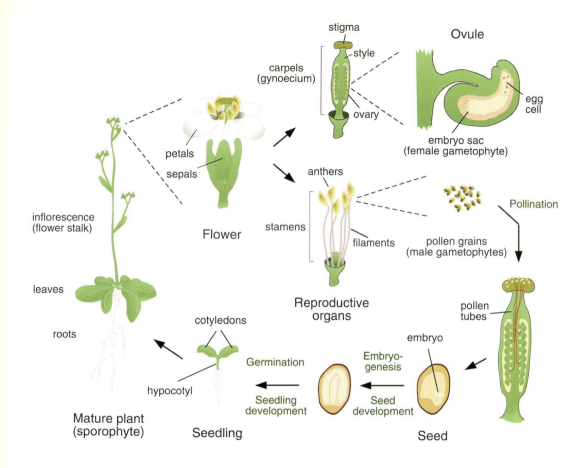

FIGURE 1.1

Stages in the life cycle of *Arabidopsis thaliana*. (Left) A flowering Arabidopsis plant
has a whorl of rosette leaves at its base and an inflorescence (flower stalk).
Cauline leaves are borne on the flower stalk, as are branches bearing flowers.
Flowers are composed of floral organs: sepals, petals, stamens, and carpels. Strip-
ping away the sepals and petals exposes the female reproductive organs (upper
figure) – the carpels or gynoecium – and male reproductive organs (lower figure)
– the stamens. Two carpels are fused to form the central gynoecium or pistil,
which is composed of the stigma, style, and ovaries. A section through the
ovaries shows locules containing rows of ovules, each bearing an embryo sac
(female gametophyte). An enlarged section through an ovule shows the embryo
sac with egg cell. The male reproductive organ is composed of an anther, which
bears pollen grains (the male gametophytes) and a supporting filament. During
pollination, pollen grains shed from anthers land on a stigma and germinate to
form pollen tubes (shown in red) that grow through the stigma papillae and
down through the style and septum (wall of the ovary). The pollen tube emerges
in the vicinity of an ovule and penetrates the ovule to fertilize the egg (and central cell) in the embryo sac. The fertilized egg develops into an embryo, and a
seed is formed from the embryo, extraembryonic, and maternal tissues. Seed germinates and gives rise to a seedling composed of cotyledons, hypocotyl, and root.
The seedling develops into a mature plant.

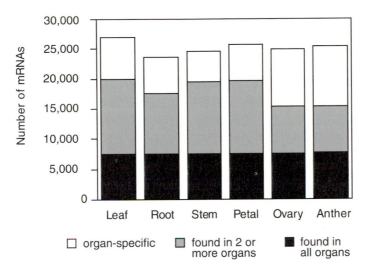

FIGURE 1.2

Number of messenger RNA species in various tobacco organs. Bars are subdivided vertically to indicate the number of mRNAs specific to the particular organ, the number found in two or more organs, and the number found in all organs. The number of RNA species is estimated from RNA driven RNA–DNA hybridization analysis using RNAs derived from the various organs. (Redrawn from Goldberg, R. B. 1988. Plants: Novel developmental processes. *Science* 240:1463. Copyright 1988 American Association for the Advancement of Science.)

are abnormalities in development. Mutations can also be used to connect developmental processes to the responsible genes. Mutations mark genes with insertions, deletions, base changes, and the like, and mutant genes can be tracked down by a variety of molecular techniques.

Genetic approaches are powerful because they require few assumptions. One simply finds mutants that affect an interesting process and then attempts to determine what genes are responsible for the mutations. Trying to understand development by exploring biochemical processes or cloning genes expressed at a given time or place in development has been less successful for several reasons. First, there is no simple way to discriminate between those genes that are important in controlling development from those that are incidentally expressed at a particular time and place. Second, there are too many genes that are specifically expressed in a particular tissue or developmental stage. Sorting through them without some criterion for selection can be an impossible task. For example, it has been estimated that 20,000 genes are expressed in tobacco embryos, and about 4,000–5,000 are uniquely expressed at the embryonic stage (Goldberg 1988) (Fig. 1.2). Third, important regulatory genes in development are thought to represent a small fraction of the genes that are uniquely expressed. Most developmental processes that have been analyzed genetically are controlled by a hierarchy of genes. A few master regulators at the top of the hierarchy control the expression of other genes in that organ or at that developmental phase. The master regulators tend to be transcription factors or cell-signaling molecules that exercise control over many other genes. Because these genes are expressed at low levels, they are usually underrepresented in cDNA libraries, even when efforts are undertaken to enrich libraries for organ-specific or developmental phase-specific genes.

Without genetics, finding genes that control various developmental processes is like searching for a needle in a haystack. The challenges will mount in the next few years as the pace of gene discovery accelerates enormously. Even now the discovery of new genes has outstripped our ability to understand what these genes do, let alone to determine their role in development. Large numbers of plant cDNAs and entire plant genomes are being sequenced (e.g., see Newman et al. 1994). The challenge of the future will be to put together the genetic puzzle of development using the vast amount of new information from various plant genome mapping and sequencing projects.

The approach that has been most successful in finding genes that control development is to uncover all the mutations that affect a particular developmental process, organ, or structure. Those who have discovered the most interesting developmental mutants have done so by grouping similar mutants in categories and examining their relationship to each other. Others have focused on mutants that are consistent with conceptual models of the process in question. Such models allow one to predict expected mutant phenotypes or to devise clever screens for other less obvious mutants. A model that has dominated thinking in modern developmental biology involves the concept of pattern formation.

PATTERN FORMATION IN DEVELOPMENT

The concept of pattern formation, which arose largely through the analysis of mutants in Drosophila, has provided a theoretical framework to classify many developmental mutants in animals (Lawrence 1992). In pattern-forming processes the basic features of the body plan, such as the development of polarity of the embryo and location of axes, are laid down early in embryogenesis. Pattern formation is driven by a hierarchical network of genes in which the earliest genes that determine the most fundamental aspects of the body plan are maternally expressed (Fig. 1.3A). The simple pattern features established through the action of the

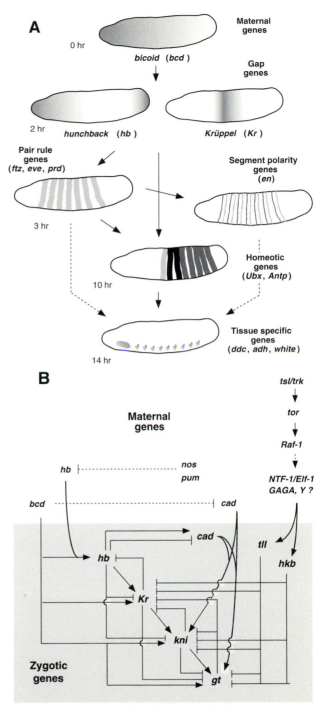

FIGURE 1.3

Pattern formation in Drosophila. (A) Drawing depicts the spatial and temporal regulation of expression of different classes of genes controlling pattern formation along the anterior–posterior axis of the developing embryo. Shading indicates the spatial location of gene expression. The control of gene expression is hierarchical, and the arrows represent control of one class of genes over another. (Redrawn from Biggin, M. D., and Tjian, R. 1989. Transcription factors and the control of Drosophila development. Trends Genet. 5:378.) (B) Circuitry controlling spatial expression of gap genes in Drosophila embryo development. Pattern of expression of maternal genes in follicle or egg establishes anterior-to-posterior pattern of expression of zygotic gap genes. Gradient of *bicoid* (*bcd*) expression differentially activates various gap genes, such as *hb, Kr, kni,* and *gt.* Boundaries of gap gene expression are sharpened by repression from neighboring gap genes. (Adapted from Rivera-Pomar, R., and Jäckle, H. 1996. From gradients to stripes in *Drosophila* embryogenesis: Filling in the gaps. Trends Genet. 12:478–483.)

maternal genes are refined by the expression of zygotic genes (i.e., genes expressed in the embryo). A common feature of animal body plans from worms to mammals is their modularity, and an early step in embryogenesis is the division of the body into modules, segments, or parasegments.

The subsequent expression of a number of genes determines the positional relationship of the segments to each other and features of the individual segments, such as their polarity. Through the action of homeotic selector genes, particular programs of development are then chosen such that each module is stamped with a special identity, such as leg segment, wing segment, and so on. The homeotic selector genes turn on constellations of other genes that define the identity of body segments by leading to the production of organs characteristic of different segments.

Pattern formation along the anterior–posterior axis of the Drosophila embryo has been studied most extensively and is an excellent example of the operation of gene networks in development. Three different maternal genetic pathways are involved: the anterior, posterior, and terminal organizer systems (St. Johnston and Nüsslein-Volhard 1992). The action of maternal genes establishes gradients of zygotically expressed transcription factors in the syncytial cytoplasm of the Drosophila embryo. *Bicoid* is a key component of the anterior organizer system. *Bicoid* RNA is synthesized in the maternal germline and accumulates at the anterior pole of the egg (Fig. 1.3A and B). BICOID protein is translated after egg deposition and activates the expression of gap genes, such as *hunchback*. Although BICOID accumulates in a concentration gradient in the embryo and activates the expression of *hunchback*, BICOID alone does not determine the spatial expression of *hunchback* (see Rivera-Pomar and Jäckle 1996). *Hunchback* is normally expressed in the anterior half and posterior quarter of the embryo (Fig. 1.3A). The localization of *hunchback* expression depends both on BICOID and on its own expression (Simpson-Brose et al. 1994). Both BICOID and HUNCHBACK activate another gap gene called *Krüppel*, which is expressed in the central region of the embryo, just behind *hunchback* (Fig. 1.3A). *Krüppel* is thought to repress the expression of *hunchback* (and *knirps*, another gap gene), establishing a boundary of *hunchback* expression (Fig. 1.3B). The spatial control of other gap genes is similar, involving the activation by maternal genes and mutual repression between companion gap genes. The spatial control of the ten pair-rule genes in Drosophila is thought to involve similar interactions at a lower tier in the zygotic gene hierarchy – control in which the pair-rule genes are activated by the gap genes (and maternal genes) and mutually repressed by neighboring pair-rule genes. The consequence of this gene circuitry and mutual repression is to turn gradients into segments in the developing embryo (Rivera-Pomar and Jäckle 1996).

Homeotic selector genes are near the lower tier of the gene hierarchy and determine segment identity. Homeotic selector genes were discovered through the unusual phenotypes of homeotic mutants. Loss-of-function homeotic mutants have the bizarre characteristic of producing normal organs in the wrong place, or, stated another way, these mutants have defects in selecting the correct organ development program in one or more segments (Lawrence 1992). For example, a consequence of *antennapedia* mutants is that adult flies have legs growing in place of antenna. (Not all segment or organ identity genes act in the same way. For example, loss-of-function *eyeless* mutants fail to form an eye and do not form another organ in its place.) The homeotic genes act as switches turning on programs of organ formation in various body segments. Homeotic genes serve in the role as major regulators because they encode transcription factors that activate the expression of constellations of effector genes further downstream in the control network. Homeotic genes encode a special class of homeodomain (homeobox) transcription factors (Gehring et al. 1994). These transcription factors have characteristic helix–loop–helix elements that allow them to bind to promoter regions of target genes.

Homeodomain-containing genes are some of the most durable in evolution and are highly conserved in all eucaryotic organisms (Scott 1994). In higher animals, homeodomain-containing homeotic genes are organized into gene clusters in an order that corresponds to the position of the body segment they define (Fig. 1.4). In mice, the homeodomain-containing homeotic genes (called *hox* genes) at the 5′-end of the cluster determine the expression of genes in the anterior end of the animal, while those at the 3′-end regulate the expression of genes in the posterior segments. The homeodomain-containing homeotic genes are found in smaller clusters in lower animals with less elaborate body plans. In higher animals, and particularly in vertebrates, the clusters are more complex and occur in multiples (Kenyon 1994).

The concept of pattern formation has served as a useful paradigm in plants. When flower development was first conceived as a pattern formation process, attention was focused on mutants that disrupt the normal pattern of floral organs (see Chap. 8). In particular, mutants in which one floral organ was formed in the place of another were interpreted to be homeotic mutants. The floral homeotic genes had the same selector function as their animal counterparts in specifying the identity of floral organs in various flower segments. Although the floral homeotic genes also encode transcription factors, the plant factors were different from the homeodomain-containing transcription factors that identify segments in animals. The flower homeotic genes encode MADs box transcription factors which, will be further discussed in Chapter 8.

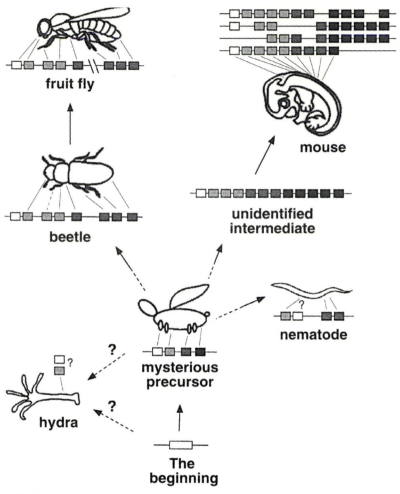

FIGURE 1.4

Homeodomain-containing genes are found in clusters in animals. Gene clusters (represented by shaded boxes) have grown larger and become more numerous during the evolution of higher animals, particularly in mammals, such as the mouse, which has multiple clusters of homeodomain genes. The arrangement of homeodomain genes within a cluster is highly conserved, and the arrangement reflects the anterior–posterior arrangement of the body segments that they specify. The ancestral form (mysterious precursor), which contains a simple cluster of homeodomain genes and gives rise to more complex animal forms, has not been identified. (From Kenyon, C. 1994. If birds can fly, why can't we? Homeotic genes and evolution. Cell 78:176. Reprinted by permission of the author and Cell Press.)

Although plants and animals are remarkably similar in their basic metabolism and cellular structure, there is no reason to think that the pattern formation processes in plants and animals must be the same. This is because plants and animals diverged about 1 billion years ago before they became multicellular organisms (Doolittle et al. 1996). The ancestors common to plants and animals had, undoubtedly, the same basic homeodomain-containing transcription factor genes that animals have used to assemble gene networks for embryonic development. In plants, one group of homeodomain-containing transcription factors, the *knotted1*-like gene family that was first identified in maize, are associated with pattern formation processes in postembryonic plant development (Jackson et al. 1994). The homeodomain-containing transcription factor genes, however, are not organized in clusters in plants and have not been shown to have homeotic selector functions. Instead, a subfamily of these transcription factors, the *knox1* genes, are expressed in meristematic tissue (Kersetter et al. 1994) and appear to play very important roles in the spatial regulation of cell proliferation. This class of transcription factors will be discussed further in Chapters 2, 5, and 6.

Despite the differences between plants and animals, plant developmental biologists have generally embraced the basic features of pattern formation, largely because the concept transcends the differences between the organisms. As stated earlier, the concept of pattern formation has been most successfully applied to flower development (see Chap. 8). There is a discussion in Chapter 3, however, about pattern formation in embryo development that illustrates the problems in transferring concepts from animals to plants without an appreciation for the differences between plant and animal systems.

IMPORTANT DIFFERENCES BETWEEN PLANT AND ANIMAL DEVELOPMENT

The similarities between plant and animal development have allowed developmental biologists to formulate certain unified theories of development, but there are also many differences between plant and animal development. Some of the differences are quite profound and are not usually discussed in the literature on animal development.

Germ Line Development

In animals, germ and somatic cell lines diverge very early in embryonic development. In mammals, the germ cell line in the female is set aside and germ cells undergo very few divisions. Because of that, germ cells in the female are spared from mutations that accumulate during divisions. The germ line is also set aside early in males; however, germ-line cells undergo many more divisions in the male, and, like somatic cells, they are more subject to mutations. In plants, it has been argued that there is no germ line at all (Walbot 1996). At least, the germ line is not set aside during embryogenesis, and the decision for germ cell differentiation is put off usually until very late in development. As will be described in Chapter 9, germ cells in plants usually derive from one of the three fundamental cell layers. The cell layer from which the germ line is derived (L2), however, is not exclusively reserved for germ cell production; rather, it gives rise to much of the vegetative part of the plant as well as germ cells.

Because the germ line is not set aside in plants, germ cells can arise from vegetative (somatic) cells that have undergone extensive divisions. In annual plants, such as maize (*Zea mays*), it has been estimated that there are about fifty divisions from zygote to zygote (Walbot 1996). (Note, however, that fifty divisions can generate an enormous number of cells. If all cells divided each division, then fifty divisions would produce 2^{50} cells.) In long-lived perennials, however, there can be many more divisions before flowers are formed. Because of this, the genetic makeup of germ cells derived from various parts of a plant might differ due to the differential accumulation of mutations.

The issue about whether there is a germ line in plants has actually been a controversy for many years. In the 1950s, Buvat (1952) proposed that the germ line in plants was composed of a group of quiescent cells in the shoot apical meristem (SAM) called the méristème d'attente, which was placed in reserve during vegetative growth. These cells in the center of the SAM were surrounded by actively dividing initial cells and carried forward by the growth of the vegetative shoot. The méristème d'attente was thought to be activated during flower development, giving rise to the reproductive organs of the flower. In morphological terms, that picture of the vegetative SAM was reasonably accurate; however, the interpretation is very different today. Indeed, there is a group of quiescent cells, called central mother cells, at the heart of the SAM (see Chap. 5). Cell lineage analysis has shown, however, that there is not a population of cells in the embryonic shoot meristem that is devoted exclusively to the formation of the flower or the floral reproductive organs. (More discussion of this matter is found in Chap. 2.)

Role of the Gametophyte

The haploid (1N) phase dominates the life cycle in some lower plants (as it does in the unicellular green alga Chlamydomonas, for example), but it plays a lesser role in the life cycles of higher plants. In higher animals, all that remains of the haploid phase are gametes. In higher plants, the haploid phase has not been so completely diminished. The gametophytes, the embryo sac and pollen grains, are haploid structures in higher plants (see Fig. 1.1). Although gametophytes are small, they are multicellular, and their development must therefore be accounted for in the life cycle of plants. One important consequence of a haploid stage is that it potentially exposes plants to the lethal or detrimental effects of recessive mutations. Polyploidy and gene duplication, however, are frequent in higher plants, so recessive mutations may not necessarily be exposed in the gametophyte. Recessive mutation exposure at the gametophyte stage may be beneficial because it may spare the species of some of its genetic load. In any case, it is important to bear in mind that plant gametophytes are multicellular entities, and mutations that affect multicellular development may not survive the gametophytic stage.

Postembryonic Development

In higher animals, most adult organs are formed during embryogenesis. For example, the human embryo appears almost as a miniature version of the adult. In higher plants, embryos do not contain any of the organs found in the adult plant. Plant embryos are composed of organs (cotyledons and axis) that make up the embryo and seedling, but not the mature plant. Plant organs are formed from shoot and root meristems during postembryonic development. Because organ formation can occur continuously in postembryonic development, the plant can adapt its body plan to meet changes in the environment. The body plans of animals are predetermined by embryonic development, while those of plants have greater plasticity in adapting to different environmental pressures (Walbot 1996).

Because most plant organs are formed in postembryonic development, plant developmental biologists are as interested in meristems as they are in embryos. The role of the SAM in shoot development will be is described in Chapter 5, and the role of the root apical meristem (RAM) in root development will be discussed in Chapter 12.

Cell Movement and the Planes of Cell Division

Plant cells have cell walls that cement them in place. As discussed earlier, plant cells and tissues do not migrate in embryonic development as they do in animals. As a result, the development of plant form is dictated by the division and expansion of otherwise immobilized cells. For that reason, plant developmental biologists often point out the patterns and planes of cell division to explain the development of plant form.

The orientation of the planes of cell divisions in a plant structure can determine how it will grow. Anticlinal divisions (i.e., divisions in which

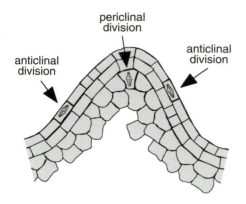

periclinal
division

anticlinal
division

anticlinal
division

FIGURE 1.5

Planes of cell division. Cell divisions in outer cell layers (epidermis and subepidermal layers) are almost exclusively anticlinal divisions in which the mitotic spindle poles are parallel to the surface (newly formed cell plate is perpendicular to the surface). In inner cell layers, divisions are more random and include periclinal divisions in which mitotic spindle poles are perpendicular to the surface (cell plate is parallel to the surface).

the cell plates are perpendicular to the surface) result in expansion of the surface, while periclinal divisions (divisions in which the cell plates are parallel to the surface) cause protrusions from the surface (Fig. 1.5). The predominant plane of division can be the defining character of a tissue. For example, epidermal tissues undergo anticlinal divisions, and in so doing, expand as a single cell layer to cover the surface of an organ or other structure. The outgrowth of a lateral root is initiated by periclinal divisions in the pericycle layer of cells in the primary root.

The division plane in plant cells is marked by the appearance of a preprophase band, which is a structure not found in animal cells. The preprophase band is a microtubule array that predicts the site where the new cell plate will be inserted into the parental walls at cytokinesis (Cleary et al. 1992). This happens even though the preprophase band is disassembled at division when the nuclear envelope breaks down. It is interesting that the plant homolog of the yeast cell division protein kinase (p34^{cdc2}) accumulates at the preprophase band, and that this protein kinase may mark the site for cell plate formation later in cell cycle (Colasanti et al. 1993).

Despite the importance of pattern and plane of cell division to the development of form in plants, the uniformity of certain plant organs or other structures does not imply that division patterns and planes are invariant. This issue will be broached again in the discussions about formation of the maize embryo in Chapter 2 and the development of the maize leaf in Chapter 6. The dilemma posed by the *tangled-1* mutant in maize is germane to this issue. In the *tangled-1* mutant, longitudinal cell divisions in the leaf are disoriented (Smith et al. 1996). As a result the cells of the leaf appear in tangled arrays instead of neat, parallel files as they do in nonmutant plants. Despite the disorientation of leaf cells, the overall shape of the leaf is fairly normal. This and other examples illustrate the fact that more global forces, rather than just the patterns and planes of cell division, help to shape organs, such as leaves.

Regeneration and Totipotency

Plants, but not most animals, have the remarkable capacity to regenerate from vegetative parts. Many terminally differentiated plant organs, tissues, and cells retain their capacity to regenerate. For example, a stem segment broken off from a prickly pear cactus (*Opuntia*) will regenerate a new plant. In tissue culture, plants can be regenerated from undifferentiated cells, generally by two different routes – by organogenesis or by somatic embryogenesis – as will be discussed in Chapter 3. In organogenesis, organs form directly without embryonic development. In somatic embryogenesis, regenerating tissue recapitulates the steps in embryonic development, although it does so in a very different environment from the zygotic embryo and in the absence of other maternal and zygotic tissue (e.g., endosperm). In higher animals, embryogenesis requires genetic input from both parents because alleles are imprinted by parental source. Maternal alleles have greater influence on the development of the embryo proper, while paternal alleles exercise greater control over the development of extraembryonic membranes. Imprinting serves to promote sexual reproduction in higher animals. In plants, imprinting also acts as a sexual reproduction checkpoint, but it has a greater impact on endosperm than embryo development (see Chap. 11) (Walbot 1996).

The ability of plants to regenerate, either by organogenesis or by somatic embryogenesis, colors one's thinking about the irreversibility of cell differentiation in plants. In a series of classic experiments, Stewart (1958) showed that *Daucus carota* (carrot) plants can be regenerated (via somatic embryogenesis) from single cells (Fig. 1.6) The inescapable conclusion from these experiments is that even single vegetative carrot cells retain their totipotency (i.e., their capacity to recapitulate development). The totipotency of differentiated plant cells means that not even differentiated plant cells are irreversibly committed to a given course of development.

The development of embryos from vegetative tissue is remarkable, but so is the fact that plants can often be propagated indefinitely in a vegetative state as nontumorous, differentiated tissue. Plants with indeterminate growth can be indefinitely propagated by cuttings. Plants with determinate growth and/or programmed senescence likewise can often be vegetatively propagated by a variety of means, by grafting, bud culture, and so on. Higher animals have finite lifetimes, as do their cells.

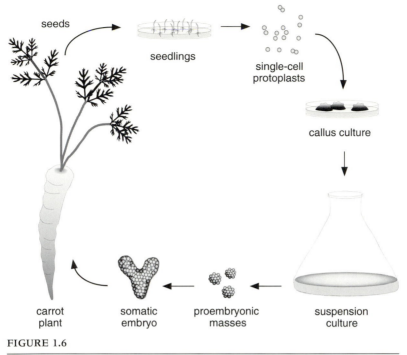

seeds

seedlings

single-cell
protoplasts

callus culture

carrot
plant

somatic
embryo

proembryonic
masses

suspension
culture

FIGURE 1.6

Regeneration of plants from single cells. Single cell protoplasts made
from hypocotyls of carrot seedlings are regenerated into calli (undifferentiated cell clumps). Calli are dispersed and grow as cell clusters in suspension culture. After a period of culture, cell clumps become proembryonic masses (PEMs), which can be induced to undergo somatic
embryogenesis by withdrawing the hormone, auxin. Somatic embryos
can then be regenerated into carrot plants. (Redrawn from Zimmerman,
J. L. 1993. Somatic embryogenesis: A model for early development in
higher plants. Plant Cell 5:1412.)

Current thinking suggests that untransformed cells in higher animals
have a replicative limit and that explanted cells are able to undergo a
certain number of divisions in culture depending on the age of the
organism at the time when the cells were explanted. There has been
considerable debate about what sets the replicative limit in animals
cells; however, it has been proposed that the limit may be set, in part,
by the loss of telomeres (Levy et al. 1992). Telomeres are specialized
structures on the ends of chromosomes, and telomeres shorten with
age. The loss of their structural integrity is thought to interfere with
ordered cell divisions and places a limit on the replicative capacity of
animal cells. The matter has not been examined extensively in plants;
however, it has been reported that telomeres in barley (*Hordeum
vulgare*) shorten during differentiation of tissues, but that they then
grow during dedifferentiation processes, such as formation of callus in
tissue culture (Kilian et al. 1995).

Table 1–1. *Plant Cell Types*

meristematic apical cells	root cap cells
parenchyma	root hairs
collenchyma	cambium
epidermis	ray cells
endodermis	cork
pericycle	phelloderm
fibers (sclerenchyma)	cork cambium
stomatal guard cells	laticifiers
stomatal subsidiary cells	secretory cells
palisade	hairs
mesophyll	egg cells
xylem vessels	synergids
tracheids	antipodal cells
phloem sieve tubes	endosperm
companion cells	aleurone
transfer cells	tapetum
gland cells	pollen generative cells
idioblasts	pollen vegetative cells
cystoliths	

Source: Lyndon, R. F. 1990. Plant Development, p. 126. London: Unwin Hyman.

Variety of Plant Organs and Cell Types

Higher animals have a greater variety of organs and cell types than do higher plants. Plants, however, produce certain organs, like leaves, in great numbers and variation. Embryos of dicots have only four organs: epicotyls or plumules (the embryonic axis above the cotyledons), cotyledons, hypocotyls (the embryonic axis below the cotyledons, but above the radicle), and the radicle (the embryonic root). Mature plants can be made up of only three vegetative organs – leaves, stems, and roots. Flowers have four organs – sepals, petals, stamens, and pistils. Other organs form during specific developmental periods, such as fruit (mature ovary).

As a consequence of having fewer, different organs, higher plants have fewer cell types than higher animals. Lyndon (1990) has listed only forty cell types in plants (Table 1.1), while hundreds of cell types are found in higher animals. In animals, organs often stamp special identity on their cells. For example, the ground cells of the liver in an animal are a distinct cell type; hepatocytes. In plants, there are fewer organs, and the organs do not always confer special cell-type characters. In this regard, it is of interest to know whether organ formation in plants is required for cell differentiation or vice versa. At least one example will be described where organogenesis and cell differentiation are uncoupled in plants. In *raspberry* mutants of Arabidopsis, organ formation during embryo development is blocked, but embryonic cell differentiation continues (Goldberg et al. 1994). Further discussion of these mutants is found in Chapter 3.

15

Because there are fewer cell types and organs in plants, the intercellular signaling network between different cells and organs is less complicated in plants than it is in animals. Animal cells bristle with receptors for the hundreds of extracellular signaling molecules and hormones. Far fewer hormones, growth factors, and signaling molecules have been described in plants. The argument can be made that there are many more, but we simply have not as yet discovered them. Five plant hormones are recognized in the classical literature – auxin, cytokinin, ethylene, gibberellic acid, and abscisic acid – and newer ones have been added to the list, such as brassinosteroids and jasmonic acid. To date, two peptide signaling molecules have been identified in plants: "systemin," which is involved in wound signaling (Pearce et al. 1991), and the product of the ENOD40 gene from soybean (*Glycine max*) (van de Sande et al. 1996). The ENOD40 gene is highly expressed in root cortical cells during early stages of nodule formation in soybean and may be involved in the induction of pericycle cells to divide. The finding suggests that there may be many other plant genes encoding peptide hormones that are involved in other plant developmental processes. Other signals, such as oligosaccharides, particularly oligouronides, appear to be unique signaling molecules in plants that are involved in wounding, pathogen attack, and perhaps normal development (Albersheim et al. 1983).

Plant cDNA sequencing projects have provided us with information about whether other components of animal cell signaling pathways are found in plants, and, if so, how many different genes are involved. It is interesting that no members of the abundant class of tyrosine kinase receptors or serpentine receptors in animal cells have been reported in plants (Meyerowitz 1997). These receptors transduce signals intracellularly through heterotrimeric G-proteins in animal cells, and although plants appear to have heterotrimeric G-proteins, they do not have the variety found in animals. Plants do have other receptors, however, for which there are no equivalents in animal cells. One is the class of surface receptors with leucine-rich repeats in the extracellular domain and serine/threonine protein kinase in the intracellular domain. Another is a class of two-component protein kinases, such as the receptor for the plant hormone ethylene to be discussed in Chapter 4.

Intercellular Communication

Given the fact that plant cells are separated by cell walls, the extent to which plant cells are interconnected seems remarkable. Plant cells are interconnected by cytoplasmic bridges called **plasmodesmata** (Robards and Lucas 1990) (Fig. 1.7). Plasmodesmata have channels with effective diameters of about 3 nm that allow for the flow of small metabolites,

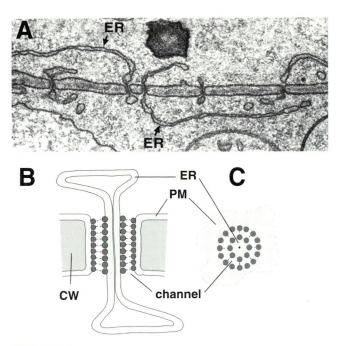

FIGURE 1.7

Structure of cytoplasmic channels (plasmodesmata) interconnecting plant cells. (A) Primary plasmodesmata pass through
the cell wall of root tip cells of the common bean (*Phaseolus
vulgaris*). Interconnection and continuity of the endoplasmic
reticulum, ER, between neighboring cells can be seen. $KMnO_4$
fixed sections observed in the transmission electron microscope. Bar = 100 nm. (B, C) Schematic model (according to
Mezitt and Lucas 1996) of a primary plasmodesma seen in
longitudinal section and transverse section along the axis of
the cytoplasmic channel. Plasma membrane, PM; cell wall,
CW. ([A] from Lucas, W. J., Ding, B., and van der Schoot, C.
1993. Plasmodesmata and the supracellular nature of plants.
New Phytol. 125:437. Reprinted by permission of Cambridge
University Press. [B] based on Mezitt, L. A., and Lucas, W. J.
1996. Plasmodesmal cell-to-cell transport of proteins and
nucleic acids. Plant Mol. Biol. 32:253.)

ions, and signaling molecules of about 1000 MW or less. In certain parts
of the plant, cells can be so well connected that they act as a syncytium
(i.e., continuous cytoplasm). Many small molecules, such as sucrose, flow
in the "symplasm" from cell to cell. Cell types vary in the number of the
plasmodesmata, and this becomes an issue in charting the paths of nutrients and metabolites into and out of the vasculature where these substances often pass through many different cell types.

The vast interconnections between plant cells do not accommodate
movement between cells by macromolecules, such as protein and
nucleic acids. The constraints on their movement would seem to create

17

barriers for the exchange of information-rich developmental signals from cell to cell. As will be discussed in the Chapter 2, however, certain proteins, such as the product of the *knotted1* gene in maize, which is a homeodomain-containing transcription factor, appear to have the capacity to dilate and to move through these channels, as well as to affect development in cells other than those in which the gene product is synthesized (Mezitt and Lucas 1996).

MODEL PLANT SYSTEMS

Throughout this book, many developmental processes will be illustrated by examples in Arabidopsis and maize. These model plants have very tractable genetic systems and have been the focus of attention for plant molecular geneticists. By specializing in the study of one or a few model plants, the plant genetics community has amassed collections of mutants and databases of sequences to allow deeper probing into questions and issues in plant development. In addition, special tools and genome maps have been produced for studying these organisms. One can access news groups, information, and databases pertaining to these organisms on the Internet.

Over the years, much genetic research has centered on maize because of the importance of corn as a crop and the impact of genetics on maize improvement (see Sheridan 1982). Hundreds of mutants have been described that affect all phases of maize development (see Fig. 1.8 for various stages in normal maize development). (Information on mutants is available through the Maize Genetics Cooperation Stock Center. The center can be contacted on the Internet at http://www.agron.missouri.edu.) In addition, high-resolution genetic maps have been constructed and marked genetic lines used in mapping are available. Maize has many advantages for genetic studies. Because maize produces separate male and female flowers, the plants are easy to cross, and hundreds of kernels can be readily scanned on a single cob. Some of the first genes recognized only by phenotype were isolated in maize (e.g., see Shure et al. 1983). The genes were identified by transposon tagging using endogenous maize transposons *Activator* (*Ac*) and *Dissociation* (*Ds*). Several other transposable element systems have been described in maize, and some are useful for tagging purposes. Mutations can be generated through chemical mutagenesis by treating maize pollen or seeds with mutagens (Fig. 1.9). Dominant mutations affecting seed characteristics (aleurone pigmentation mutants, endosperm mutants, etc.) can be assessed in M1 kernels. Recessive mutations can be unmasked by self-pollinating M1 plants, and by inspecting M2 ker-

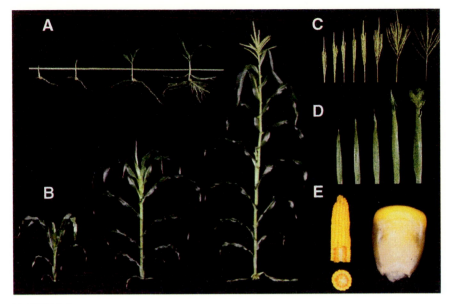

FIGURE 1.8

Development in maize (*Zea mays*). (A) Seedling development showing both shoot and root growth; (B) vegetative growth at stages V6 (6 nodes), V12 (12 nodes), and VT (tassel stage); (C) tassel development from V9 to pollen shedding stage; (D) ear shoot development; various ears from V18 plants; (E) husked cob at milk stage and enlargement of kernel at full physiological maturity. (Reproduced with permission from Iowa State University Cooperative Extension Service, Special report #48 (1993). How a corn plant develops. Available on the Internet at: http://www.ag.iastate.edu/departments/agronomy/corngrows.html)

nels and seedlings for mutations. The disadvantages of maize are its slow generation times (two to three crops per year), climatic requirements, and facilities and land needed for cultivation.

While there is a long history of maize genetics, a tremendous push has been made in the past few years to study development in Arabidopsis (Meyerowitz and Somerville 1994). Some of the disadvantages of maize are the advantages of Arabidopsis. Arabidopsis is a small plant and cultivation requires modest indoor facilities. The generation time of Arabidopsis is relatively short (~6 weeks), and individual plants produce hundreds to thousands of seeds. The Arabidopsis genome is among the smallest in higher plants, with a haploid size of about 100 megabases (mb) of DNA (Meyerowitz 1994). With a small genome size it was expected that there would be fewer problems with gene duplication. Multiple gene copies can obscure the expression of phenotype in loss-of-function mutations. (Duplicate genes within an organism or species are called paralogous genes or **paralogs**.) A surprising number of genes are present in duplicate copies in Arabidopsis, however, limiting the number of mutations which can be easily recognized.

19

Arabidopsis has other genetic advantages. The plant produces perfect flowers, is self-fertile, and tends not to open pollinate. Self-fertilization (selfing) exposes recessive mutations in homozygous form in the M2 generation following mutagenesis (Fig. 1.9). Mutagenesis is usually carried out by X-irradiating seeds or treating them with mutagens. Because the embryo in the seed is multicellular at the time of mutagenesis, the M1 plant is mosaic and heterozygous for recessive mutations. Flowers in the sectors that bear the mutation, however, will produce M2 progeny in which one quarter will be homozygous for a recessive mutation (provided that the mutation does not affect viability or recovery of gametes). Mutants can be propagated as true lines by selfing crosses. The ease in mutant selection and the analysis of thousands of progeny growing as seedlings on petri plates or in flats under lights in the lab has permitted mass screening for all kinds of developmental mutants. Mutants have been selected with alterations or defects in the development of embryos, roots, flowers, or trichomes, or that have defects in time to flowering, female fertility, phyllotaxy, or production of wax – to name just a few. (Arabidopsis mutants are available in the United States at the Arabidopsis Biological Resource Center on the Internet available at http://aims.cps.msu.edu/aims/.)

While this book focuses on model plants, such as Arabidopsis and maize, the techniques described earlier are being applied to other systems. Other plants with well-developed genetics, such as tomato (*Lycopersicon*) and rice (*Oryza*), will undoubtedly receive more attention in the future.

CLONING GENES RELEVANT TO PLANT DEVELOPMENT

A major advantage in working with model plant systems is that once interesting developmental mutants have been isolated, it is more feasible to identify the genes involved. At present, the most expedient and straightforward method is to tag mutant genes by insertional mutagenesis. A common strategy in Arabidopsis is to tag genes with T-DNA delivered to plant cells by *Agrobacterium tumefaciens.* (T-DNA is a segment of the tumor-inducing or Ti-plasmid that is transferred to plants upon inoculation with virulent strains of *Agrobacterium*.) The T-DNA is thought to be randomly inserted into the genome, although with some bias toward genes that are transcribed. Arabidopsis seeds or plants can be inoculated en masse with Agrobacterium; therefore, large collections or libraries of T-DNA insertion mutants containing, on average, a single

MAIZE

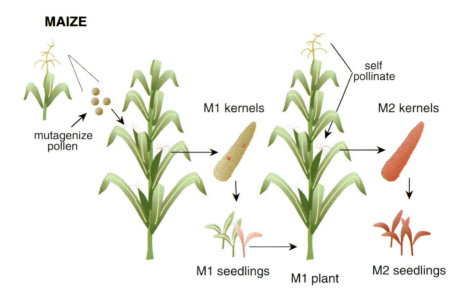

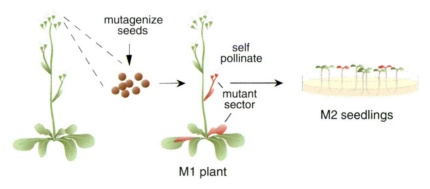

FIGURE 1.9

Generating and identifying mutants in maize (*Zea mays*) and Arabidopsis. Pollen from maize is mutagenized and used to pollinate an inbred line. Ears bear kernels that carry the mutation in heterozygous state (pinkish color kernels). Germinated seeds produce M1 seedlings in which dominant mutations can be recognized. To detect recessive mutations, M1 plants are self-pollinated (selfed) and generate ears with M2 kernels and/or M2 seedlings homozygous for the mutation (reddish color). Recessive mutations can be recognized by their phenotype and propagated as true breeding lines. Seeds from Arabidopsis are mutagenized (e.g., with the alkylating agent ethyl methane sulfonate, EMS). Because the embryo (and shoot apical meristem) is multicellular at the time of mutagenesis, the resulting M1 plants are chimeric, bearing sectors with mutations in a heterozygous state (reddish sectors). Flowers in the mutant sectors are self-fertilized and give rise to M2 seedlings of which one quarter carry recessive mutations in a homozygous state. Mutant seedlings can be recognized by their phenotype (indicated by reddish color) and propagated as true breeding lines. (Based on Neuffer, M. G. 1994. Mutagenesis. In The Maize Handbook, ed. Freeling, M., and Walbot, V., p. 214. New York: Springer Verlag.)

T-DNA insertion per genome have been developed (Feldmann 1991). Interesting mutants can be culled from those libraries, and the tagged genes identified by sequencing the plant DNA that flanks the T-DNA insertions. A number of important genes that affect plant development described in this book have been identified in this way. Other insertional mutagenesis strategies in Arabidopsis depend on heterologous transposon systems, such as the *Ac-Ds* transposable system from maize (Van Sluys et al. 1987).

Point mutations, such as those produced from chemical mutagens like ethylmethane sulfonate (EMS), are more difficult to track down because the physical change in the gene is so small. Nonetheless, genes bearing such mutations can be isolated by positional or **map-based cloning** (Tanksley et al. 1995). Positional cloning requires high-resolution maps and involves merging genetic with physical maps. An example of a high-resolution genetic map is the map of DNA markers on chromosome 4 in Arabidopsis (Fig. 1.10). The physical map of a small region of the genetic map is shown and is made up of a "contig" of yeast artificial chromosomes (YACs) (Schmidt et al. 1996). (Physical maps are generated by arranging cloned plant DNAs, such as those cloned in YACs, in order as they appear on the plant genome. Cloned DNAs are arranged by matching overlapping regions, and continuous assemblies of overlapping clones constitute a contig.) Map-based cloning is carried out by precisely determining the location of a mutant gene on the genetic map and using DNA markers to find the equivalent location on the physical map. Fewer genes have been identified by positional cloning because it is a long and labor-intensive process.

The techniques described earlier allow one to identify genes known only on the basis of phenotype. New techniques have been developed to search for mutations in specific genes without knowing the phenotype beforehand. This approach is a form of reverse genetics and is very important in understanding the function of a gene for which the sequence is known. At present, it is difficult to make directed mutations in genes within plant genomes. Therefore, the technique that has been developed uses polymerase chain reaction (PCR) methods to screen through collections of random insertion mutants to find mutations in particular genes (Bensen et al. 1995). One such collection of maize mutants developed by Pioneer Hi-Bred Seed Company is called the "Gene Machine." The collection contains a large number of F_1 plants mutagenized by a transposable element called Robertson's mutator, and the DNA from pools of these plants is screened for insertions in genes for which the sequence is known.

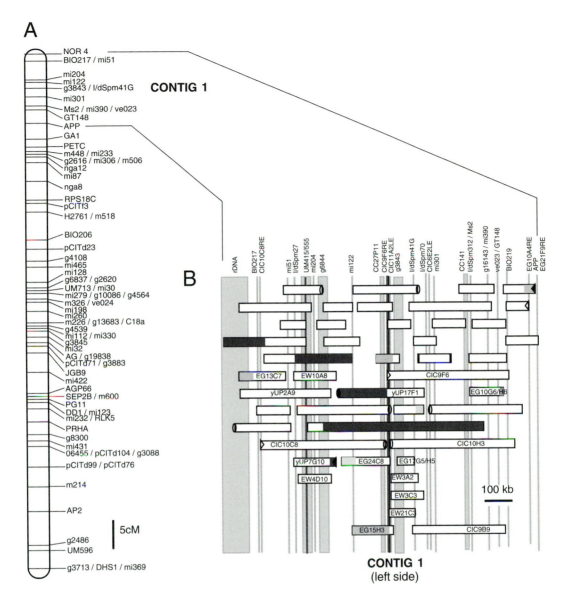

FIGURE 1.10

Genetic and physical maps of chromosome 4 in the Arabidopsis genome. (A) Genetic map showing various DNA markers including restriction fragment length polymorphism (RFLP) and simple sequence length polymorphism (SSLP) markers. Distances (in centimorgans, cM) in genetic map are determined by the frequency of recombination between markers. (B) Physical map of a small region of the upper tip of chromosome 4. Arabidopsis DNA fragments in yeast artificial chromosomes (YACs) are used to build a contig of overlapping sequences. Length of bars indicate the size of Arabidopsis DNA fragments in the individual YACs. Distances on the physical map or contig are in kilobases (kb). (Based on Schmidt, R., et al. 1996. Detailed description of four YAC contigs representing 17 Mb of chromsome 4 of *Arabidopsis thaliana* ecotype Columbia. Plant J. 9:756. From: http://genome-www.stanford.edu/Arabidopsis/JIC-contigs/Chr4_YACcontigs.html)

THE "VIRTUAL PLANT"

In the next few years, genetic studies will uncover many new genes that affect plant development, and some will be placed into developmental pathways of gene action (similar to the network shown in Fig. 1.3B). Pathways of gene action have already been assembled for several developmental processes in plants, such as time to flowering, ethylene responses, and photomorphogenesis. Many more developmental processes are currently under study. With a collection of many pathways, it may be possible to link them together in networks and to describe the operation of developmental genes for an organ or for the entire plant. In the future, one might be able to trace gene pathways through the entirety of plant development and do genetic experiments on the computer with a virtual plant. Present-day computer simulations of plant growth and morphogenesis are currently available on the internet at http://www.ctpm.uq.edu.au/Programs/IPI/ipivp.html (Somerville et al. 1997). Simulation models of the future might display the timing and action of regulatory genes and allow one to "tweak" regulatory components and observe the consequences.

However, virtual plants described by nothing more than the operation of gene pathways is a shallow representation of plant development. Development is an interplay between gene action, developmental cues and environmental signals. The interplay underlies many "nature versus nurture" arguments in plant development. The nature versus nurture polemic is brought into focus in Chapter 2, where the role of cell lineages and position effects will be discussed during plant development.

REFERENCES

Albersheim, P., Darvill, A. G., McNeil, M., Valent, B. S., Sharp, J. K., Nothnagel, E. A., Davis, K. R., Yamazaki, N., Gollin, D. J., York, W. S., Dudman, W. F., Darvill, J. E., and Dell, A. 1983. Oligosaccharins: Naturally occuring carbohydrates with biological regulatory functions. In Structure and Function of Plant Genomes, ed. Ciferri, O. and Dure, L., pp. 293–312. New York: Plenum Publishing Corporation.

Bensen, R. J., Johal, G. S., Crane, V. C., Tossberg, J. T., Schnable, P. S., Meeley, R. B., and Briggs, S. P. 1995. Cloning and characterization of the maize *An1* gene. Plant Cell 7: 75–84.

Biggin, M. D., and Tjian, R. 1989. Transcription factors and the control of Drosophila development. Trends Genet. 5: 377–383.

Buvat, R. 1952. Structure, evolution et functionnement du meristeme apical de qualques dicotyledons. Ann. Sci. Nat. Bot. Ser. 11. 13: 199–300.

Cleary, A. L., Gunning, B. E. S., Wasteneys, G. O., and Hepler, P. K. 1992. Microtubule and f-actin dynamics at the division site in living *Tradescantia* stamen hair cells. J. Cell Sci. 103: 977–988.

Colasanti, J., Cho, S. O., Wick, S., and Sundaresan, V. 1993. Localization of the functional *p34cdc2* homolog of maize in root tip and stomatal complex cells association with predicted division sites. Plant Cell 5: 1101–1111.

Doolittle, R. F., Feng, D.-F., Tsang, S., Cho, G., and Little, E. 1996. Determining divergence times of the major kingdoms of living organisms with a protein clock. Science 271: 470–477.

Feldmann, K. A. 1991. T-DNA insertion mutagenesis in *Arabidopsis:* Mutational spectrum. Plant J. 1: 71–82.

Gehring, W. J., Qian, Y. Q., Billeter, M., Furukubo-Tokunaga, K., Schier, A. F., Resendez-Perez, D., Affolter, M., Otting, G., and Wuthrich, K. 1994. Homeodomain-DNA recognition. Cell 78: 211–223.

Goldberg, R. B. 1988. Plants: Novel developmental processes. Science 240: 1460–1467.

Goldberg, R. B., de Paiva, G., and Yadegari, R. 1994. Plant embryogenesis: Zygote to seed. Science 266: 605–614.

Jackson, D., Veit, B., and Hake, S. 1994. Expression of maize *KNOTTED1* related homeobox genes in the shoot apical meristem predicts patterns of morphogenesis in the vegetative shoot. Development 120: 405–413.

Kenyon, C. 1994. If birds can fly, why can't we? Homeotic genes and evolution. Cell 78: 175–180.

Kersetter, R., Vollbrecht, E., Lowe, B., Veit, B., Yamaguchi, J., and Hake, S. 1994. Sequence analysis and expression patterns divide the maize *knotted1*-like homeobox genes into two classes. Plant Cell 6: 1877–1887.

Kilian, A., Stiff, C., and Kleinhofs, A. 1995. Barley telomeres shorten during differentiation but grow in callus culture. Proc. Natl. Acad. Sci. USA 92: 9555–9559.

Lawrence, P. A. 1992. The making of a fly: The genetics of animal design. Oxford: Blackwell Scientific Publication.

Levy, M. Z., Allsopp, R. C., Futcher, A. B., Greider, C. W., and Harley, C. B. 1992. Telomere end-replication problem and cell aging. J. Mol. Biol. 225: 951–960.

Lucas, W. J., Ding, B., and van der Schoot, C. 1993. Plasmodesmata and the suprecellular nature of plants. New Phytol. 125: 435–476.

Lyndon, R. F. 1990. Plant Development. London: Unwin Hyman.

Meyerowitz, E. M. 1994. Structure and organization of the Arabidopsis thaliana nuclear genome. In Arabidopsis, ed. Meyerowitz, E. M. and Somerville, C. R., pp. 21–36. Cold Spring Harbor: Cold Spring Harbor Press.

Meyerowitz, E. M. 1997. Plants and the logic of development. Genetics 145: 5–9.

Meyerowitz, E. M. and Somerville, C. R. 1994. Arabidopsis. Cold Spring Harbor: Cold Spring Harbor Press.

Mezitt, L. A. and Lucas, W. J. 1996. Plasmodesmal cell-to-cell transport of proteins and nucleic acids. Plant Mol. Biol. 32: 251–273.

Newman, T., de Bruijn, F. J., Green, P., Keegstra, K., Kende, H., McIntosh, L., Ohlrogge, J., Raikhel, N., Somerville, S., Thomashow, M., Retzel, E., and Somerville, C. 1994. Genes galore: A summary of methods for accessing results from large-scale partial sequencing of anonymous Arabidopsis cDNA clones. Plant Physiol. 106: 1241–1255.

Pearce, G., Strydom, D., Johnson, S., and Ryan, C. A. 1991. A polypeptide from tomato leaves induces wound-inducible proteinase inhibitor proteins. Science 253: 895–898.

Rivera-Pomar, R. and Jäckle, H. 1996. From gradients to stripes in *Drosophila* embryogenesis: Filling in the gaps. Trends Genet. 12: 478–483.

Robards, A. W. and Lucas, W. J. 1990. Plasmodesmata. Annu. Rev. Plant Physiol. Plant Mol. Biol. 41: 369–419.

Schmidt, R., West, J., Cnops, G., Love, K., Balestrazzi, A., and Dean, C. 1996. Detailed description of four YAC contigs representing 17 Mb of chromsome 4 of *Arabidopsis thaliana* ecotype Columbia. Plant J. 9: 755–765.

Scott, M. P. 1994. Intimations of a creature. Cell 79: 1121–1124.

Sheridan, W. F. 1982. Maize for Biological Research. Charlottesville, VA: Plant Molecular Biology Association.

Shure, M., Wessler, S., and Fedoroff, N. 1983. Molecular identification and isolation of the *waxy* locus in maize. Cell 35: 225–33.

Simpson-Brose, M., Treisman, J., and Desplan, C. 1994. Synergy between the hunchback and bicoid morphogens is required for anterior patterning in Drosophila. Cell 78: 855–865.

Smith, L. G., Hake, S., and Sylvester, A. W. 1996. The *tangled-1* mutation alters cell division orientations throughout maize leaf development without altering leaf shape. Development 122: 481–489.

Somerville, C., Flanders, D., and Cherry, J. M. 1997. Plant biology in the post-Gutenberg era. Everything you wanted to know and more on the world wide web. Plant Physiol. 113: 1015–1022.

St Johnston, D. and Nüsslein-Volhard, C. 1992. The origin of pattern and polarity in the Drosophila embryo. Cell 68: 210–219.

Stewart, F. C. 1958. Growth and development of cultivated cells. III. Interpretations of the growth from free cell to carrot plant. Am. J. Bot. 45: 709–713.

Tanksley, S. D., Ganal, M. W. and Martin, G. B. 1995. Chromosome landing: A paradigm for map-based gene cloning in plants with large genomes. Trends Genet. 11: 63–68.

van de Sande, K., Pawlowski, K., Czaja, I., Weineke, U., Schell, J., Schmidt, J., Walden, R., Matvienko, M., Wellink, J., van Kammen, A., Franssen, H., and Bisseling, T. 1996. Modification of phytohormone response by a peptide encoded by *ENOD40* of legumes and a nonlegume. Science 273: 370–373.

Van Sluys, M. A., Tempe, J., and Fedoroff, N. 1987. Studies on the introduction and mobility of the maize activator element in Arabidopsis thaliana and Daucus carota. EMBO J 6: 3881–3889.

Walbot, V. 1996. Sources and consequences of phenotypic and genotypic plasticity in flowering plants. Trends Plant Sci. 1: 27–32.

Zimmerman, J. L. 1993. Somatic embryogenesis: A model for early development in higher plants. Plant Cell 5: 1411–1423.

2

Cell Lineages and Positional Information

PREDICTABILITY OF CELL FATES

History is often said to predict the future. Is the same true for plant development? Can cell lineages be used to predict cell fates? In a number of simple developmental systems, such as embryo development in Shepherd's purse (*Capsella*) and root formation in Arabidopsis, the organization of cells is so regular and division patterns are so stereotyped that cell lineage patterns can easily be traced. From these examples, one might conclude that the cells follow a prescribed plan of development and that the fate of cells is predetermined.

The notion that cell lineage determines cell fate, however, has been challenged by plant developmental biologists. Although development can be very predictable in certain plant systems, cell lineage does not necessarily determine cell fate; rather, it can be argued that plant development largely relies on developmental cues and environmental signals. Plant cells do not appear to be irreversibly determined by their histories because they often assume different developmental fates when placed in a new environment. Plant development, therefore, appears to be directed by a balance of cell lineage and positional influences.

Whether or not cell lineages are prime, causal factors in plant development, it is important to be able to trace the developmental history of an organ or structure. One of the major challenges of developmental biology is to understand how cells acquire their cell fates, and cell lineage analysis can provide insight about when and where that happens. Plant systems are particularly amenable to cell lineage analysis because cells do not move about during development. There are questions that are frequently addressed when using cell lineage analysis: Are the cells in a given organ or structure clonally related? How closely related are two neighboring cells that may be very different from each other? From how many founder cells is an organ derived? It is usually the object of cell lineage analysis to mark progenitor cells at some point in development and determine the subsequent fate of the progeny cells.

MARKING CELLS WITHIN A LINEAGE

Marking cells with inherited, cell-autonomous markers allows one to trace cell lineages and to define the clonal relationships among cells. To mark single cells at a particular time in development, one can irradiate a plant or plant organ, generating mutations or chromosome breaks at low frequency. The mutations or chromosome breaks inactivate the expression of a gene, such as a pigmentation gene producing a colorless sector that stands out in the surrounding pigmented tissue. A common approach has been to knock out a pigmentation gene, which is present in heterozygous form and located near the tip of a chromosome (Fig. 2.1A). For example, loss of a chromosome tip that contains a wild-type allele required for anthocyanin or chlorophyll synthesis will mark a cell and its progeny. Cells can also be marked by irradiation to produce cytologically visible chromosome aberrations or be exposed to colchicine to produce polyploid cells. Marking cells by such drastic measures, however, may interfere with the developmental process in question.

Cell lineages can also be marked by spontaneous events such as the excision of a transposable element or transposon. More recent adaptations make use of reporter gene constructs, such as the *uidA* gene that encodes β-glucuronidase (GUS) linked through an *Ac* transposable element to the CaMV 35S promoter (35S:*Ac*:GUS) (Fig. 2.1B) (Finnegan et al. 1989). The expression of the GUS reporter gene in this construct is dependent on the excision of the *Ac* transposable element. GUS expression can be detected histologically by a chromogenic substrate that deposits a colored precipitate at the site of GUS activity. The advantage in using the transposable element is that cells can be marked at different times during development, if the excision process is not developmentally regulated.

FOUNDER CELLS AND
DEVELOPMENTAL COMPARTMENTS

Marking cells generates sectors of clonally derived cells. If cells are marked early in development and are destined to give rise to many progeny cells, then the resulting sectors will be large. If cells are marked

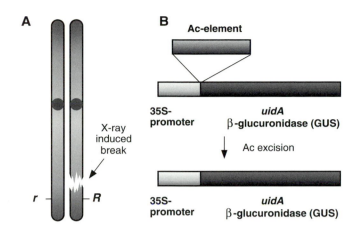

FIGURE 2.1

Markers for generating induced or spontaneous sectors for cell lineage analysis. (A) Sectors can be produced by X-ray–induced breakage of chromosomes bearing a heterozygous pigmentation gene, Rr, located near the chromosome tip. Loss of the dominant allele, R, results in a sector composed of clonally derived cells lacking pigmentation. (B) Spontaneous sectors can be produced in transgenic plants bearing a 35S:*Ac*:GUS construct. The transposable element, *Activator* or *Ac*, is located between the CaMV 35S promoter and the *uidA* or GUS coding region. Spontaneous excision of *Ac* activates expression of GUS, β-glucuronidase, which can be detected in plant tissue using histological techniques.

late in development, then fewer divisions will occur after the marking event, and the resulting sectors will be small. All cell lineages can be traced back to a single cell, which is called a **founder cell.** For sexually reproducing organisms, all cells are derived from a single-celled zygote. It is generally of interest, however, to know the number or identity of founder cells for an organ or structure at any given developmental stage, such as the time when the fate of that organ or structure is specified. (The founding cell population may be more than a single cell at that time.) Methods for determining the number of founder cells will be discussed in the next section.

During development, cells differentiate from one another by one of two general mechanisms:

1. Asymmetric division in which factors that were nonuniformly disposed in a mother cell are differentially parceled out to daughter cells;
2. Postdivisional processes in which differences in positional information between daughter cells cause one cell to develop differently from the other.

An asymmetric division will often give rise to a clone of cells with common properties, and these related cells may make up an organ or structure. Developmental compartment has been a useful term in animal systems to define a structure or region of an organism composed of clonally related cells (Lawrence 1992). The concept of a developmental compartment has had the greatest utility in animal systems where developmental compartments do not always correspond to obvious morphological structures. For example, in Drosophila embryos, body segments are divided into half segments – anterior and posterior parasegments. Parasegments cells are clonally derived and can be recognized by the localized expression of segment polarity genes.

The concept of the developmental compartment has not been as useful in plant development in defining organs, other structures, and their boundaries. In this and other chapters, we will see that boundaries of cell clones do not always demarcate the areas that we expect to be "developmental compartments." For example, as will be described in Chapter 6, the boundaries of cell clones in a dicot leaf do not lie along veins surrounding interveinal regions in a leaf. Developmental decisions that restrict cell lineages to particular organs or structures appear to be made later in plant development using positional guides and developmental cues rather than cell lineage information. As we will discuss later in this chapter, the concept of the developmental compartment is most applicable to cell layers in plants (although the terminology is not used).

DETERMINING THE NUMBER OF FOUNDER CELLS BY CELL LINEAGE ANALYSIS

As discussed earlier, the developmental history of any structure or organ can be traced back through its cell lineages to a founding cell population. It is important to be able to determine the size and identity of the founding cell population in order to understand the influences that shape the development of that structure or organ. A number of methods have been developed to determine the size of the founding cell population. Poethig (1987) described how one can determine the number of founder cells (or apparent cell number, ACN) at a given developmental stage using sector analysis. In such an analysis, plants with an appropriate genetic makeup are irradiated, marking cells at a given stage in development. The critical information needed for estimating ACN is the average fractional size of sectors produced in the structure or organ in question (Fig. 2.2). The average fractional size of sectors (i.e., size of the sector relative to the size of the organ or structure) is determined from a population of plants, organs, or structures irradiated at the same stage in devel-

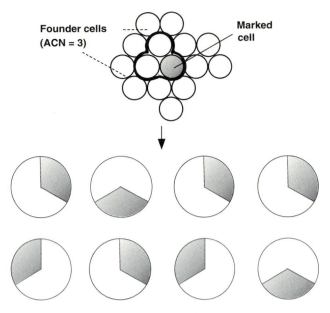

Average fractional sector size = 1/3

FIGURE 2.2

Method for determining the number of founder cells (apparent cell number, ACN) for an organ or structure based on sector size. In the example, there are three founder cells (outlined with dark lines) in the primordium for a hypothetical organ (upper part of the figure). One of the founder cells is marked (shaded cell) by the techniques described in Figure 2.1. Various sectors in organs derived from such primordia occupy, on average, one third of the organ (lower part of figure). ACN = 1/average fractional size of the sectors. Therefore, in this case ACN = 3. (Based on Poethig, R. S. 1987. Clonal analysis of cell lineage patterns in plant development. Am. J. Bot. 74:581–594.)

opment. ACN is the reciprocal of the fractional size. If the average sector size is one third of the organ, as it is in the example in Figure 2.2, then ACN = 3. This means that at the time of irradiation there were three founder cells in the primordium that gave rise to the organ. If the average size of the sectors was one tenth the size of the organ, then ACN = 10. Hence, the smaller the average fractional size of sectors, the larger the size of the founding cell population.

A number of assumptions are involved in making this estimate, and they have been addressed by Poethig (1987). The most critical is that all cells in the primordium must contribute equally to the structure in question. In studies of leaf development in tobacco where ACN has been determined, the patterns of cell growth in the leaf lamina were ascertained prior to the cell lineage analysis (Poethig and Sussex 1985).

A related method for determining the number of founding cells developed by Sturtevant (1929) for the study of Drosophila gyandromorphs is called **sector boundary analysis.** In sector boundary analysis, one marks cells very early in development and then determines the boundaries of various sectors so generated. Through the analysis of a large number of individual sectors, one can determine the number of possible sectors in an organ or structure which reflects the number of founder cells for that organ or structure. The first sector boundary analysis applied to plants was carried out by Christianson (1986), who was interested in cell lineages derived from globular embryos in cotton (*Gossypium barbadense*). Sectors in cotton were generated by the segregation of *virescent* (light green), anthocyanin, and leaf-shape markers in a semigametic seed parent. (A semigametic parent produces zygotes with unfused maternal and paternal nuclei. During early embryogenesis, the nuclei segregate, at some frequency, into maternal-haploid/paternal-haploid chimeras.)

Through a study of the sector boundaries, Christianson (1986) constructed a fate map for the cotton globular embryo and determined that there were separate cell lineages for the shoot apical meristem (SAM), each of the cotyledons, and the first and second leaves. Cotyledons appeared to be made up of fan-shaped cell clones (Fig. 2.3A) derived from eight cells arranged in a linear file in cotton embryos (Fig. 2.3B). By the same reasoning, the SAM was derived from three cells and the first and second leaves from separate cells, from two and one cell, respectively. It is a matter of debate whether this means that the first two leaves and the SAM are of separate origin. It is true that the cell lineages for the first two leaves do sort out from the rest of the shoot at a very early stage in development. Christianson argued that the first two leaves in cotton are different from the rest because they have a different phyllotaxy. (Cell lineage analysis in maize, which is described later in this chapter, also demonstrates that founder cells for the first few leaves sort out early in embryogenesis.)

Sector boundary analysis has also been applied to anther development in maize by Dawe and Freeling (1992). Maize anthers are composed of four microsporangia, the compartments in which pollen are formed. Maize coleoptiles were X-irradiated just as they emerged from the seed, and sectors lacking anthocyanin pigmentation in tassels and anthers were observed. Sectors most frequently cut anthers in half or in quarters, and sector boundaries usually lay between microsporangia (Fig. 2.4A and B). Less-frequent sector boundaries cut through individual microsporangia (Fig. 2.4C and D). Anthers therefore appear to be

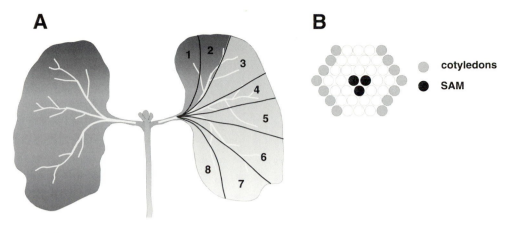

FIGURE 2.3

Sector boundary analysis in cotton (*Gossypium barbadense*). Sectors are produced from pigmentation markers in crosses involving a semigametic seed parent (see text). (A) The most frequent sector boundaries (right cotyledon) indicate that the cotyledons are composed of eight fan-shaped clones. (B) Presumptive fate map of the globular embryo in cotton based on the sector boundary analysis. Each cotyledon appears to be derived from eight cells arranged in nearly linear file. Shoot apical meristem (SAM) is clonally distinct from cotyledons at the globular embryo stage and commonly gives rise to sectors making up about a third of the shoot. The SAM is therefore thought to be derived from three cells at this stage. First and second leaves are also clonally separate at this stage (not shown). (Based on Christianson, M. L. 1986. Fate map of the organizing shoot apex in Gossypium. Am. J. Bot. 73:947–958.)

derived most frequently from primordia composed of two to four cells and less frequently from primordia composed of eight cells.

The finding that sector boundaries between microsporangia were highly reproducible suggested to Dawe and Freeling (1992) that anther development is an orderly process and that each microsporangium is a developmental compartment, composed of clonally related cells. An alternative explanation is that each microsporangium is not a strict developmental compartment; rather, the structure connecting microsporangia in the developing tassel is a large target (i.e., it is derived from many founder cells) and, therefore, random boundaries frequently lie in the connecting region. This argument was dismissed, however, because there was no morphological evidence that the connecting region is large in the developing tassel. Dawe and Freeling (1992) proposed instead that the architecture of the developing anther imposed certain constraints on the cell lineage pattern. They argued that the arrangement of cells in the founding cell population tends to influence where progeny cells will end up in a mature structure. They contended that the arrangement of the

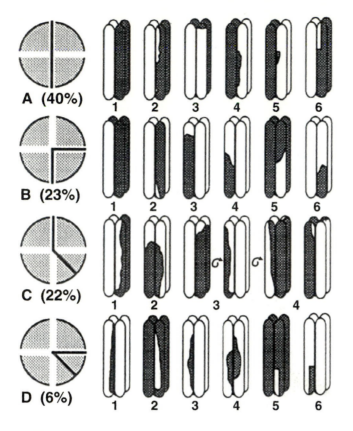

FIGURE 2.4

Sector boundary analysis used to determine the number of founding cells that give rise to anthers in maize (*Zea mays*). Plants heterozygous for several anthocyanin pigmentation genes were X-irradiated at an early seedling stage. Chimeric tassels were identified, and about one hundred chimeric anthers were characterized. Four major classes of sectors (lettered A–D) are shown with the percentage of anthers falling into that class. Representative anthers are illustrated on the right, and in the pie-shaped figures, on the left, the major sector boundaries (black lines) are shown with respect to the boundaries of the microsporangia (white lines). Curled arrow in class C anthers indicates two views of the same anther. (From Dawe, R. K., and Freeling, M. 1992. The role of initial cells in maize anther morphogenesis. Development 116:1081. Reprinted by permission of the Company of Biologists Ltd.)

small group of founder cells may represent a self-propagating "structural template" for the formation of the anther. Thus, the predictable cell lineage pattern in the developing anther may be a simple architectural phenomenon that is not based on the inherent lineage characteristics of the cells themselves.

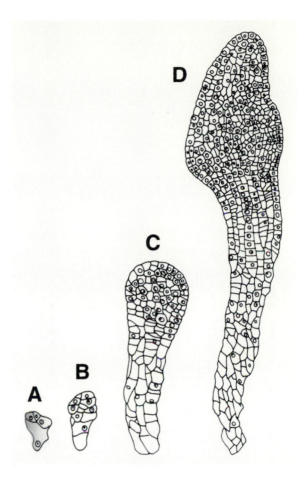

FIGURE 2.5

Early maize embryos at various days after pollination (DAP). Drawings of median longitudinal sections of embryos at 2 DAP (A), 4 DAP (B), 7 DAP (C), and 8 DAP (D). Note the lack of organization and regularity in planes of (prior) cell divisions. (From Poethig, R. S., Coe, E. H. J., and Johri, M. M. 1986. Cell lineage patterns in maize *Zea mays* embryogenesis: A clonal analysis. Develop. Biol. 117:395. Reprinted by permission of the author and Academic Press.)

CELL FATE MAPPING IN EARLY EMBRYOGENESIS

Poethig et al. (1986) used cell lineage analysis described above to produce cell fate maps for the SAM (apical initials) in maize embryos. Cell fate mapping can be straightforward in simple systems with stereotypic patterns of cell division. In maize embryos, however, cell division patterns are very irregular, and it is not possible to trace cell lineages through the analysis of histological sections (Fig. 2.5). Therefore, more general cell fate maps that were probabilistic in nature were constructed. To do so, Poethig et al. (1986) X-irradiated embryos heterozygous for dominant pigmentation markers expressed in kernels, seedlings, and/or mature plants. They irradiated maize embryos at various days after pollination (DAP) and looked for colorless sectors during subsequent plant growth.

The analysis revealed that the fate of cells in the presumptive shoot meristem was progressively restricted during development. When the embryos were irradiated early in development (4 DAP), most sectors in the shoot crossed nodal boundaries (upper panel, Fig. 2.6), indicating that individual nodes were not separate developmental compartments in the early embryo. However, sectors were eventually restricted to single nodes (8 DAP); first the lower nodes, then the upper nodes.

A fate map of the SAM was constructed in which the SAM was represented by concentric rings (lower panel, Fig. 2.6). Cells at the periphery of the SAM were assumed to give rise to the lower nodes, while those in the center of the SAM were progenitors of the upper nodes. When embryos were irradiated at 4 DAP, half of the sectors were restricted to the first three nodes, indicating at that stage that a large portion of the cell population in the SAM was devoted to the production of the lower nodes. In general, sectors were narrower in the lower nodes and much wider, often half the stem, in the upper nodes. This indicated that the founding cell population (ACN) was larger for the lower nodes (cells in the outer ring of the fate map) than it was for the upper nodes (cells in the center of the fate map). The ACN for the upper nodes was two to three at 4 DAP. At this stage, therefore, the fate of cells destined to become the upper nodes of the plant had not been sorted out. When irradiated at 7 DAP, half of the sectors were restricted to the first node, indicating that the lineage of this node is nearly separate from the other nodes at this stage.

A picture of order, not chaos, emerges from this cell lineage analysis in maize embryos. The maize embryo does not undergo a stereotypic set of divisions and the boundaries of sectors are not strict, but the fate of cells in the embryonic SAM can be defined in a general and probabilistic way. For example, a larger portion of cell population in the embryonic SAM is devoted to formation of the lower nodes. A smaller portion of the cell population gives rise to upper nodes; their fates have not been sorted out in the early embryo. As development proceeds cell fates are progressively sorted out.

The remarkable property of maize embryo development is that highly uniform corn plants arise from embryos that superficially appear to be unorganized. Even though there are few morphological markers in the early embryo, the SAM itself can be located by the expression pattern of the *knotted1* (*kn1*) gene. (*Kn1* will be discussed later in this chapter and again in several subsequent chapters.) Using immunohistochemical techniques with antibodies directed against KN1 (the protein encoded by the *kn1* gene), Smith et al. (1995) demonstrated that the SAM was located laterally along the embryo in a region at the base of

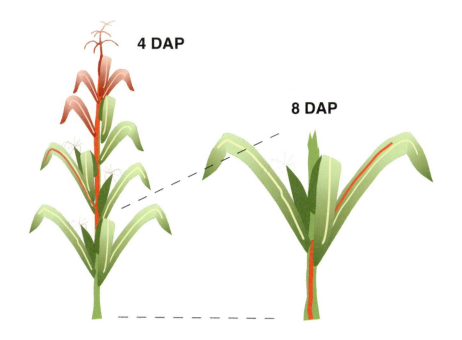

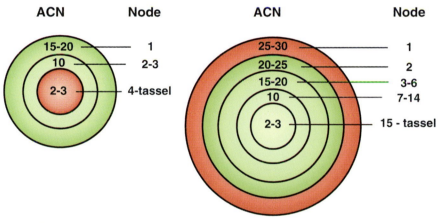

FIGURE 2.6

Sectors in mature maize plants produced by X-ray irradiation of embryos at various times after pollination. (Above) Examples of pigmented sectors, shown in red, produced by X-irradiation of embryos at 4 and 8 days after pollination DAP (see Fig. 2.5). In general, sectors were shorter and narrower in the lower part of the plant or when embryos were irradiated later. (Below) Cell fate probability maps for the embryonic SAM in transection were produced from the sector data. The map is composed of a series of concentric circles in which the inner circles represent cells giving rise to higher nodes. The apparent cell number, ACNs, for given nodes are indicated. ACN is larger for the lower nodes, and the specific cell fates are determined earlier for these nodes. (Redrawn fate maps from Poethig, R. S., Coe, E. H. J., and Johri, M. M. 1986. Cell lineage patterns in maize *Zea mays* embryogenesis: A clonal analysis. Develop. Biol. 117:397. Reprinted by permission of the author and Academic Press.)

the scutellum (see Fig. 3.6). Thus, the embryonic SAM is present and presumably functional even though it is not evident morphologically.

EXPERIMENTAL DISRUPTION OF CELL LINEAGES

A powerful tool to study the role of cell lineages in more highly ordered systems is to disrupt known cell lineages by ablating single cells and determining the consequences. Cell ablation can be carried out by using a laser microbeam that can be focused on a single cell to destroy it. Laser ablation has been employed frequently in cell lineage studies in organisms such as the nematode, *Caenorhabditis elegans,* because all the cells and their cell lineages have been identified in this organism (see Lambie and Kimble 1992).

As will be described in Chapter 12, all the cells in the Arabidopsis root tip have been identified and their cell fates described. Cell division patterns in the Arabidopsis root tips are stereotypic and the cell structure of the root tip is very uniform. Therefore, it is possible to ablate specific progenitor cells in Arabidopsis root tips and study the consequence to the cell lineage. Van den Berg et al. (1995) unexpectedly found in the Arabidopsis root that cell lineages do not dictate cell fate. For example, they found that when a cortical initial was ablated, the dead cell was simply crushed, displaced to the outside, and replaced by a pericycle cell that assumed the function of the former cortical initial. These experiments demonstrated that cell fates are not necessarily determined by cell lineages even in a highly ordered structure like the Arabidopsis root. (More discussion about cell ablation in root development is found in Chap. 12.)

POSITIONAL INFORMATION
IN ANIMAL DEVELOPMENT

If cell lineages do not determine cell fate, then what does? Current thinking is that development is an interactive process in which positional information and developmental cues have as much to do with programs of plant development as do ordered gene pathways. There is no strict distinction between positional information and developmental cues; however, positional information is usually thought to have spatial information and represents the signals that plant cells receive based upon where they are located. Developmental cues have temporal information about the developmental stage of the tissues or organs in which they are involved. We know little about positional information and developmental cues and how they operate, but signal transduction in plants is an active area of research. It is only a matter of time before some relevant examples are worked out.

The role of positional information, particularly inductive effects of tissues on each other, have been recognized for many years in animal development. The classical transplantation experiments in amphibian embryos by Spemann and Mangold (1924) demonstrated the role of the "primary organizer" (i.e., the dorsal lip of the blastopore) in the induction of neural tissue. Their experiments underscored the importance of inductive effects during tissue migrations in embryonic development. The significance of inductive effects has not been as apparent in plants because tissues do not migrate during development in plants as they do in animals.

Nonetheless, because positional information has profound effects on plant development, it is important to understand what is known about positional information in animal systems and what kinds of signals are involved. Over the years, many developmental phenomena in animals have been explained by gradients and fields of morphogenic substances diffusing over long distances (Wolpert 1981). The most interesting positional signals that have come to light, however, are ones involving short or intermediate range signals between cells and their immediate neighbors. There are now many examples of short-distance positional signaling during animal development, and one is the signaling pathway involving *wingless* (*wg*) in Drosophila (Nusse and Varmus 1992). *Wg* mutations disrupt the development of segment polarity in mutant embryos. This happens because *wg* expression is required to maintain the segmental expression of *engrailed* (*en*) during embryo development. *En* is a homeotic gene expressed in embryonic segments in a row of cells just posterior to the cells expressing *wg* (see Fig. 1.3A). *Wg* encodes a glycoprotein that appears to be secreted; however, *wg* does not work at a distance from *en* expressing cells (Fig. 2.7). Vincent and Lawrence (1994) found through the analysis of genetic mosaics that *wg* expressing cells must be adjacent to *en*-expressing cells for the maintenance of *en* expression. Thus, the secretion of the *wg* gene product provides a local signal that is effective in maintaining *en* expression only in neighboring cells. *Wg* also has long-range signaling effects; therefore, it is thought that the gene product acts as a gradient morphogen in which high local concentrations are required for *en* expression. Orthologs of *wg* have been found in other animal systems, and these genes affect a number of different developmental processes in which *wg* acts at short range or at longer range, depending on the situation and how the *wg* protein is processed (Nusse and Varmus 1992). (An orthologous gene or **ortholog** is a gene of similar sequence found in a different species.)

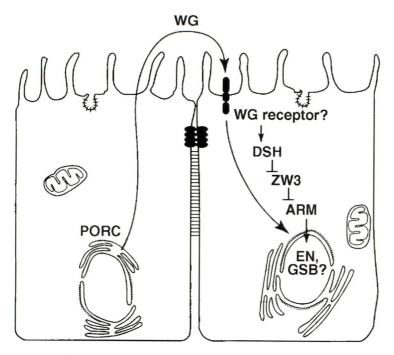

FIGURE 2.7

Short-distance signaling by the product of the *wingless* gene (WG) in Drosophila development. Production and secretion of WG by the anterior cells in each segment of the Drosophila embryo is facilitated by porcupine (PORC). WG presumably binds to unidentified receptors in cells immediately posterior to WG secreting cells. WG maintains the expression of *engrailed* (EN) through a pathway involving *disheveled* (DS), *zeste-white3* (ZW3), and *armadillo* (ARM). Signaling pathway is simplified, and it is not known whether the pathway is a single, linear pathway as depicted. (From Peifer, M. 1995. Cell adhesion and signal transduction: The Armadillo connection. Trends Cell Biol. 5:226. Reprinted by permission of Elsevier Trends Journals.)

POSITIONAL INFORMATION IN PLANT DEVELOPMENT

There are a few examples in plants of short-distance, positional signals. One interesting example involves cell differentiation signals derived from changing cell-to-cell contacts during carpel fusion in flower development. In periwinkle (*Catharanthus roseus*) two separate carpel primordia fuse to form a single carpel. The adaxial surfaces of the carpel primordia fuse and approximately 400 cells at the appressed surfaces redifferentiate from epidermal cells into parenchymal cells (Fig. 2.8). Using microsurgical transplantation techniques, Siegel and Verbeke (1989) showed that epidermal cells from surfaces of the carpels that normally do not fuse will differentiate in a similar way if juxtaposed

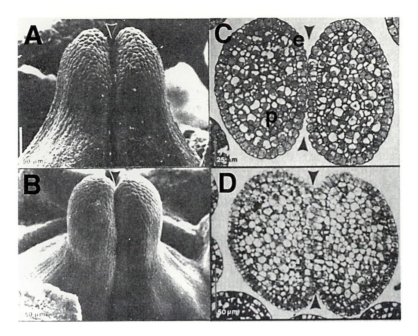

FIGURE 2.8

Fusion of carpels in developing flowers of periwinkle (*Catharanthus roseus*). (A, C) Prefusion carpels emerge as separate primordia and expand until the adaxial surfaces contact. Each carpel is surrounded by an epidermal cell layer, e. (B, D) Carpels begin to fuse, and epidermal cells at the interface (indicated by arrowheads) redifferentiate into parenchyma cells, p. (A, B) Carpels visualized by scanning electron microscopy; (C, D) sections stained with toluidine blue and viewed in the light microscope. (From Siegel, B. A., and Verbeke, J. A. 1989. Diffusible factors essential for epidermal cell redifferentiation in *Catharanthus roseus*. Science 244:580. Copyright 1989 American Association for the Advancement of Science.)

surgically. The stimulus to redifferentiate was found to be a diffusible factor(s) that could be trapped in agar blocks. When the blocks were appressed onto nonfusing regions of the carpel, the epidermal cells in this region were stimulated to undergo redifferentiation. (The diffusible factor has not yet been defined.) The cells on the two appressed surfaces rapidly develop cytoplasmic continuity and intercellular connections. This was shown by microinjecting a fluorescent tracer (Lucifer Yellow) into epidermal cells on either side of the fusion plane and demonstrating the exchange of the label after contact (Van Der Schoot et al. 1995).

Some positional information in plants may involve biophysical forces. An interesting example in this regard is the development of the regular pattern of florets in the flower head of sunflower (*Helianthus*). Hernandez and Green (1993) offered a mechanical explanation to

FIGURE 2.9

Development of disk flowers in the sunflower (*Helianthus annus*) head. (A) Sun-flower head is composed of receptacle with outer ray flowers and inner disk flowers arranged in an intersecting spiral pattern. (B) Detail of the generative zone where new florets are emerging. Florets, Fl, develop into disk flowers, DF, with floral bracts. Floret primordia appear with sinusoidal periodicity. The explanation for the pattern is that uneven expansion of the surface generates periodic deformations that serve as sites for primordia emergence. (B) Visualized by scanning electron microscopy. Bar = 100 μM. ([B] from Hernandez, L., and Green, P. B. 1993. Transductions for the expression of structural pattern: Analysis in sunflower. Plant Cell 5:1729. Reprinted by permission of the American Society of Plant Physiologists.)

describe how the sites of floret production are determined in the sun-flower head. They postulated that sites for floret formation were deter-mined by periodic deformations in the surface of the sunflower and that these deformations were caused by a nonuniform expansion of the flower head surface (Fig. 2.9). It was demonstrated that mechanical forces are involved by showing that the pattern of florets could be altered by mechanically distorting the sunflower head. Hernandez and Green (1993) concluded that deformations in the surface of the flower head switch on specific patterns of gene expression by activating stretch receptors or other positional transducers.

Further reinforcement of the idea that biophysical forces may be involved in positional signaling are studies on the localized application of the cell wall protein called "expansin." These proteins catalyze the acid-induced expansion of the cell wall and bind at the interface between cel-lulose microfibrils and polysaccharides of the cell-wall matrix (McQueen-

Mason and Cosgrove 1995). They are thought to induce extension by reversibly disrupting noncovalent bonds within the cell-wall network. Fleming (1997) isolated active expansins from cucumber (*Cucurbita*) hypocotyls and immobilized the proteins on small plastic beads. The small beads were then strategically placed on the shoot apices of tomato (*Lycopersicon*) at places where leaf primordia were scheduled to appear. The argument has been made that the surface of the shoot apical meristem (SAM) is under tension and the underlying tissue is subject to compression (Green et al. 1996). Such being the case, local application of expansins might increase the extensibility of the cell wall in the area of the bead and produce a bulge at the site of the bead. Indeed, it was found that unscheduled bulges appeared at the site of the beads and, remarkably, that these bulges developed into leaflike structures (see Chap. 5 for further discussion about this experiment). Hence, these studies suggest that there is some localized propensity for expansion on the surface of SAM that corresponds to sites of prospective leaf primordia formation. It remains to be demonstrated whether the biophysical forces in the SAM are the cause or the result of other positioning information.

Other positional information in plants is thought to be derived from diffusible chemical signals, such as gradients of plant hormones. Classic gravitropic responses in roots and phototropic responses in coleoptiles involve auxin gradients. The effects of auxin gradients on development can be demonstrated in mutants or by auxin transport inhibitors (i.e., agents that block the auxin efflux carriers). Auxin gradients result from the polarized movement of the hormone by auxin transporters. The *pin1* mutants in Arabidopsis are defective in polar auxin transport and have fused cotyledons (Okada et al. 1991). The defect blocks the development of bilateral symmetry in the embryo, a property that normally appears when the cotyledons emerge. In *pin1* mutants, cotyledons emerge as a collar, not as two lobes set across from each other (Fig. 2.10). Similar effects were found when embryos explanted from Indian mustard plants (*Brassica juncea*) were cultured in the presence of auxin transport inhibitors (Liu et al. 1993). It has been proposed consequently that gradients of auxin provide information for the establishment of normal bilateral symmetry and the division of the dorsal end of the embryo into individualized cotyledons (Liu et al. 1993).

Another example of a hormone providing positional information is found in root hair formation in Arabidopsis roots. Epidermal cells in Arabidopsis root occur in alternating cell files – some cell files form root hairs and some do not (see Chap. 12). The cells that do form root hairs are called trichoblasts, and they lie over radial walls between adjacent, underlying cortical cells. The ones that do not are called atrichoblasts,

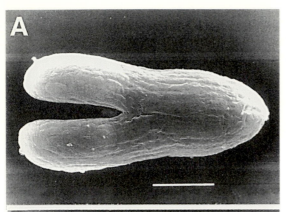

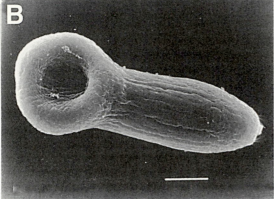

FIGURE 2.10

Effect of the *pin1*-1 mutation on embryo development in Arabidopsis. (A) Wild type; (B) *pin1*-1 embryo develops with radial symmetry and a collar rather than with bilateral symmetry and individual cotyledons. *Pin1*-1 mutants have defects in auxin polar transport in inflorescence stalks. The embryo phenotype is thought to result from the auxin transport defect. The effect can be phenocopied by auxin transport inhibitors. Bars in A and B are 50 μm. (From Liu, C.-M., Xu, Z.-H., and Chua, N.-H. 1993. Auxin polar transport is essential for the establishment of bilateral symmetry during early plant embryogenesis. Plant Cell 5:625. Reprinted by permission of the American Society of Plant Physiologists.)

and they lie over tangential cortical cell walls. As will be discussed in Chapter 12, it has been speculated that the alternating pattern of trichoblasts is due to the localized production of the hormone, ethylene (Tanimoto et al. 1995). In mutants with constitutive ethylene responses, the alternating pattern is compromised, and root hairs are formed in all cell files. On the other hand, no root hairs are produced in mutants that do not respond to ethylene. Thus, the local production of ethylene may influence the pattern of root hair production in Arabidopsis roots.

CELL LAYERS AND CHIMERAS

As discussed earlier in this chapter, the concept of developmental compartments has not been very useful in plants, except in describing cell layers. Cell layers are composed of clonally related cells and have a continuity that can be traced back through the SAM to the embryo. Angiosperms are composed of three fundamental cell layers (L1–L3) (Fig. 2.11). L1 is the outermost layer made up of the epidermis, L2 is composed of subepidermal layers, and L3 is the pith. (L2 and L3 are

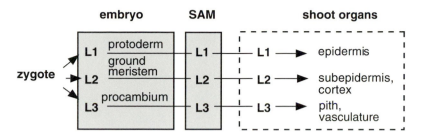

FIGURE 2.11

Continuity of cell layers from embryo to mature shoot organs. Angiosperms have three fundamental cells layers (L1–L3) that can be traced back through the shoot apical meristem, SAM, to the embryo. Cell layers are composed of clonally related cells. Organs in the shoot and primary root are made up of cells from three cell layers. For example, L1 in the early embryo is called the protoderm, which gives rise to epidermis in various organs in the mature plant.

called cell layers, but they are not necessarily one cell layer thick.) Plant organs are therefore composed of cells in cell layers that are not clonally related (except if one traces cell lineages back to the zygote or very early stages of embyogenesis). If a cell in L2 is marked early in development, then its progeny cells will produce a sector of clonally related cells within that layer (see Chap. 5) (Satina et al. 1940). The organ or organism bearing such a sector is called a **mericlinal chimera** (Fig. 2.12). Mericlinal chimeras occupy only a portion of a cell layer; therefore, they may be unstable and can be lost in subsequent growth. **Periclinal chimeras** are formed when a mericlinal chimera takes over a whole cell layer. For example, if a bud is included in a mericlinal chimera, then the bud can give rise to a branch, stem, or whole plant that is entirely chimeric.

Periclinal chimeras provide the most compelling evidence for the existence and continuity of cell layers. Cell layers are remarkably stable, as shown by the fact that periclinal chimeras can often be propagated indefinitely. Periclinal chimeras can be used to determine which cell layer exercises the most influence in specifying the development of an organ or structure. Through grafting procedures, different cell layer combinations can be produced (Tilney-Bassett 1986). For example, L1 might be derived from one grafting partner and L2 and L3 from the other. Cell layer origins in such chimeras can be identified with the appropriate cell layer markers. In tomato (*Lycopersicon*), Syzmkowiak and Sussex (1992) determined in which cell layer the *fasciated* gene must be expressed to affect the number of organs per whorl in flowers. By grafting procedures, they created chimeras between wild-type tomato plants and the *fasciated* mutant. The origin of the cell layers was determined by cell

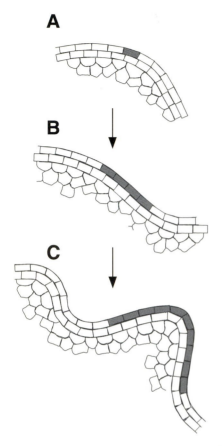

FIGURE 2.12

Formation of cell layer chimeras. (A) A mutation occurs that genetically distinguishes one cell (shaded cell) from all others in a cell layer. (B) The genetically altered cell is clonally propagated in the layer of origin (in this example, the L1 layer or epidermis), forming a mericlinal chimera that occupies only a sector in the cell layer. (C) Growth in region that contains mericlinal chimera tissue produces an organ or entire plant in which the L1 layer is genetically altered. A total cell layer chimera is called a periclinal chimera. (Redrawn from Carpenter, R., and Coen, E. S. 1995. Transposon induced chimeras show that *floricaula*, a meristem identity gene, acts non-autonomously between cell layers. Development 121:20. By permission of the Company of Biologists Ltd.)

layer–specific markers, such as *hairless* for L1 and *anthocyanin gainer* for L2. They found that the presence of *fasciated* in L3 had the most profound effect in determining the size of the vegetative meristem and number of organs produced per whorl in flowers (Fig. 2.13). They argued from this that L3 influences the number of L1 and L2 cells that are recruited into the organs of the floral meristem. The molecular nature of the factors involved in the recruitment is not yet known.

Cell layers are most intriguing because despite their stability, the cells in one layer can sometimes invade other cell layers. Dermen (1953) found in peach tree stems that although vascular tissue and pith was usually derived from L3, tissue may occasionally arise from L2. The frequency of cell layer invasion varies depending on the organ and the plant type. For example, Dawe and Freeling (1990) found that in anther development in maize the frequency of cell layer invasion was 10^{-3} per cell division. Layer invasion frequencies are thought to be much lower in dicots. Cell layer invasion has interesting consequences with respect to cell lineages and cell fates. Where it has been described,

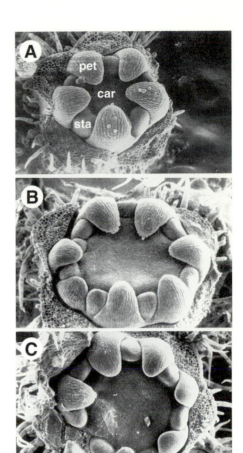

FIGURE 2.13

Floral organs produced in cell layer chimeras of tomato (*Lycopersicon esculentum*) expressing the mutation *fasciated*. Floral organs are shown at a stage when petal and stamen primordia are developing (sepals are removed). The genotypes of the cell layers are designated in order, L1 to L3: (A) nonfasciated tomato, + + +; (B) *fasciated* tomato, *f f f*; (C) L2–L3 *fasciated* chimera, + *f f*. Note that expression of *fasciated* in L2 and L3 is sufficient to confer the mutant phenotype. Petal primordium, pet; stamen primordium, sta; carpel primordium, car. (From Szymkowiak, E. J., and Sussex, I. M. 1992. The internal meristem layer L3 determines floral meristem size and carpal number in tomato periclinal chimeras. Plant Cell 9:1094–1096. Reprinted by permission of the American Society of Plant Physiologists.)

invading cells (or their progeny) take on the character of cells in the invaded layer. Given the slavish adherence of cells to particular cell layers, it is astounding that cells adopt new cell fates. The observation is more evidence that positional information and not cell lineage determines cell layer identities.

If positional information is involved in determining cell layer identities, then information must be transmitted within a cell layer and between cell layers. It can be argued that developmental information must be transmitted between cell layers because plant organs are composed of cells from all three cell layers. Therefore, in the development of an organ, signals must be able to cut across cell layers. If development of a plant organ requires a gene to be expressed in all three cell layers, then some signal must be transmitted from one layer to another to coordinate the process. The examples that follow describe novel ways in which plants might transmit information within a cell layer or from one cell layer to another.

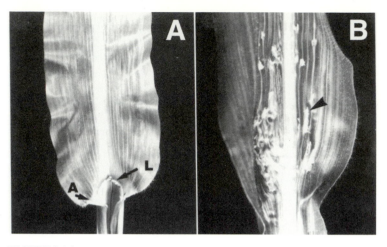

FIGURE 2.14

Knotted1 (*Kn1*) mutant in maize. (A) Wild type maize leaf has a smooth epidermis and parallel rows of lateral veins. (B) *Kn1* leaf has knots (arrowhead) over lateral veins that distort the surface of the leaf and the pattern of the veins. Auricle, A; ligule, L. (From Greene, B. A., and Hake, S. 1993. The *Knotted-1* mutants of maize: Investigating the circuitry of leaf development. Sem. Develop. Biol. 4:42. Reprinted by permission of the publisher.)

TRANSMISSION OF INFORMATION BETWEEN CELL LAYERS

As discussed earlier, developmental signals often cut across cell layers in the formation of various structures or organs. Chimeras or genetic mosaics have been very valuable in identifying circumstances in which information is transmitted between cell layers. The operation of the *knotted1* gene in maize and the *FLORICAULA* gene in snapdragons are interesting examples in which there are new ideas about the transmission of information between cell layers.

Knotted Gene

Dominant mutations in the *knotted1* (*kn1*) gene in maize produce unsightly knots or bumps on the upper surface of leaves (Fig. 2.14B). The upper surface of the leaf is normally quite smooth because the epidermis grows as a sheet by anticlinal divisions (i.e., divisions in the plane of the leaf surface). Knots produced in the mutant are initiated by periclinal divisions in the epidermal layer over lateral veins. Subsequent misorientation of divisions in subepidermal layers also contributes to the formation of knots. *Kn1* is a dominant gain-of-function mutation that leads to ectopic expression of *Kn1* at sporadic foci in leaves (to be discussed in Chapter 5). *Kn1* is not ordinarily expressed in mature leaves; only in the SAM, as described earlier in this chapter (Jackson et al. 1994).

Hake and Freeling (1986) generated genetic mosaics to determine in which cell layer *Kn1* had to be expressed in order to produce knots. Mosaics were generated by X-irradiating seedlings heterozygous for the *Kn1* (similar to Fig. 2.1A). *Kn1* is located distally on chromosome arm 1L. X-irradiation produces chromosome breaks and the loss of *Kn1* can be detected by the expression of the recessive marker, *lw*, which is an albino gene closely linked to *Kn1*. Thus, the appearance of a white sector would indicate a patch of tissue in which expression of the knotted gene was extinguished.

Hake and Freeling (1986) were most interested in sectors in which one cell layer expressed *Kn1,* and the other cell layers did not. The maize leaf blade, excepting the vasculature, is largely made up of two cell layers, L1 and L2. L2 is composed mostly of green mesophyll cells. The expression of *lw*, which produces a white sector among other green mesophyll cells, is quite obvious in L2. Such sectors are not as obvious in L1 because most cells in this layer do not contain green chloroplasts. Guard cells in L1, however, do contain chloroplasts, and, therefore, sectors not expressing *Kn1* in the L1 layer could be recognized by guard cells that lacked chloroplasts.

Sinha and Hake (1990) analyzed sections of chimeric maize leaves and made the remarkable finding that even though knots originated in the epidermis, the production of knots was determined by the genetic makeup of the subepidermal, and not the epidermal, layer. Epidermal sectors that did not express *Kn1* produced knots as long as the underlying mesophyll layer was green and expressing *Kn1*. They refined their identification of cell layers to show that knot formation required *Kn1* expression only in the middle mesophyll-bundle sheath layer.

Kn1 information must therefore be communicated from one cell layer to another, but how can that happen? The *kn1* gene has been cloned and encodes a homeodomain-containing transcription factor (Hake et al. 1989). It would seem unlikely that the gene product, the KN1 protein itself, could be transmitted cell to cell, from one cell layer to another. KN1 is not a secreted protein that can move extracellularly. Because KN1 is a moderately sized protein, it should not be able to move from cell to cell through intercellular channels either. Plasmodesmata are cytoplasmic bridges that interconnect cells, and their pores are not large enough to accommodate proteins of the size of KN1.

Quite remarkably, Lucas et al. (1995) reported that the KN1 protein may alter the structure of plasmodesmata and facilitate its own transport from cell to cell, from one cell layer to another. Their studies were spurred on by the observation described previously that KN1 acts in cells other than those in which protein was synthesized. Using *in situ* hybridization, they observed that *kn1* RNA accumulated in nonepidermal or interior cell layers (L2) of the maize SAM. Using immunolocalization techniques, however, they found that the KN1 protein was located both in L2 and in the epidermal layers of the SAM, in a cell layer in which *kn1* RNA did not accumulate. They proposed that KN1 might move from cell to cell to account for this, and they tested that hypothesis by microinjecting fluorescein-labeled KN1 (FITC-KN1) into plant cells (Fig. 2.15). For technical reasons, they were not able to inject FITC-KN1 into the cells of the maize SAM, which are quite inaccessible to microinjection. They could, however, microinject FITC-KN1 into mesophyll cells of tobacco (*Nicotiana tabacum*) leaves and found in this heterologous system that FITC-KN1 was able to migrate to neighboring cells (Fig 2.15A). KN1 was able to move and dilate the pores of plasmodesmata, which are the channels that interconnect plant cells, allowing for the movement of other macromolecules (Fig. 2.15D). It remains to be explained in mechanistic terms how KN1 is able to do this and whether these observations are relevant to developmental issues. Nonetheless, KN1 serves as a fascinating, albeit controversial, example of how cells may communicate with each other within and across cell layers in ways that could have developmental consequences.

Floricaula Gene

Another interesting example of the interactions between cell layers involves the study from Coen and co-workers of genetic chimeras produced by unstable *floricaula* (*flo*) mutants in snapdragons (*Antirrhinum majus*) (Carpenter and Coen 1995). As will be discussed in Chapter 7, the expression of the homeotic gene, *flo,* is required for the production of flowers (by conferring floral identity to the shoot meristem). *Flo* mutants produce inflorescences with an indeterminate shoot growing in place of each flower. The unstable mutant *flo-613* contains a Tam-3 transposon insertion that can excise spontaneously and reactivate the expression of *flo*. Upon excision during vegetative growth, a genetic sector (mericlinal chimera) is produced in the cell layer in which excision occurs. Excision restores the action of the *flo* gene, which then leads to the production of flowering sectors in otherwise mutant inflorescences. The cell layer in which the excision occurs can be identified by *in situ* hybridization using a *flo* gene probe (Fig. 2.16A).

The chimeras produce flowers that are not fully normal because only one layer has been restored to normal *flo* expression. The restored

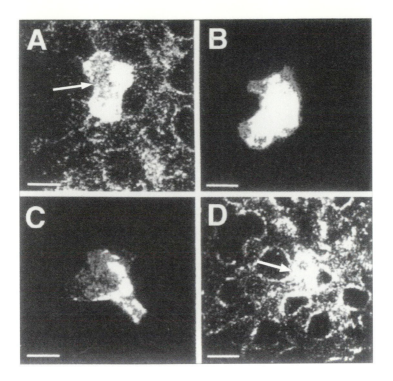

FIGURE 2.15

Cell-to-cell transport of fluorescent-tagged KN1 protein (KN1-FITC). KN1-FITC was microinjected into tobacco (*Nicotiana tabacum*) leaf mesophyll cell and the location of the fluorescent material was observed in real time using a sensitive fluorescent microscope. (A) Cell marked with white arrow was injected with KN1-FITC and observed immediately thereafter. Fluorescent-tagged protein moved into surrounding cells. (B) Cell was injected with FITC-labeled mutant form of KN1, which did not move out of the injected cell 15 minutes later. (C) Cell was injected with FITC-labeled 20 kD dextran, which remains in injected cell 60 minutes later. (D) Cell indicated with white arrow was coinjected with FITC-labeled 20 kD dextran and unlabeled KN1. FITC-labeled dextran has moved to surrounding cells 2 minutes later. Findings suggest that KN1 dilates cytoplasmic channels that interconnect cells permitting the movement of macromolecular dextran between cells. (Reprinted with permission from Lucas, W. J., et al. 1995. Selective trafficking of *KNOTTED1* homeodomain protein and its mRNA through plasmodesmata. Science 270:1981. Copyright 1995 American Association for the Advancement of Science.)

sectors fall into three classes based on the severity of their phenotypes: near wild type, intermediate, and extreme. The severity of the phenotype correlates with the layer in which the excision events have occurred. L1 chimeras are nearly wild type, L2 chimeras are intermedi-

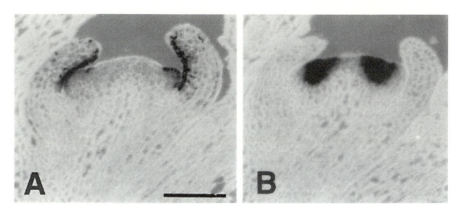

FIGURE 2.16

The effects of a regulatory gene expressed in one cell layer of a chimera on the activation of genes in other cell layers in the flowers of snapdragon (*Antirrhinum majus*). Genetic chimeras were formed in the unstable *floricaula*-613 mutant by excision of a Tam-3 transposable element activating the expression of the *flo* gene in a single cell layer. *Flo* expression is required for the expression of downstream floral homeotic genes, such as *deficiens* (*def*). *Flo* and *def* expression monitored by *in situ* hybridization with digoxigenin-labeled probes. (A) *Flo* expression is restricted to L1 in the floral bud at node 16 of L1 chimera. (B) *Def* expressed in L1, L2, and somewhat in L3 in the floral bud at node 16 of L1 chimera. (From Hantke, S. S., Carpenter, R., and Coen, C. S. 1995. Expression of *floricaula* in single cell layers of periclinal chimeras activates downstream homeotic genes in all layers of floral meristems. Development 121:33. Reprinted by permission of the Company of Biologists Ltd.)

ate, and L3 chimeras have an extreme phenotype. If excision occurred in the L2 layer, then an active *flo* gene was transmitted to progeny because the L2 layer gives rise to germ cells in the flower.

As will be discussed in Chapter 7, the expression of *flo* determines the expression of "downstream" homeotic genes that themselves determine the identity of floral organs. The authors therefore examined the expression of the floral organ identity genes, *deficiens* (*def*) and *plena* (*ple*) in sectored flowers (Hantke et al. 1995). They found that even though *flo* was expressed in one layer of the sector (Fig. 2.16A), *def* was expressed in all three layers (Fig. 2.16B). The timing of expression, however, was delayed, and the domains of expression were somewhat reduced in chimeras. Thus, the expression of *flo* in one of the cell layers signals the expression of downstream genes in all layers and leads to the production of a near normal flower. To do so must require the transmission of signals between cell layers. In this case, the signal is not known but is suspected to be the FLO protein itself.

REFERENCES

Carpenter, R. and Coen, E. S. 1995. Transposon induced chimeras show that *floricaula*, a meristem identity gene, acts non-autonomously between cell layers. Devel. 121: 19–26.

Christianson, M. L. 1986. Fate map of the organizing shoot apex in Gossypium. Am. J. Bot. 73: 947–958.

Dawe, R. K. and Freeling, M. 1990. Clonal analysis of the cell lineages in the male flower of maize. Devel. Biol. 142: 233–245.

Dawe, R. K. and Freeling, M. 1992. The role of initial cells in maize anther morphogenesis. Develop. 116: 1077–1085.

Dermen, H. 1953. Periclinal chimeras and origin of tissues in stem and leaf of peach. Am J. Bot. 40: 154–168.

Finnegan, E., Taylor, B. H., Craig, S., and Dennis, E. S. 1989. Transposable elements can be used to study cell lineages in transgenic plants. Plant Cell 1: 757–764.

Fleming, A. J., McQueen-Mason, S., Mandel, T., and Kuhlemeier, C. 1997. Induction of leaf primordia by the cell wall protein expansin. Science 276: 1415–1418.

Green, P. B., Stelle, C. S., and Rennich, S. C. 1996. Phyllotactic patterns: A biophysical mechansim for their origin. Ann. Bot. 77: 515–527.

Greene, B. A. and Hake, S. 1993. The *Knotted-1* mutants of maize: Investigating the circuitry of leaf development. Sem. Devel. Biol. 4: 41–49.

Hake, S. and Freeling, M. 1986. Analysis of genetic mosaics shows that the extra epidermal cell division in *Knotted* mutant maize plants are induced by adjacent mesophyll cells. Nature 320: 621–623.

Hake, S., Vollbrecht, E., and Freeling, M. 1989. Cloning *Knotted*, the dominant morphological mutant in maize using *Ds2* as a transposon tag. EMBO J. 8: 15–22.

Hantke, S. S., Carpenter, R., and Coen, C. S. 1995. Expression of *floricaula* in single cell layers of periclinal chimeras activates downstream homeotic genes in all layers of floral meristems. Devel. 121: 27–35.

Hernandez, L. and Green, P. B. 1993. Transductions for the expression of structural pattern: Analysis in sunflower. Plant Cell 5: 1725–1738.

Jackson, D., Veit, B., and Hake, S. 1994. Expression of maize *KNOTTED1* related homeobox genes in the shoot apical meristem predicts patterns of morphogenesis in the vegetative shoot. Development 120: 405–413.

Lambie, E. J. and Kimble, J. 1992. Genetic control of cell interactions in nematode development. Ann. Rev. Genetics 25: 411–436.

Lawrence, P. A. 1992. The making of a fly: The genetics of animal design. Oxford: Blackwell Scientific Publication.

Liu, C.-M., Xu, Z.-H., and Chua, N.-H. 1993. Auxin polar transport is essential for the establishment of bilateral symmetry during early plant embryogenesis. Plant Cell 5: 621–630.

Lucas, W. J., Bouche-Pillon, S., Jackson, D. P., Nguyen, L., Baker, L., Ding, B., and Hake, S. 1995. Selective trafficking of *KNOTTED1* homeodomain protein and its mRNA through plasmodesmata. Science 270: 1980–1983.

McQueen-Mason, S. J. and Cosgrove, D. J. 1995. Expansin mode of action on cell walls: Analysis of wall hydrolysis, stress relaxation, and binding. Plant Physiol. 107: 87–100.

Nusse, R. and Varmus, H. E. 1992. *Wnt* genes. Cell 69: 1073–1087.

Okada, K., Ueda, J., Komaki, M. K., Bell, C. J., and Shimura, Y. 1991. Requirement of the auxin polar transport system in early stages of Arabidopsis floral bud formation. Plant Cell 3: 677–684.

Peifer, M. 1995. Cell adhesion and signal transduction: The armadillo connection. Trends Cell Biol. 5: 224–229.

Poethig, R. S. 1987. Clonal analysis of cell lineage patterns in plant development. Am. J. Bot. 74: 581–594.

Poethig, R. S., Coe, E. H. J., and Johri, M. M. 1986. Cell lineage patterns in maize *Zea mays* embryogenesis: A clonal analysis. Dev. Biol. 117: 392–404.

Poethig, R. S. and Sussex, I. M. 1985. The cellular parameters of leaf development in tobacco *Nicotiana tabacum:* A clonal analysis. Planta 165: 170–184.

Satina, S., Blakeslee, A. F., and Avery, A. G. 1940. Demonstration of the three germ layers in the shoot apex of *Datura* by means of induced polyploidy in periclinal chimeras. Am. J. Bot. 27: 895–905.

Siegel, B. A. and Verbeke, J. A. 1989. Diffusible factors essential for epidermal cell redifferentiation in *Catharanthus roseus.* Science 244: 580–582.

Sinha, N. and Hake, S. 1990. Mutant characters of knotted maize leaves are determined in the innermost tissue layers. Devel. Biol. 141: 203–210.

Smith, L. G., Jackson, D., and Hake, S. 1995. Expression of *knotted1* marks shoot meristem formation during maize embryogenesis. Devel. Gen. 16: 344–348.

Spemann, H. and Mangold, H. 1924. Uber Inducktion von Embryonalanlagen durch Implantation artfremder Organisatoren. Arch Mikr. Anat. Entw. Mech. 100: 599–638.

Sturtevant, A. H. 1929. The claret mutant type of *Drosophila simulans:* A study of chromosome elimination and cell lineage. Z. Wiss. Zool. 135: 323–356.

Szymkowiak, E. J. and Sussex, I. M. 1992. The internal meristem layer L3 determines floral meristem size and carpal number in tomato periclinal chimeras. Plant Cell 4: 1089–1100.

Tanimoto, M., Roberts, K., and Dolan, L. 1995. Ethylene is a positive regulator of root hair development in *Arabidopsis thaliana.* Plant J. 8: 943–948.

Tilney-Bassett, R. A. E. 1986. Plant Chimeras. Baltimore: Edward Arnold.

van den Berg, C., Willemsen, V., Hage, W., Weisbeek, P., and Scheres, B. 1995. Cell fate in the Arabidopsis root meristem determined by directional signalling. Nature 378: 62–65.

Van Der Schoot, C., Dietrich, M. A., Storms, M., Verbeke, J. A., and Lucas, W. J. 1995. Establishment of a cell-to-cell communication pathway between separate carpels during gynoecium development. Planta 195: 450–455.

Vincent, J.-P. and Lawrence, P. A. 1994. Drosophila *wingless* sustains engrailed expression only in adjoining cells: evidence from mosaic embryos. Cell 77: 909–915.

Wolpert, L. 1981. Positional information and pattern formation. Phil. Trans. Royal Soc. Lond. 295 B: 441–450.

3

Embryogenesis

EARLY EVENTS IN EMBRYOGENESIS

Embryogenesis has been a focal point of research in animal development. In many animals, embryogenesis is the time when the basic features of the body plan unfold and organ systems are formed. Unlike in animals, development in plants is largely postembryonic. Adult organs are often not even present in the plant embryo. They are formed later in development from the shoot and root meristems. Plant embryos are simple, and are composed of two major organ systems – the axis and cotyledons (Fig. 3.1). The axis is made up of tissue that will give rise to structures of the seedling and the adult plant. In the mature embryo, the axis consists of the epicotyl, shoot apical meristem (SAM), hypocotyl, radicle, and the root apical meristem (RAM). (In plants, such as Arabidopsis, the epicotyl is not highly developed in the embryo and consists only of the SAM. In other plants, such as bean, the epicotyl or plumule is made up

FIGURE 3.1

Mature Arabidopsis embryo. Embryo is cleared and nuclei are stained with propidium iodide and visualized by confocal laser scanning microscopy. Shoot apical meristem (SAM) and root apical meristem (RAM) are regions densely populated with nuclei. Hypocotyl, hyp; cotyledon, cot. (From Clark, S. E., Running, M. P., and Meyerowitz, E. M. 1995. *CLAVATA3* is a specific regulator of shoot and floral meristem development affecting the same processes as *CLAVATA1*. Development 121:2059. Reprinted by permission of the Company of Biologists Ltd.)

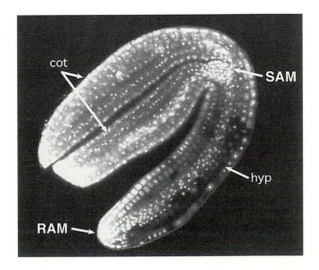

of the SAM and several foliage leaves.) Cotyledons are terminal structures that contain food reserves for the embryo and seedling. Embryogenesis involves not only the morphogenic events that give rise to the embryo, but also the events that prepare the embryo for desiccation, dormancy, and germination. This chapter deals largely with the morphogenic events in the embryo. Chapter 11 will focus on events leading to seed production and germination.

Embryo development begins with fertilization of the egg and zygote formation. (To simplify the discussion, embryo development in dicotyledonous angiosperms will be discussed with emphasis on the Capsella form of embryogenesis observed in Arabidopsis.) Fertilization in plants is a double fertilization process in which two sperm cells from the pollen (male gametophyte) fuse with two nuclei in the embryo sac (female gametophyte). One sperm cell nucleus fuses with the egg cell nucleus; the other fuses with the two polar nuclei in the central cell (see Chap. 10). The fertilized egg cell becomes a zygote and the central cell gives rise to the triploid endosperm. The egg arises from the cellularization of the egg nucleus in the megasporocyte. Plant egg cells are small (50–100 µM) compared, for example, with the eggs of invertebrate animals that do not develop in a maternal environment. Plant embryos are nourished by maternal sources, and food is stored in embryonic and extraembryonic structures, such as the endosperm.

The development of polarity is thought to be one of the earliest and, perhaps, the most fundamental steps in the establishment of the body plan. The polarity of the Arabidopsis embryo is determined quite early because the zygote begins to elongate along the presumptive embryo axis before the first division (Figs. 3.2 and 3.3). The embryo axis is aligned along the chalaza-micropyle axis of the ovule, suggesting that the maternal tissue influences the orientation of the axis (Laux and Jürgens 1997). In support of the idea that maternal tissue may influence the orientation of the embryo axis is the finding, to be discussed later, that a maternal mutation appears to affect development along the apical-basal axis in Arabidopsis embryos (Ray et al. 1996). Recall that maternal genes are key in setting up the anterior–posterior axis in Drosophila embryo development.

As the zygote elongates, cortical microtubules are laid down transversely at the end of the zygote distal to the micropyle (Webb and Gunning 1991). The first two divisions are transverse, first producing a filament of two, then four cells (Figs. 3.2 and 3.3). The first division itself is asymmetric, which gives rise to two cells of different sizes (i.e., a small apical cell and a larger basal cell) and different cell fates (Laux and Jürgens 1997). For example, the asymmetric division partitions the capacity

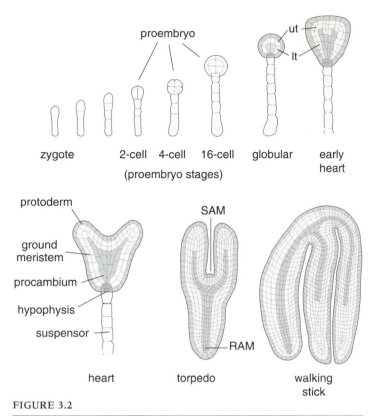

proembryo

ut

lt

zygote   2-cell   4-cell   16-cell   globular   early
heart

(proembryo stages)

protoderm

ground
meristem

procambium

hypophysis

suspensor

SAM

RAM

heart   torpedo   walking
stick

FIGURE 3.2

Diagrammatic representation of embryo development in Arabidopsis. Early stages are named according to the number of cells
in the proembryo; later stages by the shape of the embryo. At the
globular stage, transverse divisions along the apical-basal axis
divide the proembryo into an upper tier, ut, and lower tier, lt, of
cells. In addition, signs of tissue differentiation (histogenesis) are
evident at this stage, and cells distinctive to the protoderm, ground
meristem, and procambium (indicated by shading) are formed.
The hypophysis is derived from the uppermost cell of the suspensor. The root apical meristem (RAM) and shoot apical meristem
(SAM) become evident at the early and late heart stage, respectively; however, they are pointed out here at the torpedo stage.

to accumulate mRNA from the *ARABIDOPSIS THALIANA MERISTEM
LAYER1* (*ATML1*) gene into the smaller apical cell, but not into the larger
basal cell (Lu et al. 1996). The smaller distal cell will become the proembryo. (**Proembryo** is a term given to describe the embryo up to the
stage when the protoderm can be recognized, which is usually the globular stage.) The basal cell, which is attached to the maternal tissue,
develops into the suspensor, a structure that anchors the growing
embryo to maternal tissue.

At the four-cell proembryo stage, divisions in both longitudinal and
transverse planes give rise to the globular embryo (Figs. 3.2 and 3.3). No

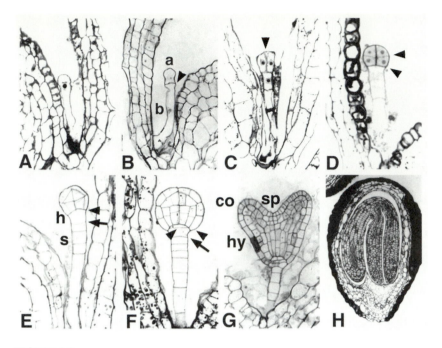

FIGURE 3.3

Arabidopsis embryos at various stages of development. (A) Zygote. (B) One-cell proembryo stage in which the single apical cell, a, is destined to become the proembryo and the basal cell, b, the suspensor. (C) Two-cell stage. (D) Eight-cell stage. Arrowheads point out the division of the embryo into upper and lower tiers. (E) Early globular stage. The uppermost cell derived from the basal cell is the hypophysis, h, and remaining cells form the suspensor, s. (F) Midglobular stage. (G) Heart stage. (H) Walking stick or mature embryo stage. Light microscope sections. (From Mayer, U., Buettner, G., and Jürgens, G. 1993. Apical-basal pattern formation in the Arabidopsis embryo studies on the role of the *gnom* gene. Development 117:155. Reprinted by permission of the Company of Biologists Ltd.)

organs form at the globular stage; however, there are clear signs of histogenesis or differentiation of embryonic cells. As discussed in the previous chapter, angiosperms are composed of three fundamental cell layers (L1–L3), and the presumptive layers can be distinguished at this stage: (1) protoderm cells (L1) that give rise to the epidermis; (2) ground meristem cells (L2) that accumulate storage proteins and oils during embryo development and which give rise in the embryo to the subepidermal and cortical parenchyma cells, and (3) procambium cells (L3) that are the progenitors to vascular tissue in the embryo. The cells in different layers can be recognized not only by their appearance, but by their distinctive patterns of cell division. For example, at the sixteen-cell proembryo stage, divisions are anticlinal in the protoderm, which is a pattern that is carried forward in the L1 through later stages in embryo development and in the developing plant. To signal the end of the globular stage, the embryo forms rudimentary organs: the cotyledons and axis system. Formation of organs gives shape and form to the globular embryo, and the subsequent stages describe the changing shape of the embryo: globular → heart → torpedo → walking stick (or bent cotyledon). Shape changes are largely associated with the emergence of the cotyledons.

The basal cells of the four-celled filamentous embryo form the suspensor (Figs. 3.2 and 3.3). The suspensor anchors the embryo to the maternal tissue and is thought to function in providing nutrition to the embryo. The suspensor, surprisingly, has no direct connection via plasmodesmata to the cells of the embryo; however, in certain species the suspensor cell adjacent to the proembryo often forms elaborate cell wall outgrowths characteristic of transfer cells.

A crucial event in embryo development is the formation of the shoot and root apical meristems (SAM and RAM, respectively). These structures are the postembryonic centers of plant development. Organization of the RAM can be seen near the beginning of the heart stage when the hypocotyl elongates and the cotyledons begin to form (Scheres et al. 1994). The RAM has a dual origin, derived from cells of the proembryo and from the upper cell of the suspensor (Fig 3.3). Formation of the SAM becomes apparent in histological sections at the late heart stage, later than when the RAM forms. The SAM forms in the crotch between the emerging cotyledons and is composed of small, cytoplasmically dense cells. New cell-type specific markers, however, have made it possible to identify SAM progenitors at even earlier stages in embryo development (Chap. 5). In the organization of the SAM, cells are recruited from the three fundamental cell layers, resulting in cell lineages in L1–L3 of the shoot that can be traced back to the early embryo.

COMPLEXITY OF GENE EXPRESSION IN THE EMBRYO

Plant embryos are morphologically simple, but molecularly complex. Surprisingly, an enormous number of genes are expressed in plant embryos. The complexity of RNAs (number of genes represented by RNAs) in plant embryos has been estimated through RNA driven RNA–DNA hybridization studies (Goldberg et al. 1989). Embryonic RNAs in *Nicotiana tabacum* (tobacco) are derived from about 20,000 genes, about the same number of genes that are expressed in highly differentiated plant organs, such as roots leaves, anthers, etc. (see Fig. 1.2). That is quite amazing because the situation is different from animals in which patterns of gene expression become more complex during development. In plants, the RNAs that are stage-specific are a modest fraction of the total; however, they represent a large number of genes. For example, in cotton, about 10 percent of the approximately 15,000 different mRNAs in midmaturation phase embryos are unique to that stage (Galau and Dure 1981).

Some of the genes expressed in a stage-specific manner during embryogenesis have been cloned as cDNAs by differential screening or by other methods. In general, most of the cDNAs represent genes that are highly expressed, such as the seed storage protein genes (Perez-Grau and Goldberg 1989). These genes show temporal as well as spatial patterns of gene expression during seed development. Further discussion about stage-specific gene expression in the embryo during seed development is found in Chapter 11. The stage-specific cDNAs have been useful as markers of cell differentiation. An example is a cDNA that represents the Kunitz trypsin inhibitor (Jofuku and Goldberg 1989). RNA accumulates in the ground meristem tissue at the micropylar pole of the globular embryo (where the sperm tube enters the embryo sac, see Chap. 9) before obvious cell differentiation occurs. The expression of the gene is one of the first molecular indicators of polarity in the transition from spherical globular embryo to a bilaterally symmetrical, heart-shaped embryo. Kunitz trypsin inhibitor RNA later accumulates at the outer surface of the cotyledons; at maturation, the mRNA accumulation pattern moves as a wave toward the inner surface of the cotyledons.

GENETICS OF EMBRYOGENESIS

As discussed in Chapter 1, the concept of pattern formation has been a dominant paradigm in embryonic development. In Drosophila and mammalian embryos, the basic features of the body plan, such as polarity of the embryo, are laid out early in development. The embryo is then divided into discrete modules or segments and the general properties of the segments, such as their individual polarity and relationship to each other, are determined. The principles of pattern formation in embryogenesis were discovered through the analysis of mutants in animal systems, and similar mutants have been sought in plants.

Two types of embryo mutants have been pursued in plants: embryo-lethal mutants and mutants that affect patterning in the embryo. The distinction between the two groups of mutants is important. Embryo-lethal mutants are blocked in any process necessary for embryo viability and development. Such mutants have been collected and analyzed without any prior assumptions about the underlying plan of plant development. On the other hand, pattern formation mutants have been identified that are consistent with various hypotheses about pattern formation in plant embryo development. The important genes in pattern formation are master regulators, at the top of the regulatory pyramid, that control fundamental aspects of the embryo body plan. The master regulators are few in number, and the mutations affecting master regulators should be rare among all developmental mutants. From the frequency at which embryo-lethal mutants appear in Arabidopsis, Jürgens (1991) estimated that 4000 genes are required for normal embryogenesis; however, only forty to fifty genes affect pattern formation in embryogenesis. A vast number of genes represented by embryo-lethal mutants may be required for normal embryogenesis, but they may not be master regulators of the process.

EMBRYO-LETHAL MUTANTS

Arabidopsis

Franzmann et al. (1995) identified and analyzed a large number of embryo-lethal mutants in Arabidopsis. They estimate that 500 genes (out of the 4,000 predicted by Jürgens) can be easily mutated to give an embryo lethal phenotype in Arabidopsis. The mutants were blocked at various stages of embryogenesis and were maintained as heterozygotes. Some embryo-lethal mutants have been useful in understanding development, but others have not. The question is whether these mutants have defects in important developmental regulators or in simple metabolic functions. A case in point is *biol* in Arabidopsis. This mutant was first recognized because it arrests at the early heart stage in embryogenesis. *biol*, however, was later shown to be a biotin auxotroph by the fact that the arrested embryo could be rescued by biotin. The explanation for the fact that the biotin-requiring embryo was arrested at a specific developmental stage is not clear. It could be that the embryo simply runs out of biotin at a certain developmental stage. On the other hand, the embryo may have special biotin requirements at the arrested stage. Whatever the reason, *biol* shows that caution must be exercised in interpreting the role of embryo-lethal mutations in development. If *biol* simply exhausts the supply of biotin, then the process that governs the arrest in the mutant may not be important as a specific regulator of development.

Another group of embryo-arrested mutants called *raspberry* mutants have been isolated by Yadegari et al. (1994). These mutants are blocked in the globular stage of embryo development and produce protuberances in the outer cell layers, giving the embryo a raspberry appearance (Fig. 3.4B). (The protuberances are actually not an abnormality of mutant embryos because wild type embryos produce similar protuberances at

61

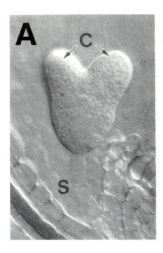 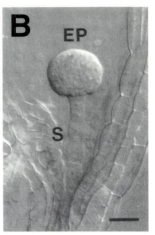

FIGURE 3.4

Embryo of *raspberry* (*rasp*) mutant in Arabidopsis. (A) Wild type and (B) *rasp* embryos at equivalent stages of development. Mutant embryos in selfed *rasp/+* Arabidopsis plants were compared with wild type embryos in the same seed pod or silique. Cotyledon, C; suspensor, S; proembryo or embryo proper, EP. Light microscopy with Nomarski optics. Bar = 26 μm. (From Yadegari, R., et al. 1994. Cell differentiation and morphogenesis are uncoupled in Arabidopsis *raspberry* embryos. Plant Cell 6:1719. Reprinted by permission of the American Society of Plant Physiologists.)

the late globular stage. In normal embryos, however, this stage is transient, and so the protuberances are often overlooked in the descriptions of embryogenesis.) *Raspberry* embryos are blocked at a stage before organ formation. Yadegari et al. (1994) raised the question whether histogenesis (cell-specific differentiation) occurs in the absence of organ formation. Although morphological differentiation of cells to form ground meristem and procambium cells was not observed in *raspberry* embryos, clear evidence of cell-specific expression was found using cell-specific probes. For example, an epidermal cell marker (a lipid transfer protein RNA probe) in wild type embryos, hybridized exclusively with the outer layer cells in the *raspberry* embryos. Probes for several storage protein RNAs were also examined. In wild type embryos, RNAs hybridizing to these probes accumulated starting at the late torpedo stage in epidermal cells and in ground meristem, but not in the procambium. In *raspberry* embryos, these same RNAs also accumulated in a similar temporal and spatial pattern (i.e., in the outer cell layers, but not in the inner cell layers, where procambium cells would ordinarily be found). The authors therefore concluded that the cells in *raspberry* embryos undergo normal cell differentiation (with normal timing) in the absence of organ formation. This

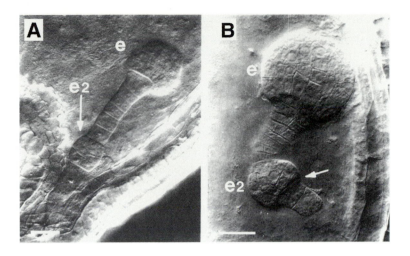

FIGURE 3.5

Twin mutant in Arabidopsis develops two embryos in tandem. (A, B) Proembryos, e, at the globular stage. Cells of the suspensor form an additional embryo, e2. (From Vernon, D. M., and Meinke, D. W. 1994. Embryogenic transformation of the suspensor in twin, a polyembryonic mutant of Arabidopsis. Develop. Biol. 165:571. Reprinted by permission of the author and Academic Press.)

finding addresses a classical issue in development – whether morphogenesis precedes histogenesis.

Given the intimate connection between the embryo and suspensor, it is interesting to ask what effects embryo-lethal mutations have on the development of the suspensor. The suspensor attaches the embryo to maternal tissue and is thought to play an important role early in embryogenesis in the transport of nutrients to the embryo and in providing growth regulators to the embryo (Yeung and Meinke 1993). The suspensor can be quite elaborate in some species, but it degenerates late in embryogenesis in most cases and is not present in the mature seed. In *raspberry* embryos, described earlier, the suspensor enlarges and expresses genes, in a cell layer–specific fashion, that are normally expressed only in embryos. Yeung and Meinke (1993) suggested that development of the suspensor may be negatively controlled by the embryo in Arabidopsis. When development of the embryo is blocked, the suspensor enlarges. Mutations have been found that compromise the repressive control of the embryo over the suspensor (Vernon and Meinke 1994). A mutant called *twin* produces two, or sometimes three embryos in tandem by transformation of the cells in the suspensor (Fig. 3.5). The embryos are viable, but they have a number of developmental defects. The formation of viable embryos from the suspensor plays havoc with pattern formation models in which the cell fates for the proembryo and suspensor are partitioned in the first division of the zygote. (Pattern formation will be discussed further in the next section.)

Maize

Embryo-lethal mutants have also been described in maize (*Zea mays*). Clark and Sheridan (1991) described two types of mutants: *dek* mutants (defective embryonic lethals), in which both embryo and endosperm development are inhibited and *emb* mutants in which only embryo development is blocked. They identified 51 *emb* mutants that were blocked at different stages of embryo development (Fig. 3.6). From this they argued that there are many temporally ordered, gene-regulated steps in embryonic development. Surprisingly, none of the mutants were early mutants, polarity mutants or mutants blocked at early division stages, suggesting that these early events may be maternally controlled. While all of the mutations were embryonic lethals, none seemed to have global effects on the entire embryo. Most appeared to affect specific organs, like the suspensor or the scutellum.

PATTERN MUTANTS

Jürgens and co-workers (1991) took a very different genetic approach to the analysis of embryogenesis in Arabidopsis by seeking pattern mutants rather than embryo-lethal mutants. They proposed a model for the body plan in the Arabidopsis embryo, and then identified mutants with defects in establishing or carrying out the basic features of the plan. Unlike embryo-lethal mutants, pattern formation mutants might not interfere with the progress of embryogenesis, but they might produce seedlings with recognizable phenotypes that reflect the defects in the embryo (Jürgens 1994). The authors theorized that the Arabidopsis embryo was assembled from two superimposed segmentation patterns: one along the apical-basal axis and another along the radial axis (Fig. 3.7). The apical-basal axis could be divided into five segments that give rise to discrete organs or structures in the seedling: the shoot apical meristem (SAM), cotyledons, hypocotyl, radicle, and root apical meristem (RAM) (Laux and Jürgens 1997). These segments roughly correlated with domains set out by transverse divisions in the early Arabidopsis embryo (Scheres et al. 1994). The first transverse division of the proembryo at the four-cell stage created an upper tier of cells and a lower tier (Fig. 3.2). The upper tier gives rise to most of the cotyledons and the SAM, while the lower tier gives rise to the basal part of the embryo, the basal part of the cotyledons, hypocotyl, and root. (More will be discussed about the domains in the early embryo in a later section on cell fate maps.)

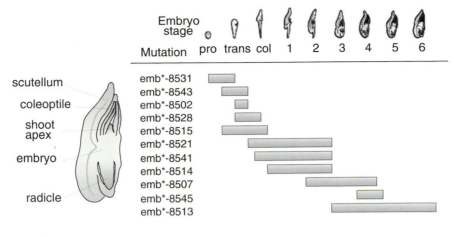

FIGURE 3.6

Embryo-lethal mutants in maize (*Zea mays*). Drawing of maize embryo and scutellum at stage 6 (left) and various developmental stages (above). Shaded bars indicate developmental stages at which various embryo-lethal (*emb*) mutants are blocked in development. (Redrawn from Clark, J. K., and Sheridan, W. F. 1991. Isolation and characterization of 51 embryo-specific mutations of maize. Plant Cell 3:939. Reprinted by permission of the American Society of Plant Physiologists.)

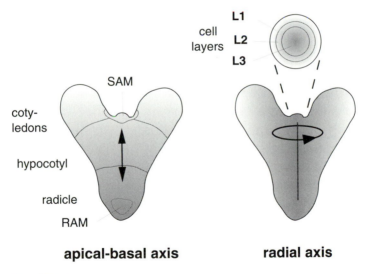

apical-basal axis   **radial axis**

FIGURE 3.7

Body plan and segmentation pattern proposed for Arabidopsis embryos. Segmentation along the apical–basal axis divides the embryo into shoot apical meristem (SAM), cotyledons, hypocotyl, radicle, and root apical meristem (RAM) regions. Segmentation across the radial axis divides the embryo into fundamental cell layers, L1–L3. (Based on Mayer, U., et al. 1991. Mutations affecting body organization in the Arabidopsis embryo. Nature 353:402–407.)

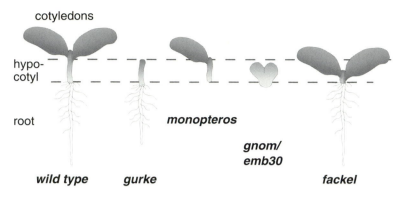

cotyledons

hypo-
cotyl

root **monopteros**

**gnom/
emb30**

wild type   **gurke**   **fackel**

FIGURE 3.8

Seedling phenotypes of some Arabidopsis embryonic pattern mutants. With the exception of *gnom/emb30,* mutants are interpreted to have deletions of various segments along the apical–basal axis of the embryo. *Gnom/emb30* is thought to have defects in establishing an apical-basal axis. Apical segments, shoot apex, and cotyledons are deleted in *gurke.* Basal segments, radicle, and hypocotyl do not develop in *monopteros.* Midsegment or hypocotyl fails to develop in *fackel.* (Based on Mayer, U., et al. 1991. Mutations affecting body organization in the Arabidopsis embryo. Nature 353:402–407. Adapted from Goldberg, R. B., de Paiva, G., and Yadegari, R. 1994. Plant embryogenesis: Zygote to seed. Science 266: 605–614.)

Several kinds of mutations that affect patterning in Arabidopsis embryo development were identified by Jürgens as well as others. One group of mutants was interpreted to represent polarity or apical-basal axis mutants; another group was thought to represent deletions of various segments along the apical-basal axis, giving rise to seedlings that were missing one or more organs or structures (Fig. 3.8). For example, the mutant *gurke* lacked the apical segment of the embryo and the mutant *fackel* was missing the middle region or hypocotyl. Along the radial axis, the embryo was thought to be segmented into cell layers, L1–L3, as discussed in Chapter 2. Pattern mutants along the radial axis might be missing a layer or a layer could be misidentified.

POLARITY OR APICAL–BASAL AXIS MUTANTS

Short Integument Mutants

As discussed in Chapter 1, the earliest features of the body plan in Drosophila embryo development, such as embryo polarity, are determined by maternal genes (rather than by zygotic genes). Very few maternal genes affecting embryo development have been found in plants, and because plants are able to develop from somatic embryos or from zygotes fertilized *in vitro* (both discussed in later sections), there is no reason to suspect that maternal genes play an essential pattern formation role in embryogenesis. Ray et al. (1996) argued nonetheless that as a maternal gene, *SHORT INTEGUMENT1* (*SIN1*) influences pattern formation in Arabidopsis embryos. *Sin1* mutants affect ovule formation and flowering time (see Chap. 9), but defects in embryos that develop in ovules of *sin1* mutants were also found. Defects ranged from embryos with funnel-shaped cotyledons to masses of unorganized tissue; however, most of the defects appeared to be abnormalities in development along the apical–basal axis. The phenotype of maternal effect mutations depends on the genotype of the maternal tissue, not on the zygotic tissue. In the case of the *sin1* mutants, a *sin1/sin1* or *sin1/+* embryo that was nursed by a *sin1/sin1* maternal sporophyte had morphogenic defects. A *sin1/sin1* embryo, however, was normal when nursed by a *sin1/+* sporophyte. It remains to be seen whether *SIN1* is really required for (or even contributes to) pattern formation in normal embryogenesis or whether the defect in the development of maternal tissues in the mutant ovule imposes some limitation on the development of the embryo.

Gnom/emb30 Mutants

Gnom/emb30 (*gn/emb30*) was considered to be a key pattern formation mutant because it is a zygotic mutation (due to the expression of the mutation in zygotic tissue and not in maternal tissue) that appears to affect one of the earliest steps in establishing the embryo body plan (Mayer et al. 1991). The mutant was originally described as a segment deletion mutant lacking the RAM. Through the analysis of additional alleles, however, it was argued that *gn/emb30* was defective in establishing apical–basal polarity of the embryo (Mayer et al. 1993). The mutant embryos were nearly round in more severe alleles, having failed to establish either apical or basal poles (Fig. 3.8). When *gn/emb30* was analyzed further, a defect was found very early in development, at the first division of the embryo. Recall that the first division is asymmetric producing a smaller, cytoplasmically dense apical cell that gives rise to the proembryo and a larger, vacuolated basal cell that forms the suspensor. In *gnom,* the first division was less asymmetric, producing two cells of nearly equal size. *GN/EMB30* therefore appeared to act very early in development.

The finding that the mutation appears to affect the same regulatory pathway and is **epistatic** to other segment deletion mutants such as *monopteros* was consistent with the observation that *gn/emb30* acts very early. (**Epistasis analysis** can be used to determine the order of genes on the same regulatory pathway and is performed by analyzing double mutants. The analysis requires that mutations in the different genes have distinguishable phenotypes based on where they block the pathway and are null mutants or are sufficiently severe so as to disrupt information flow on the pathway. A mutation is said to be epistatic to another if the mutation is in a gene that acts "upstream" to the other. Loss-of-function mutations in an "upstream" gene will mask the effects of mutations in downstream genes in a double mutant.) Even though *gnom* is epistatic to other embryo pattern mutants and appears to act very early in development, *gnom* is a zygotic, and not maternal, gene. This indicates that genes expressed in the zygote affect some of the first embryonic divisions.

GN/EMB30 has been cloned (Shevell et al. 1994) and, quite surprisingly, encodes a protein with sequence similarity to the *Gea2p* gene in yeast. *Gea2p* in yeast functions on the secretion pathway as a guanine nucleotide exchange factor involved in vesicle transport associated with the Golgi complex. In Arabidopsis, *GN/EMB30* affects the development of the embryo, but it is expressed throughout the plant in seedlings, root, leaf, stem, inflorescence, and flower. *Gea2p* is required for cell growth in yeast. It is possible that *GN/EMB30* serves a very basic role in plants; however, plant cells derived from homozygous *gnom* mutants survive and divide in cell culture (Shevell et al. 1994). *GN/EMB30,* therefore, is apparently not required for basic cell growth. Although the function of *GN/EMB30* is not yet known, it has been speculated that it is also involved in a secretion role in plants. A secretory function may be consistent with the effects of the *gnom* mutation on embryogenesis and might have a major impact on various aspects of plant development that results from its influence on cell shape, cell expansion, and so on.

An important issue is whether *GN/EMB30* is really a key regulator in pattern formation in Arabidopsis embryonic development or whether it simply functions to stabilize the apical–basal axis in the developing embryo. The discovery that the *GN/EMB30* does not encode a gene product with obvious master gene regulatory functions has raised some doubts about the importance of *GN/EMB30* to the control of Arabidopsis embryogenesis. The concern is whether *GN/EMB30* is a "housekeeping gene" in the pattern formation process or whether it is a master regulator. It may be that our notion of what constitutes a master regulator in a pattern formation process is too narrow-minded and that *GN/EMB30* may be a regulator of other processes that generate patterns in ways that we presently do not understand. It is a still premature, however, to judge the relevance of *GN/EMB30* to pattern formation until we know more about the function of the gene product.

SEGMENT DELETION MUTANTS

Monopteros Mutants

Monopteros (*mp*) is a segment deletion mutant that lacks the basal region of the embryo, the hypocotyl and RAM (Fig. 3.8) (Mayer et al. 1991). The name of the mutant is derived from the fact that the cotyledons are variously arranged and more severe alleles have a single cotyledon. The *mp* phenotype appears early in development, although somewhat later than *gnom*. In wild type embryos, the proembryo is composed of two tiers of cells at the eight-cell stage (Fig. 3.9A). The upper tier gives rise to the epicotyl and cotyledons. The hypocotyl, radicle, and root arise from the lower tier. The uppermost cell of the suspensor becomes the hypophysis that, along with cells of the lower tier, gives rise to the essential elements of the RAM. During the later globular stages of development, cells of the upper tier remain isodiametric and do not divide in any preferential direction. Cells of the lower tier, however, elongate along the axis of polarity and divide to produce precisely ordered cell files.

In *mp* embryos, all cells in the proembryo behave as upper tier cells (Fig. 3.9 D–F) (Berleth and Jürgens 1993). The lower tier cells remain isodiametric and do not form linear cell files. In addition, the uppermost cell of the suspensor fails to form the hypophysis. A stack of somewhat flattened cells accumulates at the top of suspensor instead. The failure of the hypophysis to differentiate may not be a primary defect in the mutant. The primary defect appears to be in the elongation and/or divisions in the lower tier of proembryo cells. The defect in hypophysis formation may be the consequence of a failure to produce a signal from the lower tier of cells.

Why are the cotyledons misplaced in *mp*? Mayer et al. (1991) argued that this, too, may be a secondary consequence of developmental defects in the basal region. The cotyledon leaves in Arabidopsis normally emerge directly opposite each other. Mayer et al. (1991) speculated that the basal region is needed to establish bilateral symmetry reflected in the positioning of the cotyledons. The cotyledons show less positional symmetry in more severe alleles (i.e., those with greater truncation of the basal region). There is also the possibility that the developing basal region may exercise control over the production of hormones or the establishment of hormone gradients. Interesting obser-

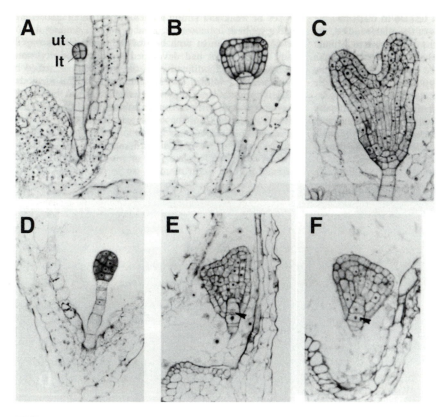

FIGURE 3.9

Arabidopsis embryos in wild type (A–C) and *monopteros* (*mp*[T313b]) embryos (D–F). (A) Eight-cell stage in which embryo has been divided into an upper tier (ut) and lower tier (lt) of cells; (B) early heart stage in which primordial cell layers become evident in the proembryo; (C) late heart stage. (D) Globular stage in *mp* embryo; (E, F) early heart stage. Note the lack of development of the lower part of the embryo in the *mp* mutant, and the piling up of flattened cells (arrowheads) above the suspensor. Light microscope sections stained with toluidine blue. (From Berleth, T., and Jürgens, G. 1993. The role of the *monopteros* gene in organising the basal body region of the Arabidopsis embryo. Development 118:583. Reprinted by permission of the Company of Biologists Ltd.)

vations on the role of hormone gradients in determining the position of cotyledon emergence are described later.

An intriguing question was whether *mp* mutants, which are defective in basal region formation, had lost the ability to generate roots. Berleth and Jürgens (1993) unexpectedly found that the mutants still retained the capacity to form adventitious roots. To demonstrate this, they cut off the basal part of the hypocotyl and placed the embryos on auxin-containing medium. Roots emerged in most of the embryos with weak *mp* alleles; however, the efficiency of root formation declined with allele strength. Although *mp* mutants were unable to organize the basal region of the embryo from which roots ordinarily develop, they did not lose the capacity to produce roots.

Gurke Mutants

In *gurke* (*gk*), one of the segment deletion mutants mentioned earlier, defects were found in the development of the apical end of the embryo (Fig. 3.8). Alleles representing a continuum of phenotypes all affected the development of the apical region, and the apical end and some of the central region of the embryo failed to develop at all in the most severe alleles (Torres-Ruiz et al. 1996). The defect was traced back to the early heart stage when the embryo undergoes a transition from radial to bilateral symmetry (Figs. 3.2 and 3.3). The transition occurs when anticlinal divisions in the upper tier of cells flatten the top of the embryo. Mutants were normal up to the globular stage, but the divisions required for the development of cotyledon primordia were misoriented or delayed in the mutant. Cotyledons were not formed in severe *gk* alleles. Cotyledons are derived from both the apical domain of the embryo and the central domain (Scheres et al. 1994); therefore, *GK* may control the formation of the central as well as the apical domain. Only in weaker alleles that formed rudimentary cotyledons was a distinctive group of cells characteristic of the SAM formed. Thus, *gk* affects the development at the apical (and possibly central) domains of the embryo, including the formation of cotyledon primordia and the development of the SAM region.

A number of other segmentation mutants are less well described (Mayer et al. 1991). For instance, another type of pattern mutant is a deletion associated with a duplication, such as *doppelwurzel*. In this mutant, the apical meristem is deleted and replaced with a mirror image of basal meristem. There are other types of pattern mutants, such as those in which there is a duplication of cotyledons or a transformation of cotyledons into shoots. Pattern formation mutants that affect the shape, but not the identity of organs, may also be informative.

RADIAL AXIS MUTANTS

Knolle Mutants

Few mutants have been described with defects in the radial axis (i.e., with defects in the formation of one of the three tissue layers – protoderm, procambium, or ground meristem). One mutant thought to be in this category is *knolle* (*kn*), which appears to lack an epidermis or L1 (Mayer et al. 1991). In *kn*, the defect is evident at the early globular stage. With regard to the role of *KN* in radial pattern formation in embryo development, the question is whether *kn* forms an epidermis at all or whether the epidermis is simply not recognizable. As discussed earlier, one of the distinguishing marks of the epidermis is the tangential or anticlinal divisions that allow this layer to grow as a sheet. Divisions in the outer cell layer in *kn* are not uniform, disrupting this cell layer and the layers below. To address the question whether *kn* has an epidermis, Lukowitz et al. (1996) examined whether an epidermal marker in *fusca* (*fus*) mutants could be expressed in *kn* embryos. *Fus* mutants (described further in Chap. 4) accumulate large quantities of the pigment anthocyanin in the epidermis of the embryo. The analysis of double *kn fus* mutants showed that anthocyanin was accumulated in some of the outer cells. This suggested that *kn,* indeed, has epidermal cells; however, the cell layer is disrupted by misoriented and abnormal divisions.

If *kn* has an epidermis, albeit disrupted, then is *kn* a pattern formation mutant? Certainly *kn* mutations cause abnormalities in embryonic development, but is *KN* a regulator of other genes and a controller of pattern formation? *KN* was cloned and found to encode a product similar to syntaxins (Lukowitz et al. 1996). Syntaxins are involved in secretion and are thought to be plasma membrane targets for secretory vesicles. At a molecular level, therefore, *KN* does not fulfill our expectations as a master regulator of genes in a pattern formation process. Like *GN*, *KN* could be involved in secretion, which indicates that secretion is very important in early steps of embryonic development. It should be cautioned again, however, that the actual function of *KN* is not yet known.

SUMMARY ABOUT PATTERN FORMATION IN EMBRYOGENESIS

In summary, embryo development in Arabidopsis is clearly a pattern formation process. One can see patterns emerge as the embryo develops. Cell layers appear and regions of the embryo begin to specialize according to their presumptive fate. It is not yet clear, however, that the present set of mutants represent the key regulators in the pattern formation process. Jürgens (1994) argued that interesting pattern formation mutants might survive to the seedling stage, and that they could be recognized by distinctive seedling phenotypes. It could be that the interesting pattern formation mutants are embryo lethal and buried among the hundreds of embryo-lethal mutants that are so difficult to sort out on the basis of phenotype.

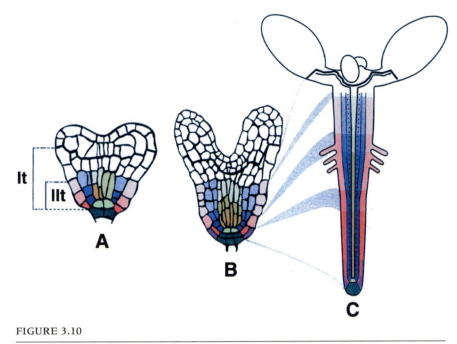

FIGURE 3.10

Fate map of the Arabidopsis embryo focusing on root and hypocotyl regions. Lower tier, lt, of cells is divided by transverse divisions at an early heart-shaped stage (A) into an upper lower tier (ult) and a lower lower tier (llt). (B) Late heart-shaped stage and (C) seedling. (From Scheres, B., et al. 1994. Embryonic origin of the Arabidopsis primary root and root meristem initials. Development 120:2486. Reprinted by permission of the Company of Biologists Ltd.)

CELL FATE MAPS IN EMBRYO DEVELOPMENT

As discussed in Chapter 2, cell division patterns in the maize embryo are very irregular, yet a predictable embryo with well-defined embryonic organs is formed during embryogenesis. In addition, Poethig et al. (1986) demonstrated that embryo cell fate maps can be constructed and overlaid on embryos despite their irregular cell organization and patterns of cell division. Organizational principles other than regular patterns of cell division obviously come to play in shaping the maize embryo.

As described in the previous sections, the organization of cells and the pattern of cell divisions are much more regular in Arabidopsis. Scheres et al. (1994) followed the fate of cells in heart-shaped embryos in Arabidopsis, focusing on root and hypocotyl cell lineages (Fig. 3.10). Whether or not roots and hypocotyls were developmental compartments with well-defined boundaries was at issue. To carry out their cell lineage analysis, Scheres et al. used a transgene marker consisting of a CaMV 35S promoter-GUS construct interrupted by an *Ac* transposable element (Fig. 2.1B). The *Ac* element infrequently excises from the

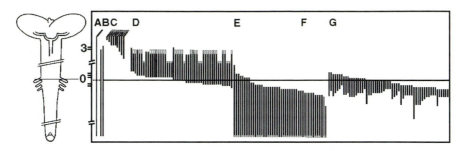

FIGURE 3.11

Representation of sector endpoints in a population of Arabidopsis seedlings. Sectors are generated by excision of the *Ac* transposon from a 35S:*Ac*:GUS construct introduced into transgenic plants. Black-ended bars indicate instances where the endpoints of sectors were exactly determined. White-ended bar indicates sectors in which endpoints were estimated. Sectors are classified in several different categories (A–G). (A) Sector that demonstrates lower tier/upper tier division of embryo. Sectors represent (B) root and hypocotyl, (C) cotyledon shoulders, (D) hypocotyl, (E) root with apical ends of sectors overlapping basal ends of hypocotyl sectors, (F) root with sectors that do not overlap basal ends of hypocotyl sectors, and (G) intermediate zone. From Scheres, B., et al. 1994. Embryonic origin of the Arabidopsis primary root and root meristem initials. Development 120:2483. Reprinted by permission of the Company of Biologists Ltd.)

35S:*Ac*:GUS construct, activating the expression of the GUS marker in the cells where the event occurs and in any progeny cells.

Scheres et al. (1994) proposed that the four tiers of ground meristem cells at the heart stage of development give rise to the bulk of the root and hypocotyl (Fig. 3.10). During the early globular stage, the embryo is divided into an upper tier and lower tier of cells, which are derivatives of the first transverse division of the initial proembryo cell. The lower tier is further subdivided by transverse divisions into an upper lower tier (ult) and a lower lower tier (llt). The llt tier is two cells high and consists in cross-section of an inner core of procambium cells surrounded by a layer of ground meristem and protoderm cells. At the late heart stage, the llt is further subdivided into four tiers, and the most basal protoderm cell becomes a RAM initial recognized by periclinal divisions giving rise to the lateral root cap.

Scheres et al. (1994) found that the hypocotyl, intermediate zone, root, and RAM could be traced back to the four tiers of llt cells. They found that the largest embryonic sectors spanned the basal region of the cotyledons where they join the axis, through the hypocotyl, the root, and to the root cap columella (Fig. 3.11). The authors proposed that this sector arose from the first division within the embryo demarcating the upper and lower halves of the embryo. The next largest sectors in the lower part of the embryo spanned the root and hypocotyl and were interpreted to represent the separation of cell lines that occurs in the divisions that separate the ult from the llt cells. The next smaller sectors defined four sections of the root-hypocotyl region (Figs. 3.10 and 3.11). This pattern was consistent with the proposition that these sectors arise from the four tiers of cells in the basal half of the heart-shaped embryo.

A critical observation in the study by Scheres et al. was that the sector boundaries were not as sharp as the morphological boundaries (Fig. 3.11). They only approximated the organ or zone boundaries, indicating that cell lineages correlated generally with body plan patterns, but not exactly. Even though division patterns in the formation of the embryo and root were highly regular (see Chap. 12), these patterns did not mark the precise boundaries of developmental compartments. It appears in Arabidopsis, as in maize, that there are other forces at work beyond cell lineages that define the boundaries of organs.

IN VITRO FERTILIZATION AND DEVELOPMENT OF PLANT EMBRYOS

Aside from the requirement for nutrients, embryos are quite independent of their maternal hosts and become more so as they mature. Embryos can be explanted at various stages of development and cultured in tissue culture. In fact, it is common practice to rescue embryos in crosses where certain incompatibilities block normal embryo development. In most cases, younger embryos are more difficult to culture and have more stringent medium requirements. Embryos acquire greater autonomy as they mature, and they can usually survive in simple, but rich, medium. As mentioned earlier, the lack of maternal mutations affecting embryo development is also an indicator of the independence of embryo development.

Kranz and Lörz (1993) made a significant breakthrough in fertilizing egg cells from maize *in vitro* and regenerating the eggs into fertile plants. The feat was accomplished through the development of techniques to isolate egg protoplasts. Single egg protoplasts, manually dissected from ovule tissue, were mixed with sperm nuclei released from ruptured pollen grains. The nuclei were introduced into egg protoplasts by electrofusion (Fig. 3.12). The fused cells proceeded to divide and in many cases formed embryo-like structures from which a coleoptile developed. It was reported that development did not require exogenous plant hormones, but did require "feeder" cells. The fact that maize eggs can be fertilized and develop *in vitro* argues that maternal signals and the physical envi-

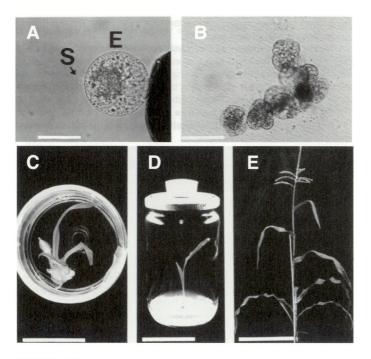

FIGURE 3.12

Fusion of maize egg and sperm cell protoplasts *in vitro,* and development of zygotes into fertile maize plants. (A) Maize egg, E, and sperm protoplast, S, on electrode prior to electrofusion. Bar = 50 μM. (B) Multicellular clusters 5 days after fusion. Bar = 100 μM. (C) Plantlet 35 days after fusion. Bar = 2 cm. (D) Plant 39 days after fusion. Bar = 6 cm. (E) Flowering plant 99 days after fusion. Bar = 50 cm. (From Kranz, E., and Lörz, H. 1993. *In vitro* fertilization with isolated, single gametes results in zygotic embryogenesis and fertile maize plants. Plant Cell 5:741. Reprinted by permission of the American Society of Plant Physiologists.)

ronment of the gametophyte are not required for embryo development. These experiments, however, do not exclude the possible role of maternal effect genes because development was initiated with an egg that undoubtedly had already been subjected to maternal influences.

SOMATIC EMBRYOGENESIS

A unique property of plants (discussed in Chap. 1) is their capacity to reproduce vegetatively and to initiate embryonic development from somatic tissues. The process is called **somatic** or **asexual embryogenesis** and involves the production of embryos from nonzygotic tissue (Fig. 1.6) (for review, see Zimmerman 1993). Somatic embryos are similar to zygotic embryos, except that somatic embryos do not form suspensors. Somatic embryos, however, appear to pass through similar developmental stages as their zygotic counterparts (i.e., globular, heart, torpedo stages). In some plants, somatic embryos arise on foliar surfaces, such as in the orchid, *Malaxis paludosa.* Somatic embryogenesis can also be induced in tissue culture and has the promise of providing developmental biologists with easy access to large quantities of plant embryos. Embryogenesis has always been difficult to study in plants because embryos are small and inaccessible in maternal tissues.

Somatic embryogenesis is an important form of plant propagation; therefore, efforts to induce embryogenesis have been described for almost all economic plants. Few generalizations arise out of a comparison of the conditions required for somatic embryogenesis other than that the process can usually be induced in a variety of plant species by tissue culture manipulations. These usually involve changing media, establishing different regimes of hormone types or concentrations, and/or controlling other culture conditions, such as cell density, nutrients, or illumination. The remarkable feature of somatic embryogenesis is that these simple manipulations can usually set a dramatic series of events in motion that give rise to the embryo. Plants appear to be loaded with a preset embryogenic program that can be triggered by specific conditions. Zygotic embryogenesis is set off by fertilization, and in tissue culture the process is initiated by other means.

The best-studied system for somatic embryogenesis is carrot (*Daucus carota*) cell suspension cultures. When the auxin, 2,4-dichlorophenoxyacetic acid (2,4-D), is withdrawn from carrot suspension cultures, cell clumps undergo somatic embryogenesis and form embryos (Fig. 3.13). Under the best conditions, millions of embryos in a suspension culture can be obtained. With the availability of so many embryos at similar stages in development, one might imagine that cell suspension cultures would be the system of choice for studying embryo development. The somatic embryo system, however, has presented some unexpected complications.

The difficulties can be represented by the early studies of Sung and Okimoto (1981), who used two-dimensional gel electrophoresis analysis to look for changing patterns of gene expression during somatic embryogenesis in carrot suspension culture cells. They unexpectedly found very few differences in protein patterns between somatic embryos and proliferating cells. These findings contradicted the conclusions described earlier in this chapter from the studies that showed that ~1,000 genes are uniquely expressed at different stages of development in zygotic embryos. There are several explanations for this inconsis-

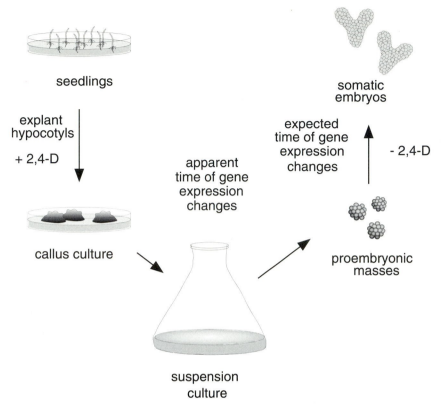

seedlings

explant
hypocotyls

+ 2,4-D

callus culture

apparent
time of gene
expression
changes

suspension
culture

somatic
embryos

expected
time of gene
expression
changes

- 2,4-D

proembryonic
masses

FIGURE 3.13

Somatic embryogenesis in carrot (*Daucus carota*) cultures. Hypocotyls are explanted from seedlings and placed on callus induction medium containing the auxin, 2,4-D. Calli are dispersed and grown as cell clusters in suspension culture. Continual cell culture leads to the formation of proembryonic masses, PEMs, which can be induced to form somatic embryos by withdrawing 2,4-D. Most gene expression changes appear to occur in the formation of the PEMs rather than at the time when somatic embryos are induced.

tency. One is that the gene products from genes involved in embryo development might not be abundant and not easily detected in two-dimensional protein gel electrophoresis or in cDNA libraries. Another possible explanation is that the embryos may not be as synchronized in development as they appear, particularly in later stages, and that differences at various stages may be dampened by asynchrony. Yet another possible explanation is that gene constellations involved in somatic embryogenesis may already be activated in proliferating cells in culture. There is some information on the latter point.

Embryogenic carrot cell suspension cultures are produced by explanting sections of seedling hypocotyls into culture medium containing 2,4-D (Fig. 3.13). 2,4-D stimulates provascular cells to divide and form clumps of small proliferating isodiametric cells (Guzzo et al. 1994). After further culture, cell clumps competent to undergo embryogenesis are called proembryonic masses (PEMs) (de Vries et al. 1988). Wilde et al. (1988) examined the changes in gene expression between PEMs and embryos. They found few differences in the two-dimensional gel electrophoresis patterns of translation products from RNAs isolated from PEMs and torpedo-shaped embryos. It was concluded from this that many of the genes expressed in the embryo were already active in PEMs. If that is the case, then many of the interesting molecular events that normally accompany embryo formation may have taken place earlier in the cell culture, rather than at the time when embryogenesis is hormonally induced.

Others have examined earlier stages in the establishment of cell cultures in the effort to identify markers for the development of embryogenic cells. Schmidt et al. (1997) used a sensitive technique, called **differential display** (Liang and Pardee 1992), to look for genes that might be differentially expressed in "competent cells" (i.e., cells that had undergone a transition between somatic and embryogenic states). They enriched the cell population for embryogenic cells and identified by differential display a cDNA representing a gene that was differentially expressed in competent cells. The sequence of the cDNA (called *SERK*) predicted that it would encode a leucine-rich repeat containing receptorlike protein kinase. (Receptorlike protein kinases are discussed with respect to the *CLAVATA* gene in Chap. 5 and with respect to the *SRK* gene in Chap. 10). The expression of the gene was correlated with the appearance of competent cells in primary cultures derived from hypocotyl explants. In established cell suspension lines, the *SERK* promoter was linked to the firefly luciferase gene so that cell clumps expressing *SERK* could be identified. By tracking the development of the cells with videotape, Schmidt et al. (1997) found that *SERK* was expressed in most of the cell clusters giving rise to embryos, and that *SERK* expression continued until the globular stage of embryonic development.

Other investigators have examined the developmental events during somatic embryogenesis once the process has been induced (by withdrawal of 2,4-D). Temperature-sensitive cell lines have been isolated that are defective in their ability to progress through embryogenesis at restrictive temperature, and one of these lines, *ts11*, produces misshapen heart stage embryos (Giuliano et al. 1984). It was found, however, that the defect in *ts11* could be overcome by the addition of conditioned media from wild type cells. In the effort to find the factor(s) that restore development, De Jong et al. (1992) fractionated conditioned medium.

In doing so, they found an acidic endochitinase that would restore the developmental capacity of the cells at restrictive temperature, although not efficiently. They speculated that the endochitinase may hydrolyze a substrate in plant cell walls, such as an oligomer N-acetylglucosamine, and release a compound that is required for somatic embryogenesis. Such a substrate has not yet been identified; however, they found that a lipo-oligosaccharide from *Rhizobium,* NodRlv-V(Ac, C18:4) was capable of rescuing somatic embryogenesis in *ts11*. The search for a natural substrate in plant cell walls is underway.

The finding that other factors, such oligosaccharides, may be involved in somatic embryogenesis is important to understand. The program of events that transpires during somatic embryogenesis is too elaborate to be regulated by a single hormone (in this case the withdrawal of single hormone, auxin, or 2,4-D). The perturbation caused by the withdrawal of the hormone must set a chain of events in motion that involves the action of more specific growth factors and biophysical forces that confer both positional and developmental information. A challenge in the future is to understand the hormone effects, to identify the further steps in the process and to discover the factors and forces that regulate them.

REFERENCES

Berleth, T. and Jürgens, G. 1993. The role of the *monopteros* gene in organising the basal body region of the Arabidopsis embryo. Development 118: 575–587.

Clark, J. K. and Sheridan, W. F. 1991. Isolation and characterization of 51 embryo-specific mutations of maize. Plant Cell 3: 935–952.

Clark, S. E., Running, M. P., and Meyerowitz, E. M. 1995. *CLAVATA3* is a specific regulator of shoot and floral meristem development affecting the same processes as *CLAVATA1*. Development 121: 2057–2067.

De Jong, A. J., Cordewener, J., Lo Schiavo, F., Terzi, M., Vandekerckhove, J., Van Kammen, A., and De Vries, S. C. 1992. A carrot somatic embryo mutant is rescued by chitinase. Plant Cell 4: 425–433.

de Vries, S. C., Booij, H., Meyerink, P., Huisman, G., Wilde, H. D., Thomas, T. L., and van Kammen, A. 1988. Acquistion of embryonic potential in carrot cell-suspension cultures. Planta 176: 196–204.

Franzmann, L. H., Yoon, E. S., and Meinke, D. W. 1995. Saturating the genetic map of Arabidopsis thaliana with embryonic mutations. Plant Journal 7: 341–350.

Galau, G. A. and Dure, L. 1981. Developmental biochemistry of cottonseed embryogenesis and germination: Changing messenger RNA populations as shown by reciprocal heterologous complementary DNA/mRNA hybridization. Biochemistry 20: 4169–4178.

Giuliano, G., LoSchiavo, F., and Terzi, M. 1984. Isolation and developmental characterization of temperature-sensitive carrot cell variants. Theor. Appl. Genet. 67: 179–183.

Goldberg, R. B., Barker, S. J., and Perez-Grau, L. 1989. Regulation of gene expression during plant embryogenesis. Cell 56: 149–160.

Goldberg, R. B., de Paiva, G., and Yadegari, R. 1994. Plant embryogenesis: Zygote to seed. Science 266: 605–614.

Guzzo, F., Baldan, B., Mariani, P., Lo Schiavo, F., and Terzi, M. 1994. Studies on the origin of totipotent cells in explants of *Daucus carota* L. J. Exp. Bot. 45: 1427–1432.

Jofuku, K. D. and Goldberg, R. B. 1989. Kunitz trypsin inhibitor genes are differentially expressed during the soybean life cycle and in transformed tobacco plants. Plant Cell 1: 1079–1094.

Jürgens, G. 1994. Pattern formation in the embryo. In Arabidopsis, ed. Meyerowitz, E. M. and Somerville, C. R., 297–312. Cold Spring Harbor: Cold Spring Harbor Press.

Jürgens, G., Mayer, U., Ruiz, R. A. T., Berleth, T., and Misera, S. 1991. Genetic analysis of pattern formation in the Arabidopsis embryo. Development 27.

Kranz, E. and Lörz, H. 1993. *In vitro* fertilization with isolated, single gametes results in zygotic embryogenesis and fertile maize plants. Plant Cell 5: 739–746.

Laux, T. and Jürgens, G. 1997. Embryogenesis: A new start into life. Plant Cell 9: 989–1000.

Liang, P. and Pardee, A. B. 1992. Differential display of eukaryotic messenger RNA by means of the polymerase chain reaction. Science 257: 967–971.

Lu, P., Porat, R., Nadeau, J. A., and O'Neill, S. D. 1996. Identification of a meristem L1 layer-specific gene in Arabidopsis that is expressed during embryonic pattern formation and defines a new class of homeobox genes. Plant Cell 8: 2155–2168.

Lukowitz, W., Mayer, U., and Jürgens, G. 1996. Cytokinesis in the Arabidopsis embryo involves the syntaxin-related *KNOLLE* gene product. Cell 84: 61–71.

Mayer, U., Buettner, G., and Jürgens, G. 1993. Apical-basal pattern formation in the Arabidopsis embryo studies on the role of the *gnom* gene. Development 117: 149–162.

Mayer, U., Ruiz, R. A. T., Berleth, T., Misera, S., and Jürgens, G. 1991. Mutations affecting body organization in the Arabidopsis embryo. Nature 353: 402–407.

Perez-Grau, L. and Goldberg, R. B. 1989. Soybean seed protein genes are regulated spatially during embryogenesis. Plant Cell 1: 1095–1110.

Poethig, R. S., Coe, E. H. J., and Johri, M. M. 1986. Cell lineage patterns in maize *Zea mays* embryogenesis a clonal analysis. Dev. Biol. 117: 392–404.

Ray, S., Golden, T., and Ray, A. 1996. Maternal effects of the *short integument* mutation on embryo development in Arabidopsis. Devel. Biol. 180: 365–369.

Scheres, B., Wolkenfelt, H., Willemsen, V., Terlouw, M., Lawson, E., Dean, C., and Weisbeek, P. 1994. Embryonic origin of the Arabidopsis primary root and root meristem initials. Development 120: 2475–2487.

Schmidt, E. D. L., Guzzo, F., Toonen, M. A. J., and de Vries, S. C. 1997. A leucine-rich repeat containing receptor-like kinase marks somatic plant cells competent to form embryos. Development 124: 2049–2062.

Shevell, D. E., Leu, W. M., Gillmor, C. S., Xia, G., Feldmann, K. A., and Chua, N. H. 1994. *EMB30* is essential for normal cell division, cell expansion, and cell adhesion in Arabidopsis and encodes a protein that has similarity to Sec7. Cell 77: 1051–1062.

Sung, Z. R. and Okimoto, R. 1981. Embryonic proteins in somatic embryos of carrot. Proc. Natl. Acad. Sci. U.S.A. 78: 3683–3687.

Torres-Ruiz, R. A., Lohner, A., and Jürgens, G. 1996. The *GURKE* gene is required for normal organization of the apical region in the Arabidopsis embryo. Plant J. 10: 1005–1016.

Vernon, D. M. and Meinke, D. W. 1994. Embryogenic transformation of the suspensor in twin, a polyembryonic mutant of Arabidopsis. Dev. Biol. 165: 566–573.

Webb, M. C. and Gunning, B. E. S. 1991. The microtubular cytoskeleton during development of the zygote, proembryo and free-nuclear endosperm in *Arabidopsis thaliana (L) Heynh.* Planta 184: 187–195.

Wilde, H. D., Nelson, W. S., Booij, H., de Vries, S. C., and Thomas, T. L. 1988. Gene expression programs in embryonic and nonembryonic carrot cultures. Planta 176: 205–211.

Yadegari, R., de Paiva, G. R., Laux, T., Koltunow, A. M., Apunya, N., Zimmerman, J. L., Fischer, R. L., Harada, J. J., and Goldberg, R. B. 1994. Cell differentiation and morphogenesis are uncoupled in Arabidopsis *raspberry* embryos. Plant Cell 6: 1713–1729.

Yeung, E. C. and Meinke, D. W. 1993. Embryogensis in angiosperms: Development of the suspensor. Plant Cell 5: 1371–1381.

Zimmerman, J. L. 1993. Somatic embryogenesis: A model for early development in higher plants. Plant Cell 5: 1411–1423.

4

Seedling Development

No stage in the life of a plant is more vulnerable than the seedling. The tender shoot of the seedling must burrow up through the soil and emerge, only to be buffeted by the storms of life above ground (Fig. 4.1). For survival, the seed must sense suitable conditions to germinate, and the seedling must respond to changing environmental conditions. The seedling is equipped with an array of sensors and mechanisms that allow it to respond to light, gravity, water conditions, and so on. The seedling is a transition form between embryonic and postembryonic development. In seedlings, such as Arabidopsis seedlings, postembryonic shoot development is initiated by activation of the shoot apical meristem (SAM) and appearance of the first true leaves. (In seedlings,

FIGURE 4.1

Soybean seedlings emerging from the soil. Soybean (*Glycine max*) is an epigeous plant in which the cotyledons emerge from the soil with the sprout. Note that cotyledons, cot, are closed and held in a folded position by the curved apical hook at the top of the hypocotyl, hyp. (From Scott, W. O., and Aldrich, S. R. 1970. Modern Soybean Production, p. 15. Champaign IL: S & A Publications.)

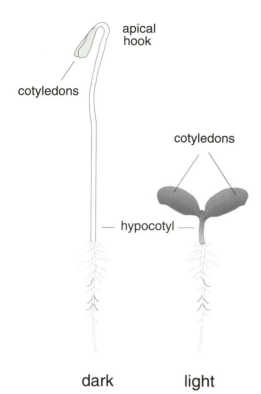

FIGURE 4.2

Comparison of Arabidopsis seedlings grown in the dark and light. Illustration demonstrates that dark-grown seedlings are bleached, tall, and lanky. Etiolated seedlings have long hypocotyls, pronounced apical hooks (at early stages), and unexpanded, unopened cotyledons. Light-grown seedlings have short hypocotyls, lack an apical hook, and have green cotyledons that are opened and have expanded.

such as bean seedlings, the SAM is activated and the first true leaves are formed during embryonic development.)

Seedling development is of particular interest to Arabidopsis geneticists because it is a stage of development when it is easy to screen large numbers of plants for mutant phenotypes. For example, hundreds of Arabidopsis seedlings can be germinated on a single petri dish, where they can be screened for traits such as hypocotyl elongation, root growth, trichome formation, and so forth. Many mutants with altered responses to environmental cues can also be detected at this stage. Mutants that affect light perception, hormone response, and other responses have been identified at the seedling stage.

PHOTOMORPHOGENESIS

Light is a major environmental factor influencing the course of seedling development. It is also an important germination signal in some species. Seeds that germinate beneath the surface of the soil undergo a form of development in the dark called **skotomorphogenesis** (dark development). Seedlings grown in the dark are bleached, lanky, and said to be etiolated (Fig. 4.2). In an etiolated Arabidopsis seedling, the hypocotyl elongates and pushes the folded cotyledons upward through the soil. (Arabidopsis and soybean are epigeous plants in which the cotyledons emerge from the soil.) When a seedling emerges to the surface of the soil and is exposed to light, development switches from a program of skoto-morphogenesis to **photomorphogenesis,** and seedlings begin to de-eti-olate (i.e., they green and begin to grow as normal light-grown seedlings). In the light, elongation of the hypocotyl stops, cotyledons green and enlarge, the SAM is activated, and the first true leaves emerge.

Photomorphogenesis or de-etiolation has been intensively studied as a developmental process and offers a context in which to investigate light signal transduction. Photomorphogenic mutants have been isolated and shown to have defects in light perception or light signal transduction. Two very different types of mutants have been found: light insensitive mutants and mutants that undergo photomorphogenesis in the dark. The discovery of the latter class of mutants has radically changed our thinking about photomorphogenesis and the role of light in plant development.

LIGHT-INSENSITIVE PHOTOMORPHOGENIC MUTANTS – *HY* MUTANTS

Koornneef and co-workers (1980) isolated a group of light-insensitive, photomorphogenic mutants in Arabidopsis called *long hypocotyl* or *hy* mutants. In wild type seedlings, hypocotyls elongate rapidly in the dark, but slowly in the light. *Hy* seedlings fail to perceive the light, and their hypocotyls elongate rapidly as though they are growing in the dark (Fig. 4.3). *Hy* mutants stand out in large populations of mutagenized seedlings because they are much taller than their cohorts. Most of the *hy* mutants have defects in photoreceptors, but not in other steps in the light-signaling pathway. The *hy* mutants have therefore been key in identifying light receptors in photomorphogenesis.

In higher plants, different photoreceptors sense light in different regions of the spectrum. In Arabidopsis, photomorphogenesis can be activated by far-red, red, blue, UV-A, and UV-B light. The best-studied photoreceptors are the phytochromes. In Arabidopsis seedlings, phy-tochromes sense red and far-red light. Because photomorphogenesis can be activated by light in different regions of the spectrum, it is thought that the different photoreceptors funnel their signals into the same pho-tomorphogenic pathway in seedlings (Fig. 4.4). Because of this, mutants that affect different photoreceptors have the same phenotype at the seedling stage – long hypocotyls in the light. Mutants with defects in a particular photoreceptor have a *hy* phenotype when illuminated with a

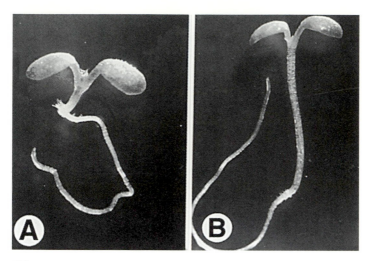

FIGURE 4.3

Comparison in growth of wild type and *long hypocotyl5, hy5,* Arabidopsis seedlings. (A) Light-grown wild type seedling has short hypocotyl and opened, expanded cotyledons. (B) *hy5* mutant seedling has long hypocotyl when grown in the light, characteristic of growth in the dark. Seedlings are 6 days old. (From Ang, L. H., and Deng, X. W. 1994. Regulatory hierarchy of photomorphogenic loci: Allele-specific and light-dependent interaction between the *HY5* and *COP1* loci. Plant Cell 6:617. Reprinted by permission of the American Society of Plant Physiologists.)

narrow band of light characteristic of the action spectrum for that photoreceptor. These mutants, however, grow normally in full spectrum white light because signaling from other photoreceptors can compensate for the defect in the *hy* mutant. Mutants with defects "downstream" in the common segment of the photomorphogenic response pathway or with pleiotropic defects in many different photoreceptors can be recognized by their *hy* phenotype in full-spectrum white light.

The major photoreceptors involved in photomorphogenesis – the phytochromes – are composed of an apoprotein and a light-absorbing chromophore. Phytochrome apoproteins are encoded by a family of genes, and the best-studied phytochrome is the red-light absorbing form, phytochrome B (PHYB) in Arabidopsis. PHYB is biologically inactive in its red-absorbing form. When exposed to red light, PHYB undergoes a conformational change and is converted to a far-red, light-absorbing, biologically active form. Subsequent exposure to far-red light returns PHYB back to its red-light absorbing, biologically inactive form. The interconversion of forms acts as a switch that can render PHYB biologically active or inactive.

One of the light-insensitive *hy* mutants from the Koornneef collection, originally called *hy3*, is deficient in PHYB, and photomorphogene-

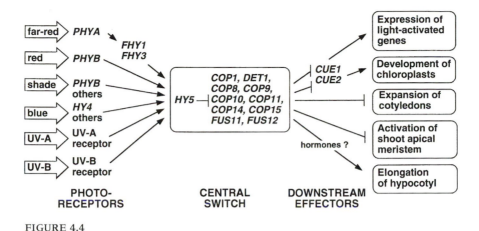

FIGURE 4.4

Regulatory pathway for the control of photomorphogenesis in Arabidopsis seedlings. Different wavelengths of light are perceived by different photoreceptors. Most loss-of-photomorphogenic-function mutants (particularly *hy* or *phy* mutants) have defects in specific photoreceptors. Light signals from different photoreceptor systems (although not yet demonstrated for the UVA and UVB photoreceptors) converge on a central processor that involves *hy5*, which appears to repress the combined action of *cop/det/fus* gene products. *Cop/det/fus* gene products act as negative regulators of photomorphogenesis. Some of the products are known to act as a complex. The pathway involves two negative operator steps. (Activation is indicated by an arrow, ↓, while inactivation is represented by ⊥.) Mutations that inactivate *COP/DET/FUS* genes constitutively activate photomorphogenesis. (Adapted from McNellis, T. W., and Deng, X.-W. 1995. Light control of seedling morphogenetic pattern. Plant Cell 7:1749–1761; Wei, N., and Deng, X.-W. 1996. The role of the *COP/DET/FUS* genes in light control of Arabidopsis seedling development. Plant Physiol. 112:871–878.)

sis in *hy3* is insensitive to red light (Nagatani et al. 1991; Somers et al. 1991). The gene responsible for the mutant has been cloned and found to encode the apoprotein of PHYB (Reed et al. 1993). (The gene involved in the *hy3* mutant has been renamed *PHYB* to reflect its function.) Another *hy* mutant, originally called *hy8*, is insensitive to far-red light. The gene responsible for this mutant has been cloned and found to encode the apoprotein of phytochrome A (PHYA). It has been shown through the analysis of light responses in *phyA* and *phyB* mutants that the two phytochromes have different but overlapping functions (Reed et al. 1994). Some of the other *hy* mutants, such as *hy1*, *hy2*, and *hy6*, appear to encode functions involved in the production of the phytochrome chromophore (Chory 1992). These mutations affect all the phytochromes, and photomorphogenesis in these mutants is blocked in both red and far-red light.

Another *hy* mutant, originally called *hy4*, is insensitive to blue light (Ahmad and Cashmore 1993). *HY4* has also been cloned and found to encode the apoprotein of a putative blue-light receptor. Only one of the

FIGURE 4.5

Wild type and *de-etiolated1*-1, *det1*-1, Arabidopsis seedlings grown in the dark. (Left) Wild type seedling shows typical etiolated stature with longer, slender hypocotyl. (Seedling does not have an apical hook at this late stage of growth.) (Right) Dark-grown *det1*-1 seedling has characteristics of a seedling grown in light with shortened hypocotyl, and opened and expanded cotyledons. Seedlings were grown for 7 days in the dark. (From Pepper, A., et al. 1994. *DET1*, a negative regulator of light-mediated development and gene expression in Arabidopsis, encodes a novel nuclear-localized protein. Cell 78:109. Reprinted by permission of the author and Cell Press.)

hy mutants, *hy5*, does not encode a photoreceptor component, but it does appear to be involved in a step downstream in the photomorphogenic response pathway, which will be discussed shortly.

CONSTITUTIVE LIGHT RESPONSE MUTANTS – *DET* AND *COP* MUTANTS

Because photomorphogenesis is a developmental program that is activated when seedlings are exposed to light, it is natural to think that light drives the photomorphogenic process. Quite surprisingly that does not appear to be the case! Chory et al. (1989) discovered a fascinating set of mutants that undergo photomorphogenesis in the dark. These mutants, called de-etiolated or *det* mutants, have short hypocotyls and open cotyledons in the dark (Fig. 4.5, right), which are characteristics of seedlings grown in the light. (*Det* mutants are not green because certain steps in chlorophyll biosynthesis require light.) Similar mutants were isolated by Deng and Quail, and these were called *constitutive photomorphogenic* (*cop*) mutants (Deng and Quail 1992; Hou et al. 1993). *Cop* mutants are similar to the *det* mutants. *Cop1* has short hypocotyls, open and expanded cotyledons and chloroplastlike differentiation in cotyledon plastids in the dark (Deng and Quail 1992; Hou et al. 1993). *Cop* and *det* mutants are highly pleiotropic, but they have similar phenotypes. One group of *cop* mutants, *cop2-cop4,* are less pleiotropic and more specifically affect cotyledon opening and enlargement.

In addition to the effect of *det1* on seedling morphogenesis, the mutation relieves the light requirement for certain photoregulated genes (Chory et al. 1989). In *det1* mutants, the light-regulated genes that encode ribulose bisphosphate carboxylase small (*RBCS*) and large subunit (*rbcL*) are expressed in the dark. (*RBCS* and *rbcL* are located in the nuclear and chloroplast genomes, respectively.) *DET1* has been cloned and encodes a novel protein that is located in the nucleus (Pepper et al. 1994). DET1 is thought to control the cell type-specific expression of light-regulated promoters, however; DET1 itself does not appear to be a DNA-binding protein and its function is not yet known.

Although *det* and *cop* mutations constitutively activate photomorphogenesis, they are recessive or loss-of-function mutations. The properties of the mutants are consistent with the idea that *DET* and *COP* genes normally suppress photomorphogenesis in the dark and that loss-of-function mutations in the gene relieve the repression. The action of the *DET* and *COP* genes can be represented by a pathway of gene action with two negative operators (Fig. 4.6). Such pathways operate logically like simple pathways with only positive operators. In such pathways, activation of photoreceptors elicits a positive response (Fig. 4.6A), and there is no response without activation of photoreceptors (Fig. 4.6B). In pathways with negative operators, however, loss-of-function mutations in intermediate components (e.g., *DET* and *COP* gene) produce a constitutive response (Fig. 4.6C) (i.e., the response occurs without activation of photoreceptors; therefore, it is constantly "on"). The fact that photomorphogenesis is under negative control by *DET* and *COP* suggests that photomorphogenesis is a "default" pathway in seedling development that must be actively repressed in the absence of light to allow for skotomorphogenesis. In keeping with that line of thinking, it has been proposed that skotomorphogenesis is an evolutionary adaptation in angiosperms that allows them to compete in conditions of low light or darkness (McNellis and Deng 1995).

Do *DET* and *COP* genes, indeed, act in a single photomorphogenic pathway along with *HY* genes, and, if so, where do they act? The issue has been addressed by generating double mutants and performing **epistasis analysis** (Chory 1992; Ang and Deng 1994). (Epistasis analysis involving pattern formation mutants was described in Chap. 3. In the example in Chap. 3, epistasis analysis was carried out with two mutants,

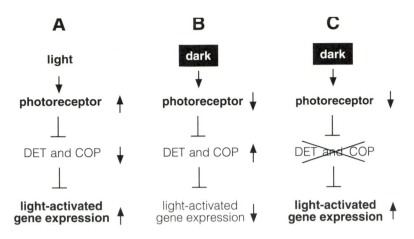

FIGURE 4.6

Operational logic of the photomorphogenesis pathway. (A) Light stimulates light-activated gene expression. The pathway has two negative operators, ⊥, but acts like a pathway with only positive operators. (B) As expected, in the dark, light-activated genes are not expressed. (C) In *DET* or *COP* mutants, however, light-activated genes are expressed even in the dark. (Upward-pointing arrows indicate activation; downward-pointing arrows indicate inactivation.)

both of which are loss-of-function mutants. If the mutants are null or severe alleles and act on the same regulatory pathway, then the phenotype of the double mutant should be similar to the phenotype of the mutation that lies furthest "upstream" on a pathway. As it relates to a constitutive mutant, such as *det1*, and a loss-of-function mutant, such as a *hy* mutant, photomorphogenesis in the double mutant will be constitutive if the constitutive mutant acts downstream on a pathway of gene action. In this case, the "downstream" constitutive mutant is epistatic to the "upstream" loss-of-function mutant.) Using this approach, it was found that the constitutive *det* and *cop* mutants are epistatic to the *hy* mutants; therefore, *det* and *cop* mutants appear to act downstream in the photomorphogenic response pathway (Fig. 4.4) (McNellis and Deng 1995). By similar reasoning, *hy5* appears to act upstream from all the *det* and *cop* mutants that have been tested and downstream from the other *hy* mutants. *Hy5* mutants are insensitive to broad-band illumination, which indicates that the *hy5* mutation blocks light signals from all the photoreceptors. It is argued that signals from the various photoreceptors converge and negatively regulate the machinery (the "central switch") that represses photomorphogenic development (Fig 4.4) (McNellis and Deng 1995).

To date, ten different *det* and *cop* mutants have been found that act downstream from *hy5* (Wei and Deng 1996). Because *det* and *cop* mutants have similar phenotypes, they have been difficult to sort out from each other and to determine whether they act on the same regulatory pathway. In the case of *cop1* and *det1*, however, **synthetic lethality** between weak alleles has provided some evidence that the genes act on the same pathway (Ang and Deng 1994). (Synthetic lethality is a severe defect that occurs when the gene products from two weak alleles interact.) Severe alleles of *cop1* and *det1*, on their own, are lethal after the seedling stage. Double mutants constructed between two weak alleles, *cop1*-6 and *det1*-1, showed synthetic lethality, and the double mutants die at the adult stage. Because the mutations tend to interact, it has been argued that all of the constitutive photomorphogenic mutants act at the same point on the same pathway. How can that be? One possibility is that all the gene products may form a complex with each other.

Several of the *COP* genes have been cloned, and *COP1* encodes a novel protein that has domains similar to other signaling proteins and transcription factors (Deng et al. 1992). COP1 contains a zinc finger-binding motif, a potential coiled-coil region, and a domain with multiple repeats similar to those found in the β subunit of trimeric G-proteins (G_β). (The G_β region is similar to a domain in a subunit of the TFIID transcription factor complex in Drosophila.) *COP9* encodes a small protein that is found in a heterogeneous array of supercomplexes in dark-grown seedlings (Wei et al. 1994). Upon exposure to light, the supercomplexes containing COP9 break down and resolve into smaller, more homogenous complexes of about 560 kD (Chamovitz et al. 1996). The complexes are 12 nm particles composed of twelve subunits and one of the subunits has been identified as COP11. A translational fusion between COP9 and β-glucuronidase or GUS (COP9-GUS) has been used to determine the subcellular location of the complexes because COP9 is only found in complexes. Complexes that contain COP9-GUS are located in the nucleus of root cells of seedlings grown in either light or dark.

The subcellular location of COP1 has been determined in a similar way (von Armin and Deng 1994). In hypocotyl cells, the location of COP1 depends on light conditions (Fig. 4.7). In the dark, the COP1-GUS fusion is largely nuclear, but in the light, it is located in the cytoplasm. In root cells, the complex remains in the nucleus even in the light. The nuclear localization of COP1-GUS in hypocotyl cells appears to depend on the presence or integrity of COP9-containing complexes. In *cop9* mutants grown in the dark, COP1-GUS fails to accumulate in nuclei. COP1, however, is not a component of the COP9-containing complexes (Chamovitz et al. 1996). It has been proposed that in the dark, COP1 is located in the nucleus, where it associates with the

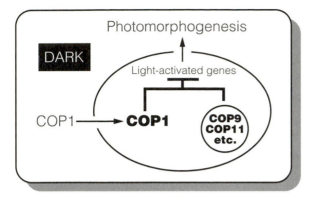

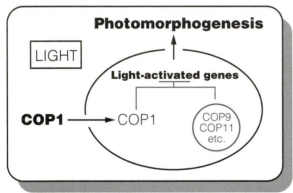

FIGURE 4.7

Change in location in the dark and light of CONSTITUTIVE PHOTOMORPHO-GENESIS1 (COP1) protein in Arabidopsis hypocotyl cells. Localization of the COP1 protein is indicated by the location *COP1*-GUS fusion. COP1 is located in the nucleus in dark grown-seedlings and in the cytoplasm in light-grown seedlings. An explanation for the change in location of COP1 is that it is transported to the nucleus in the dark where it may interact with the COP9-containing complex to repress light-activated genes involved in photomorphogenesis. Just the opposite is thought to happen in the light. COP1 is transported out of the nucleus, relieving the repression of light-activated genes, which contribute to photomorphogenic development when expressed. (Adapted from Wei, N., and Deng, X.-W. 1996. The role of the *COP/DET/FUS* genes in light control of Arabidopsis seedling development. Plant Physiol. 112:871–878.)

supramolecular complexes formed by COP9-containing complexes and suppresses photomorphogenic gene expression. In the light, COP1 is located outside of the nucleus abrogating its suppressive action.

FUSCA MUTANTS

Because *det* and *cop* mutants were selected on the basis of their constitutive photomorphogenic phenotypes, it was quite surprising when it was found that nearly all of the pleiotropic mutants were allelic to *fusca* (*fus*) mutants. *Fus* mutants are a class of seedling-lethal mutants that accumulate anthocyanins during embryo development (Castle and Meinke 1994). For example, *fus1* is allelic to *cop1*, *fus2* to *det1*, and so forth. The allelism between the *det, cop,* and *fus* mutants in Arabidopsis represent an example of the complexity and the overlapping action of genes in seedling development. The *fus* mutants were originally described as embryo-lethal mutants, but, in fact, most of the *fus* mutants are blocked in seedling development (Fig. 4.8). The *fus* mutants accumulate anthocyanin pigments in the cotyledons of the developing seedlings (the term **fusca** means dark purple color). Anthocyanin biosynthesis in plants is tightly regulated by light and other environmental and hormonal influ-

FIGURE 4.8

Comparison of Arabidopsis wild type and *fusca6*-1 (*fus6*-1) mutant. (A) Wild type and (B) *fus6*-1 after 2 weeks growth in the light. The *fus6*-1 mutant is arrested at the seedling stage and has accumulated dark anthocyanin pigment in its cotyledons. (From Castle, L. A., and Meinke, D. W. 1994. A *FUSCA* gene of Arabidopsis encodes a novel protein essential for plant development. Plant Cell 6:28. Reprinted by permission of the American Society of Plant Physiologists.)

ences. The accumulation of anthocyanin pigment is not the cause of the developmental arrest, but it appears to be an independent effect of the *fus* mutation. This can be shown by the fact that a double *fus6* and anthocyanin biosynthesis mutant (*transparent testa*, *tt3* or *ttg*) fails to accumulate anthocyanin, but still suffers from the developmental defects (Castle and Meinke 1994). The pleiotropic character of the *fus* mutants indicates a defect in some basic mechanism that affects a number of regulated processes. The allelism between *cop/det/fus* mutants suggests that part of the machinery that represses photomorphogenesis may also have a more general function in seedling development.

DOWNSTREAM EVENTS IN PHOTOMORPHOGENESIS

What are the downstream targets of the COP/DET/FUS machinery (Fig. 4.4)? A number of key effector genes are typically used to monitor photomorphogenic development, such as the *CAB* genes encoding the chlorophyll A/B binding proteins, which are major chlorophyll binding proteins found in chloroplast membranes. Light activates the transcription of *CAB* genes in etiolated seedlings as part of a program of chloroplast development in greening tissue (see the review by Chory and

Susek 1994). CAB gene expression has been used to identify mutations in genes encoding transactivating factors that positively regulate effector genes in photomorphogenic development. To do so, Li et al. (1995) developed a construct that allowed both positive and negative selection of *CAB* gene expression. The construct (*CAB3* promoter:*ADH*) consisted of the promoter from the *CAB3* gene in Arabidopsis linked to the *alcohol dehydrogenase* gene (*ADH*). *ADH* expression can be selected against by using allyl alcohol that is converted by *ADH* to toxic acrolein. Thus, mutants that fail to produce ADH upon exposure to light can be selected in the presence of allyl alcohol. (The construct also contains another *ADH* promoter linked to the *uidA* reporter gene or *ADH* promoter:GUS. The two-promoter system reduces the likelihood that cis-acting promoter mutations will be picked up in the screen because such mutations would require two independent mutations, one in each promoter.)

The selection scheme was found to be robust and netted recessive mutants that defined two new loci, *cab underexpressed1* and *2* (*cue1* and *cue2*) (see Fig. 4.4) (Li et al. 1995). In the light, *cue* mutants accumulated less RNA from genes encoding chloroplast components, such as *CAB* RNA (from the endogenous *CAB* genes), RBCS subunit, and psbA (chloroplast gene encoding D1 protein from photosystem II). In wild type plants, *CAB* genes were expressed rather uniformly across the leaf, whereas *CAB* expression was only observed in the bundle sheath cells around veins in the *cue1* mutant. Light-grown *cue1* plants reflect this pattern of expression in that plastid development and chlorophyll was confined to paraveinal regions of the leaf (bundle sheath cells around veins), whereas the interveinal regions were pale. Such leaves have a reticulate pattern of veins on a pale green background. (The expression of another light-regulated gene, chalcone synthase [*CHS*], was monitored in the *cue* mutants. *CHS* is normally expressed in epidermal cells, and its expression was unaffected in the *cue1* mutant.) *CUE1* action therefore appears to be involved in directing the action of the photomorphogenic machinery toward plastid development in mesophyll cells of the leaf.

BRASSINOSTEROIDS AND THE CONTROL
OF PHOTOMORPHOGENESIS

Surprisingly, plant hormones called brassinosteroids are also required to suppress photomorphogenic development in the dark. These hormones are related to mammalian steroid hormones, but they were once thought to be minor hormones redundant in action to other major hormones such as gibberillic acid and cytokinin. It is now realized that brassinosteroids play a major role in Arabidopsis development and that

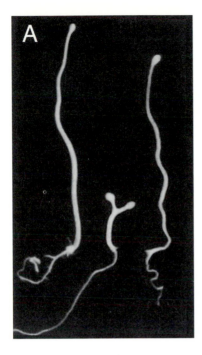

Brassinolide

FIGURE 4.9

Comparison of wild type and *de-etiolated2* (*det2*) mutant seedlings in Arabidopsis grown in the dark. (A) Wild type on left displays typical etiolated features, long hypocotyl, and unopened cotyledons. In center, *det2* mutant shows de-etiolated characteristics, short hypocotyl, and open, expanded cotyledons. Phenotype of *det2* mutant on right is restored nearly to wild type conditions by application of 10^{-6} M brassinolide. Seedlings are 10 days old. (B) Structure of brassinolide. (Reprinted with permission from Li, J., et al. 1996. A role for brassinosteroids in light-dependent development of Arabidopsis. Science 272:400. Copyright 1996 American Association for the Advancement of Science.)

a number of mutants in the *det/cop/fusca* group can be rescued by the action of brassinosteroids.

The first constitutive photomorphogenic mutant that was reported to involve brassinosteroids was *det2* (Li et al. 1996). *Det2* was originally identified by its photomorphogenic development in the dark (i.e., the inhibition of hypocotyl elongation, the opening of cotyledons, and the precocious development of true leaves in the dark) (Fig. 4.9A, center). Although the phenotype of *det2* in the dark had many features in common with *det1*, double mutant analysis suggested that *det2* was on a different pathway from *det1*, *cop1*, and *cop9* (Chory 1992). Furthermore, *det2* also showed pronounced growth defects in the light. These mutants were dwarfed and darker green than wild type, with reduced apical dominance and fertility. The phenotype in *det2* could be rescued by brassinosteroids

(Fig. 4.9A, right). *Det2* was cloned by map-based cloning and found to encode a protein similar to a key enzyme involved in steroid synthesis in mammals, steroid 5α reductase (Li et al. 1996), which suggests that *det2* was defective in a step in brassinosteriod biosynthesis. It was concluded from this that without an intact brassinosteroid synthesis pathway, photomorphogenesis is not fully suppressed in the dark.

Other brassinosteroid mutants have been identified with the same phenotype, either by their constitutive photomorphogenesis in the dark or dwarfism in the light. One such mutant, called *constitutive photomorphogenesis and dwarfism* (*cpd*), could also be rescued by supplying brassinosteroids (Szekeres et al. 1996). This mutant was cloned by T-DNA tagging and was found to encode a novel cytochrome P450 similar to other steroid hydroxylases, which is another step in the biosynthesis of brassinosteroids. Another group of dwarfed mutants, called *cabbage* mutants (*cbb1-cbb3*), was also found to respond to brassinosteroids (Kauschmann et al. 1996). *Cbb* mutants were identified by their stunted shoot axis, compact rosette structure, and dark green leaves.

Brassinosteroids promote cell elongation. These hormones, therefore, can rescue a number of mutants in which elongation of the hypocotyl has been inhibited in the dark. These include the photomorphogenic mutants described earlier, including *det1, cop1, fus4–fus9, fus11, fus12,* and, in addition, the *diminuto* (*dim*) mutant, which is defective in cell-elongation processes independent of light signaling (Szekeres et al. 1996). Because brassinosteroids rescue such a broad range of photomorphogenic mutants on the *det1* pathway, it has been argued that the *det1* and *det2* pathways must interact. Brassinosteroids do not rescue the short hypocotyl phenotype of other mutants, such as gibberillic acid (GA) mutants. This suggests that GA works in an independent pathway that is also required to promote hypocotyl elongation in the dark. (More discussion of GA mutants will be found in Chap. 5.)

ETHYLENE MUTANTS AND THE TRIPLE RESPONSE

Another important role for plant hormones in seedling development is the response that occurs when the shoot of an elongating seedling hits an obstacle as it grows toward the soil surface (see Fig. 4.1). This stress redirects the growth of the hypocotyl to development of girth rather than length, adding "muscle" to the seedling and helping it to push away obstacles. In its encounter with an obstacle, the tendency of the shoot to grow upward (negative geotropism) is also partially relieved, and the shoot grows diageotropically (not directly upward). This

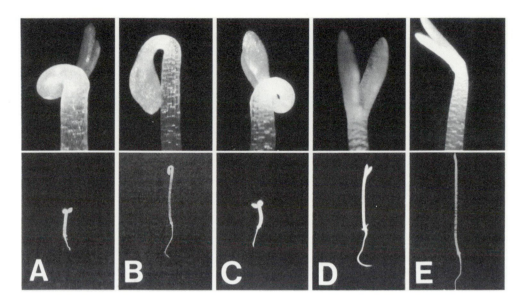

FIGURE 4.10

Effects of ethylene on growth of Arabidopsis wild type and triple response mutant seedlings. Lower panels show whole seedlings; upper panels are higher magnifications detailing the apical region of the seedlings. Three-day-old seedlings have been grown in the dark in the presence or absence of ethylene (10 µl l⁻¹). Wild type seedling grown (A) in the presence of ethylene and (B) in the absence. Note that ethylene inhibits hypocotyl elongation and exaggerates coiling of the apical hook, which are characteristics of the triple response. (C) *Ethylene overproducer1*-1 (*eto1*-1) mutant grown in the absence of ethylene. Note that *eto1*-1 displays a constitutive triple response in the absence of added ethylene. (D) *Hookless1*-1 (*hls1*-1) grown in the absence of ethylene. Mutant does not form an apical hook. (E) *Ethylene insensitive2*-1 (*ein2*-1) mutant grown in the presence of ethylene. Note that the mutant does not display a triple response in the presence of ethylene. (From Guzmán, P., and Ecker, J. R. 1990. Exploiting the triple response of Arabidopsis to identify ethylene-related mutants. Plant Cell 2:515. Reprinted by permission of the American Society of Plant Physiologists.)

response, which aids the shoot in finding alternative routes to the surface, is mediated, in part, by the hormone ethylene.

The ethylene-mediated responses described earlier constitute part of the so-called triple response. The triple response includes inhibition of stem elongation, radial swelling of the stem, and absence of a normal geotropic response (Crocker et al. 1913). Another telltale response to ethylene is the exaggerated coiling of the apical hook (Fig. 4.10A). The hook holds the cotyledons in a downward "tuck" position, and in the absence of ethylene (or when the ethylene response is blocked), the exaggerated bend of the hook is relieved (Fig 4.10B). Ethylene is thought to produce an imbalance in auxin distribution across the axis in the hook region, leading to unequal growth on the sides (and coiling) of the apical hook.

Ethylene-insensitive mutants in Arabidopsis have been isolated that fail to show the triple response in the presence of exogenous ethylene (Fig. 4.10E) (Bleecker et al. 1988; Guzmán and Ecker 1990). In addition, constitutive responders have also been isolated (i.e., mutants that display a triple response even in the absence of exogenous ethylene). Guzmán and Ecker classified the constitutive mutants into two groups based on their response to ethylene synthesis inhibitors. Those mutants in which constitutive ethylene responses were blocked by ethylene synthesis inhibitors were considered to be ethylene overproducers, lacking some control in the ethylene biosynthesis pathway. Those mutants that were unaffected by ethylene synthesis inhibitors were considered to be constitutive responders.

Several of the ethylene mutants have been cloned and tentatively identified by sequence comparison. Some of the genes encode protein kinases, which indicates that ethylene responses probably involve a protein phosphorylation cascade (Ecker 1995). The order of action of components in the ethylene signal transduction pathway has been examined by constructing double mutants and carrying out the epistasis analysis described earlier. Through this analysis, it has been possible to order most of the components into a linear pathway (Fig. 4.11). The most "upstream" component of the pathway is *ETR1* (or *EIN1*). *ETR1* encodes a very unusual protein that is similar to two-component protein kinases in bacterial systems or to the *SLN1* component in the osmoregulation pathway in yeast (Chang et al. 1993). Two-component protein kinases are composed of a "sensor" with histidine kinase activity and a "response regulator" which contains an aspartate group that is phosphorylated by the protein kinase. Both components of what is normally a two-component system in bacteria are located in the C-terminal region of the ETR1 protein.

Because *ETR1* acts furthest upstream in the ethylene response pathway, it had been suspected that its gene product might also be an ethylene receptor. This was supported by the discovery that when *ETR1* from Arabidopsis is expressed in yeast, ethylene bound to the yeast with high affinity and in a saturable manner (Schaller and Bleecker 1995). The N-terminus of ETR1, with hydrophobic (possible membrane spanning) domains, appears to be the receptor region (Fig. 4.11). This was demonstrated by showing that C-terminal truncations of ETR1 leaving only the three membrane-spanning domains still conferred upon yeast the ability to bind ethylene. *Etr1* mutations are dominant, although it is not entirely clear why. The mutations, however, are thought to be domi-

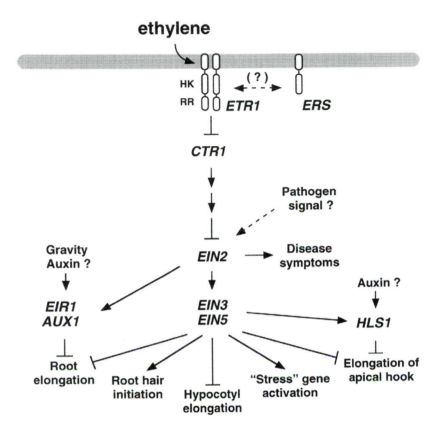

FIGURE 4.11

Model for ethylene action. *ETHYLENE RESISTANCE1* (*ETR1*) encodes an ethylene receptor that binds ethylene and initiates the signal transduction process. Other genes, such as *ERS,* encode proteins that are similar to ETR1 and may interact with ETR1 (Hua et al. 1995). ETR1 is a two-component-like protein kinase, composed of a histidine kinase (HK) and response regulator (RR), that in response to ethylene may directly or indirectly inactivate the product of the *CONSTITUTIVE ETHYLENE RESPONSE1* (*CTR1*) gene, which acts downstream from ETR1 (Chang et al. 1993). CTR1 is related to RAF-like protein kinases, and through its downstream intermediates CTR1 negatively regulates *ETHYLENE INSENSITIVE2* (*EIN2*). EIN2 positively regulates *EIN3* and *EIN5*, which in turn activate the expression of a wide number of events associated with ethylene action. *ETHYLENE INSENSITIVE RESPONSE1* (*EIR1*) and *AUXIN RESISTANT1* (*AUX1*) are involved in gravity and auxin responses in the root and appear to function on another pathway. *HOOKLESS1* (*HLS1*) is thought to mediate an auxin gradient response in the apical hook. *EIN6* and *EIN7* have not been fully characterized. Like photomorphogenic signaling, the ethylene response pathway involves two negative operators. Loss-of-function mutations in *CTR1* or in genes immediately downstream are predicted to activate ethylene responses constitutively. (Redrawn from Ecker, J. R. 1995. The ethylene signal transduction pathway in plants. Science 268:667–675.)

nant because they lie in hydrophobic regions of the protein and may either (1) poison the action of other subunits if ETR1 acts in a complex or (2) lock ETR1 into a constitutively active conformation (Chang et al. 1993). The ETR1 protein is known to exist as a dimer associated with membranes (see Bleecker and Schaller 1996). An additional gene with sequence similarity to *ETR1* has been found and is called *ETHYLENE RESPONSE SENSOR* (*ERS*) (Hua et al. 1995). *ERS* differs from *ETR1* in that *ERS* lacks the response regulator region (Fig. 4.11). To demonstrate that ERS is involved in ethylene responses, a mutation similar to that in *etr1*-4 was made in ERS and introduced into transgenic plants. The mutated gene conferred dominant ethylene insensitivity similar to the effect of the *etr1*-4 mutation. The mutation demonstrates that ERS may be a part of the ethylene response pathway and that it may interact with ETR1 in sensing ethylene.

The Arabidopsis *CTR1* gene, which encodes the next component downstream in the pathway, has also been cloned and identified (Kieber et al. 1993) (Fig. 4.11). *CTR1* encodes a protein with a domain similar to RAF-type protein kinases. RAF-type protein kinases are serine/threonine protein kinases that are components of mitogen-activated protein (MAP) kinase cascades in other organisms. The *ctr1* mutation is a recessive mutation and confers a constitutive ethylene response. This suggests that *CTR1* negatively regulates downstream components in the pathway either directly or indirectly. That is to say, the activity of *CTR1* inactivates the function of components further downstream. *CTR1* is thought to be a negative controller of *EIN2* function (via several unknown steps) (Fig. 4.11). With respect to its upstream interactions, if CTR1 interacts directly with ETR1, then the dominant ethylene insensitivity of *ETR1* might be explained if ETR1 constitutively activated CTR1, which, in turn, would inhibit ethylene responses (Chang et al. 1993).

It is interesting that the pattern of logical operators in the ethylene response pathway is similar to that in the photomorphogenic pathway (Fig 4.6), even though the pathway components are different. In either the ethylene or light response pathways, perception of the stimulus activates a receptor. In turn, the activated receptor directly or indirectly turns off the activity of the midstream components. In the photomorphogenic pathway, light activation of the appropriate photoreceptor inactivates the function of the *COP* and *DET* components. In the ethylene response pathway, ethylene activation of *ETR1* (and associated components) turns off the function of *CTR1*. The inactivation of the midstream components of the pathways promotes the action of the downstream components and, from there, the response. Because the midstream components turn off or repress the response, loss-of-funtion mutations that inactivate the midstream components, such as *det* and *cop* mutations in the photomorphogenic pathway and *ctr1* mutations in the ethylene response pathway, constitutively activate their respective responses (Fig 4.6).

REFERENCES

Ahmad, M. and Cashmore, A. R. 1993. *HY4* gene of *A. thaliana* encodes a protein with characteristics of a blue light photoreceptor. Nature 336: 162–165.

Ang, L. H. and Deng, X. W. 1994. Regulatory hierarchy of photomorphogenic loci: Allele-specific and light-dependent interaction between the *HY5* and *COP1* loci. Plant Cell 6: 613–628.

Bleecker, A. B., Estelle, M. A., Somerville, C., and Kende, H. 1988. Insensitivity to ethylene conferred by a dominant mutation in *Arabidopsis thaliana.* Science 241: 1086–1089.

Bleecker, A. B. and Schaller, G. S. 1996. The mechanism of ethylene perception. Plant Physiol. 111: 653–660.

Castle, L. A. and Meinke, D. W. 1994. A *FUSCA* gene of Arabidopsis encodes a novel protein essential for plant development. Plant Cell 6: 25–41.

Chamovitz, D. A., Wei, N., Osterlund, M. T., von Arnim, A. G., Staub, J. M., Matsui, M., and Deng, X. W. 1996. The COP9 complex, a novel multisubunit nuclear regulator involved in light control of a plant developmental switch. Cell 86: 115–121.

Chang, C., Kwok, S. F., Bleecker, A. B., and Meyerowitz, E. M. 1993. *Arabidopsis* ethylene-response gene *ETR1:* Similarity of product to two-component regulators. Science 262: 539–544.

Chory, J. 1992. A genetic model for light-regulated seedling development in Arabidopsis. Development 115: 337–354.

Chory, J., Peto, C., Feinbaum, R., Pratt, L., and Ausubel, F. 1989. *Arabidopsis thaliana* mutant develops as a light-grown plant in the absence of light. Cell 58: 991–999.

Chory, J. and Susek, R. E. 1994. Light signal transduction and the control of seedling development. In Arabidopsis, ed. Meyerowitz, E. M. and Somerville, C. R., 579–614. Cold Spring Harbor: Cold Spring Harbor Press.

Crocker, W., Knight, L. I., and Rose, R. C. 1913. A delicate seedling test. Science 37: 380–381.

Deng, X.-W., Matsui, M., Wei, N., Wagner, D., Chu, A. M., Feldmann, K. A., and Quail, P. H. 1992. *COP1,* an Arabidopsis regulatory gene, encodes a novel protein with both a Zn-binding motif and a Gβ-protein homologous domain. Cell 71: 791–801.

Deng, X. W. and Quail, P. H. 1992. Genetic and phenotypic characterization of cop-1 mutants of *Arabidopsis thaliana.* Plant J. 2: 83–95.

Ecker, J. R. 1995. The ethylene signal transduction pathway in plants. Science 268: 667–675.

Guzmán, P. and Ecker, J. R. 1990. Exploiting the triple response of Arabidopsis to identify ethylene-related mutants. Plant Cell 2: 513–524.

Hou, Y., von Arnim, A. G., and Deng, X.-W. 1993. A new class of *Arabidopsis* constitutive photomorphogenic genes involved in regulating cotyledon development. Plant Cell 5: 329–339.

Hua, J., Chang, C., Sun, Q., and Meyerowitz, E. M. 1995. Ethylene insensitivity conferred by Arabidopsis ERS gene. Science 269: 1712–1714.

Kauschmann, A., Jessop, A., Koncz, C., Szekeres, M., Willmitzer, L., and Altmann, T. 1996. Genetic evidence for an essential role of brassinosteroids in plant development. Plant J. 9: 701–713.

Kieber, J. J., Rothenberg, M., Roman, G., Feldmann, K. A., and Ecker, J. R. 1993. CTR1, a negative regulator of the ethylene response pathway in Arabidopsis, encodes a member of the RAF family of protein kinases. Cell 72: 1–20.

Koornneef, M., Rolff, E., and Spruit, C. J. P. 1980. Genetic control of light-inhibited hypocotyl elongation in *Arabidopsis thaliana* L Heynh. Z. Pflanzenphysiol. 100: 147–160.

Li, H. M., Culligan, K., Dixon, R. A., and Chory, J. 1995. *CUE1:* A mesophyll cell-specific positive regulator of light-controlled gene expression in Arabidopsis. Plant Cell 7: 1599–1610.

Li, J., Nagpal, P., Vitart, V., McMorris, T. C., and Chory, J. 1996. A role for brassinosteroids in light-dependent development of Arabidopsis. Science 272: 398–401.

McNellis, T. W. and Deng, X.-W. 1995. Light control of seedling morphogenetic pattern. Plant Cell 7: 1749–1761.

Nagatani, A., Chory, J. and Furuya, M. 1991. Phytochrome-B is not detectable in the hy3 mutant of Arabidopsis, which is deficient in responding to end-of-day far-red light treatments. Plant Cell Physiol. 32: 1119–1122.

Pepper, A., Delaney, T., Washburn, T., Poole, D., and Chory, J. 1994. *DET1*, a negative regulator of light-mediated development and gene expression in arabidopsis, encodes a novel nuclear-localized protein. Cell 78: 109–116.

Reed, J. W., Nagatani, A., Elich, T. D., Fagan, M., and Chory, J. 1994. Phytochrome A and phytochrome B have overlapping but distinct functions in Arabidopsis development. Plant Physiol. 104: 1139–1149.

Reed, J. W., Nagpal, P., Poole, D. S., Furuya, M., and Chory, J. 1993. Mutations in the gene for the red/far-red light receptor phytochrome B alter cell elongation and physiological responses throughout Arabidopsis development. Plant Cell 5: 147–157.

Schaller, G. E. and Bleecker, A. B. 1995. Ethylene-binding sites generated in yeast expressing the Arabidopsis *ETR1* gene. Science 270: 1809–1811.

Scott, W. O. and Aldrich, S. R. 1970. Modern soybean production. Champaign IL: S & A Publications.

Somers, D. E., Sharrock, R. A., Tepperman, J. M., and Quail, P. H. 1991. The *hy3* long hypocotyl mutant of Arabidopsis is deficient in phytochrome b. Plant Cell 3: 1263–1274.

Szekeres, M., Nemeth, K., Koncz Kalman, Z., Mathur, J., Kauschmann, A., Altmann, T., Redei, G. P., Nagy, F., Schell, J., and Koncz, C. 1996. Brassinosteroids rescue the deficiency of CYP90, a cytochrome P450, controlling cell elongation and de-etiolation in Arabidopsis. Cell 85: 171–182.

von Armin, A. G. and Deng, X.-W. 1994. Light inactivation of Arabidopsis photomorphogenic repressor COP1 involves a cell-specific regulation of its nucleocytoplasmic partitioning. Cell 79: 1035–1045.

Wei, N., Chamovitz, D. A., and Deng, X. W. 1994. Arabidopsis COP9 is a component of a novel signaling complex mediating light control of development. Cell 78: 117–124.

Wei, N. and Deng, X.-W. 1996. The role of the *COP/DET/FUS* genes in light control of Arabidopsis seedling development. Plant Physiol. 112: 871–878.

5

Shoot Development

"Mighty oaks from tiny acorns grow." So the old adage goes. The development of the shoot (i.e., the aerial portion of the plant consisting of stems, branches, leaves, etc.) can be quite a wonder, especially in large trees. It is particularly impressive when one realizes that the primary growth of the shoot represents tip growth from microscopic meristems such as the shoot apical meristem (SAM) and axillary bud meristems. In plants such as Arabidopsis, these meristems are composed of about one hundred cells. It has not been easy to study SAMs in biochemical terms because they are small and usually inaccessible. Ordinarily, the operation of SAMs has been inferred from changes in the formation of the shoot and its lateral organs. Only through genetics and techniques such as *in situ* hybridization has it been possible to probe directly into the molecular workings of the SAMs.

Meristems are remarkable self-renewing structures. They house proliferative or stem cells that divide and give rise to the primary shoot, and they produce a variety of lateral organs, such as leaves, branches, tendrils, and thorns. The constant differentiation of cells in organ formation is balanced by the continuous replenishment of cells through cell divisions. The capacity to balance cell division and cell differentiation is one of the remarkable features of shoot meristems.

During vegetative growth, SAMs give rise to repeating units of the shoot called **phytomers** (Fig. 5.1) (Evans and Grover 1940). Phytomers are usually produced through reiterations of a process in which leaf primordia emerge from the SAM. Although successive phytomers are often composed of the same organs, phytomers usually differ in internode length, leaf size and shape, axillary bud potential, and so on, depending on their location within the shoot. The time interval between the production of successive leaf primordia is called a **plastochron** (Sharman 1942), which is an important measure of developmental time. (The

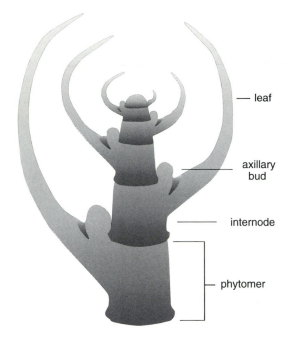

— leaf

axillary
bud

internode

phytomer

FIGURE 5.1

Diagrammatic representation of the repeating phytomer structure at the apex of a
vegetative shoot. A phytomer consists of a
leaf, axillary bud, and subtending internode. Although phytomers are repeating
structures, they can vary in size, internode
length, and kinds of lateral organs produced depending on developmental stage
and their location in the plant.

expression of developmental time in plastochrons normalizes for differences in growth conditions.) A typical phytomer consists of a node to
which a leaf is attached, a subtending internode, and an axillary bud at
the base of the leaf. The axillary bud has a meristem that is similar to the
SAM and can give rise to indeterminate structures, such as branches and
leaves, as well as to determinate structures. Plant shoots can be single and
unbranching or multiple and highly branched, depending on whether
axillary buds are active and whether they produce determinate or indeterminate structures. (In most of the discussion that follows, the tip
meristems of the main shoot and those derived from axillary buds will be
treated similarly. The functionality of these meristems with respect to
organ formation is similar; however, these meristems differ with respect
to apical dominance and origin. The SAM of the main shoot is formed
embryonically, whereas the others are formed postembryonically.)

SAM ORGANIZATION – LAYERS AND ZONES

As discussed in Chapter 2, angiosperm shoots are composed of three fundamental cell layers that can be traced back through the SAM to the
embryo. One can get a better grasp of the continuity of cell layers by
understanding the topography and operation of the SAM. The anatomy
of the SAM is defined in terms of layers and radial zones (Fig. 5.2B).
SAMs of both angiosperms and gymnosperms are stratified in two distinct

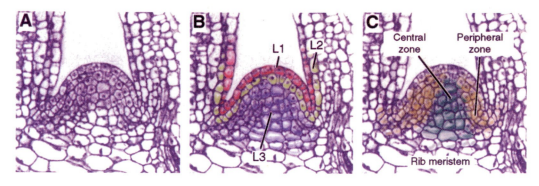

FIGURE 5.2

Organization of the shoot apical meristem (SAM) into layers and zones. (B) SAM is organized in three fundamental cell layers. The outer layer in dicots constitutes the tunica and is composed of L1 and L2, whereas the inner layer or corpus is made up of L3. (C) SAM is also organized radially into zones: a central zone composed of central mother cells and a peripheral zone composed of rapidly dividing apical initials. Lateral organs (leaf primordia) form on the flanks of the SAM in an area sometimes referred to as the morphogenic zone. The initial cells below the apical dome constitute the rib or file meristem and contribute to the elongation of the stem. Median longitudinal section of Arabidopsis shoot apex falsely colorized to show layers and zones. (Photo from Medford, J. I., et al. 1992. Normal and abnormal development in the Arabidopsis vegetative shoot apex. Plant Cell 4:635. Reproduced by permission of the American Society of Plant Physiologists.)

layers: an outer tunica layer and an inner corpus layer (see the review by Kerstetter and Hake 1997). The tunica-corpus arrangement reflects the patterns of cell divisions within the layers; tunica cells undergo anticlinal cell divisions and, therefore, have a sheetlike appearance, whereas the corpus is composed of cells that largely undergo periclinal divisions. Through the cell-lineage analysis that will be discussed shortly, it has been shown that SAMs in angiosperms are, in fact, composed of three clonally related layers, L1–L3 (which are not necessarily one cell layer thick) (Fig. 5.2B). In dicots, two layers, L1 and L2, constitute the tunica, and L3 makes up the corpus. The outermost layer is the L1 layer, or epidermis, which gives rise to a cell layer that covers all organs. L1 is a sheet of cells because almost all divisions in this layer are anticlinal, driving the expansion of the layer in two dimensions. In the middle layer, or L2, divisions are oriented less regularly along the anticlinal plane; these cells give rise to subepidermal tissue, the procambium, and part of the ground meristem (the cortex and sometimes part of the pith). The innermost layer is the pith meristem or L3. Divisions in this layer are largely periclinal, giving rise to the rest of the ground meristem and the pith.

Evidence that the three fundamental layers have separate cell lin-

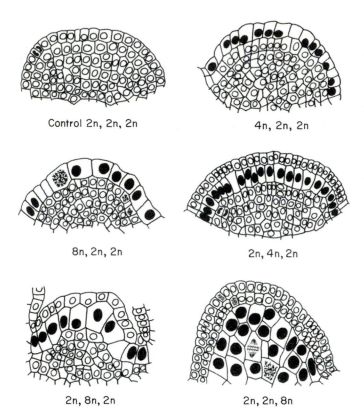

Control 2n, 2n, 2n

4n, 2n, 2n

8n, 2n, 2n

2n, 4n, 2n

2n, 8n, 2n

2n, 2n, 8n

FIGURE 5.3

Drawings of sections of periclinal chimeras of *Datura*. Shoots were treated with colchicine to induce polyploidy in various layers. Polyploid cells were recognized cytologically as cells with enlarged nuclei with multiple nucleoli (represented here as darkly shaded nuclei). Polyploid cells occupied one of three fundamental cell layers, L1, L2, or L3. The ploidy numbers of layers in the various examples are indicated for the various shoots. ($2n =$ diploid.) (Based on Satina et al. 1940. From a reproduction in Steeves, T. A., and Sussex, I. M. 1989. Patterns in Plant Development, 2d ed., p. 79. Cambridge, UK: Cambridge University Press. Reprinted by permission of the publisher.)

eages is derived from the analysis of periclinal chimeras as discussed in Chapter 2. In periclinal chimeras, one of the cell layers in the meristem differs genetically from the other layers. In fruit trees, where such chimeras have horticultural value, they usually arise by spontaneous or induced mutations and become part of a lateral shoot. Periclinal chimeras are perpetuated by the pattern of cell division in all subsequent lateral shoots. Satina et al. (1940) demonstrated the existence of layers by generating chimeras in Jimsom weed (*Datura*) using colchicine to induce the formation of polyploid cells (Fig. 5.3). Polyploid cells have enlarged nuclei with multiple nucleoli and can be recognized cytologically. The polyploid cells occupy one or more of three distinct layers in the chimeras, demonstrating that all cells in the shoot are organized into three fundamental layers derived from the SAM. The modern equivalents of these experiments are ones in which various cell markers have been shown to be expressed in single layers. For example, Lu et al. (1996) found a gene (*ARABIDOPSIS THALIANA MERISTEM LAYER1* or *ATML1*, mentioned in Chap. 3) that encodes a homeodomain protein which is expressed exclusively in the L1 layer. The gene was expressed in the L1 layer in the embryo up to the torpedo stage, and again later in the L1 layer of the SAM. These observations do not show the continuity of the L1 layer through development as did the experiments of Satina et al. (1940); however, they do demonstrate the uniqueness of cell layers in formative structures, such as embryos and meristems.

SAMs are also organized radially into zones (Fig. 5.2C) (Vaughn 1952). The SAM consists of a central zone made up of **stem cells** surrounded by a peripheral zone of apical initial cells. Used in this context, the term stem cell refers to undifferentiated, self-renewing cells that give rise to various cell populations in an adult organism. In animal systems, blood cells, intestinal crypt cells, epidermis, and spermatocytes are produced from undifferentiated stem cells in the adult organism (Gilbert 1994). The central zone contains **central mother cells,** which are large, vacuolated cells that divide infrequently. They are surrounded by initials, which are undifferentiated, cytoplasmically rich cells that divide frequently, unlike the central mother cells. At the flanks of the meristem is a morphogenic zone where organ primordia are formed (Steeves and Sussex 1989). Initials in the peripheral zone generate cells used in organ formation. Initials that lie below the central zone constitute the rib meristem. Their division (and elongation) leads to growth of the stem. Central mother cells are thought be progenitors of the rapidly dividing initials in the peripheral zone, but that fact is very difficult to prove. As discussed in Chapter 1, the central mother cells were once considered to be part of the méristème d'attente, which are cells placed in reserve during vegetative growth for later germ cell development. The common assumption now, however, is that cells of the SAM are derived from infrequent divisions of the central mother cells giving rise to apical initials that undergo extensive proliferation.

Dermen (1945) addressed the issue of whether the stem cells in the SAM were permanent or were turned over and replaced by other cells during development. He did so by examining the permanence of mericlinal chimeras of polyploid cells induced by colchicine treatment. Dermen found that the longevity of sectors varied, some disappearing after a period of growth. This would indicate that the progenitor cell population might be transient and subject to turnover. The issue is still unresolved despite Dermen's findings. Dermen's results could be questioned on the grounds that polyploidy might not have been induced in the true central mother cells because they divide so infrequently. It is more likely that polyploidy was not induced in the central mother cells, but in the apical initials that have considerable, but not indefinite, proliferative capacity. Questions about the identity and permanence of the true stem cells in the SAM are particularly relevant to those who want to produce stable transgenic plants by biolistically transforming cells in the SAM (i.e., bombarding with the gene gun).

The organization of the SAM has been largely described in terms of morphology, patterns of division, and cell lineages; however, there have been other attempts to understand the organization of the SAM with respect to patterns of activities of genes involved in meristem functions. Of particular interest is the pattern of expression of the maize *knotted1* gene (*kn1*), which encodes a homeodomain-containing transcription factor described earlier in Chapter 2. The gene is expressed in tissues with meristematic activity, and the pattern of expression in the maize meristem has been examined by Smith et al. (1992). They found that the gene is expressed in the meristem, but not in organ primordia that are being formed. The most interesting observation was that *kn1* expression faded at the position where the next leaf primordium would appear and could be used to predict the site of lateral organ formation. The expression pattern of *kn1* therefore allows one to recognize regions of prospective morphogenic activity that are not obvious by morphological features alone (Kerstetter and Hake 1997).

SAM MUTANTS

Mutants That Fail to Maintain Proliferative Cells in the Meristem

The operation of the SAM is so important to the understanding of shoot development that there has been considerable interest in mutants that affect the formation of the SAM. One of the most fascinating mutants is *shoot apical meristemless* (*stm*) in Arabidopsis that was identified by Barton and Poethig (1993). *Stm*-1 fails to organize a recognizable SAM during embryogenesis, and true leaves do not emerge as expected at the seedling stage (Fig. 5.4). These mutants, however, produce seemingly normal seedlings with cotyledons and hypocotyls demonstrating that the organization of the embryonic SAM is not really required for the formation of these organs. That finding might not really be surprising given the cell lineage analysis of Christianson (1986), which concluded that the SAM is derived from different cell lineages than cotyledons in cotton embryos (recall the discussion in Chap. 2).

A weaker allele in Arabidopsis, *stm-2*, also failed to organize an embryonic SAM, although the mutant developed a SAM at the seedling stage. Shoots were produced in *stm-2*, but they were determinate structures that terminated after producing just a few leaves. No inflorescences were formed from these shoots; however, inflorescences were

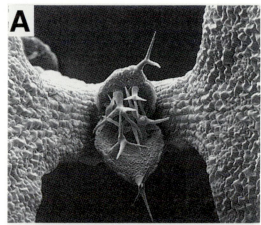

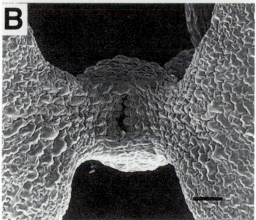

FIGURE 5.4

Comparison of shoot apical meristem (SAM) regions in wild type and *shoot meristemless* (*stm*-1) seedlings. (A) First true leaves, covered with trichomes, emerge from the functional SAM in wild type seedling. (B) Similar region in *stm*-1 mutants shows no evidence of leaf or shoot development. Seedlings are 1 week old and visualized by scanning electron microscopy. Bar = 100 µM. (From Long, J. A., et al. 1996. A member of the KNOTTED class of homeodomain proteins encoded by the *STM* gene of Arabidopsis. Nature 379:66. Reprinted by permission of the author and Nature.)

occasionally produced from meristems in the axils of leaves (Clark et al. 1996). Thus, it appears from the weaker *stm*-2 allele that *STM* is also needed for the maintenance of the SAM, and not just for its formation.

The stronger *stm-1* allele produced a root and appeared to have a normal root apical meristem (RAM). The defect in *stm-1*, therefore, is not a general lesion that affects meristems per se. An interesting question is whether *stm-1* can produce leaves at all. Adventitious leaves formed, albeit infrequently, from *stm-1* shoots regenerated in culture, although the leaves were not produced from a normal shoot and were not arranged properly (Barton and Poethig 1993). Some of the weakest *stm* alleles produced several leaves, arranged normally; however, inflorescence development terminated prematurely in the weak alleles (Endrizzi et al. 1996). In intermediate alleles, the embryonic SAM appeared to be consumed in the production of a single, large leaf primordium that split later in development to give rise to basally fused leaves. Although strong alleles did not form leaves, leaf primordia were produced in several cases, but they were encased by the cotyledon petioles fused at

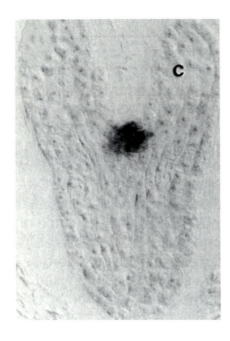

FIGURE 5.5

Localized expression of *SHOOT MERISTEMLESS* (*STM*) gene in shoot apical meristem (SAM) region of wild type Arabidopsis embryo at torpedo stage. *STM1* expression detected by *in situ* hybridization. Cotyledon, c. (From Long, J. A., et al. 1996. A member of the KNOTTED class of homeodomain proteins encoded by the *STM* gene of Arabidopsis. Nature 379:68. Reprinted by permission of the author and Nature.)

their edges. From the series of alleles, it appeared that the meristem in *stm* was depleted of proliferative cells and unable to restrain the differentiation of cells in the center of meristem. In the strongest alleles, it was thought that cell differentiation consumed proliferative cells at the earliest stages in meristem development. In weaker alleles, the proliferative cell population in the center of the meristem was maintained longer, but did ultimately succumb to the pressures to differentiate. *Stm* was therefore interpreted to be defective in maintenance of the meristem and not in its formation, even though strong alleles did not appear to form a SAM at all.

STM was cloned by Long et al. (1996) and used as a hybridization probe to detect *STM* expression during embryonic development in normal (i.e., nonmutant) plants. *STM* was expressed only in those cells predicted to form the embryonic SAM. *STM* was first expressed in only 1–2 cells in early to midglobular stage embryos. In mid- to late-globular stage, *STM* RNA was found in cells in a stripe across the top half of the embryo. In heart-shaped embryos, *STM* was expressed in the crotch between the cotyledons where the presumptive SAM forms (Fig. 5.5). In older plants, *STM* was expressed in the SAM and axillary bud meristems. In shoot meristems, *STM* expression faded rapidly at sites of leaf primordia formation, and *STM* was not expressed in the newly forming leaves. *STM* was also expressed in the inflorescence meristem, but disappeared as floral buds were initiated. *STM* was expressed again in the floral meristem, but ceased during floral organ formation.

Long et al. (1996) found that *STM* encodes a transcription factor that belongs to the same class of homeodomain-containing transcription factors as encoded by the *knotted1* (*kn1*) gene in maize described in Chapter 2, another gene that affects meristematic activity. Recall that *Kn1* mutants in maize produce knots on maize leaves (Fig. 2.14). Knots are thought to be foci of meristematic activity caused by the ectopic expression of *kn1*. In normal maize plants, the expression pattern of *kn1* is similar to *STM* in Arabidopsis in that both genes are active in meristematic tissue (Jackson et al. 1994). It has been proposed that *kn1* expression is involved in meristem functions and, perhaps, in maintaining the meristem in an undifferentiated state. Like *STM, kn1* is expressed in undifferentiated cells in the meristem. As expected for two genes with similar functions, the overexpression of *kn1* has effects opposite to the underexpression of *STM*. Overexpression of *kn1* results in ectopic formation of shoot meristems. When *kn1* was expressed under the control of the strong CaMV 35S promoter (35S:*kn1*) in transgenic tobacco plants, a number of the transformants formed ectopic shoot meristems that resulted in a "shooty" phenotype. These *kn1*-expressing plants produced epiphyllous inflorescences on the surface of leaves on renewal branches (formed on the lower nodes of the plant after flowering) (Fig. 5.6), and many shoots arose from the junction between the blade and petiole.

Another Arabidopsis mutant that fails to maintain a functional SAM is *wuschel* (*wus*). *Wus* mutants form defective SAMs that prematurely terminate growth after having produced only a few leaves. Unlike the dome-shaped SAM in wild type, the terminated meristems in *wus* were flat and very thin (Fig. 5.7B). Laux et al. (1996) interpreted the defect in *wus* as a depletion of proliferative or stem cells in the meristem. From the depleted meristems, however, leaf primordia and secondary shoot meristems were ectopically initiated. During vegetative growth, the formation of secondary meristems leads to the production of hundreds of rosette leaves. During inflorescence development, *wus* mutants display a "stop-and-go" mode of growth where flower stalks progressively arise out of clusters of cauline leaves (Fig. 5.7C) (Laux et al. 1996).

It was reasoned that the depletion of SAM in *wus* results from heavy recruitment of the proliferative cell population for organ formation. If that is the case, then *WUS* might normally restrain cell differentiation, maintaining the pool of proliferative cells. Double mutants between strong *stm* alleles and *wus* indicated that *stm* was epistatic to *wus* (Endrizzi et al. 1996). In double mutants between intermediate *stm-2* and *wus*-1 alleles, however, *wus*-1 enhanced the expression of *stm-2*, producing a novel phenotype. *Stm* and *wus* mutants, therefore, have similar effects, but they do not appear to lie on the same genetic pathway.

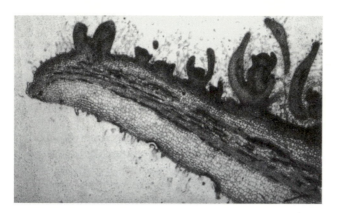

FIGURE 5.6

Epiphyllous shoots on leaf of transgenic tobacco plant (*Nicotiana tabacum*). The maize *knotted1* gene (*kn1*) was constitutively expressed by the CaMV 35S promoter (35S:*kn1*). Inflorescence shoots are produced on the adaxial surfaces of leaves on renewal branches of plants that have finished flowering. Leaf section viewed in the light microscope. (From Sinha, N. R., Williams, R. E., and Hake, S. 1993. Overexpression of the maize homeobox gene, *KNOTTED-1*, causes a switch from determinate to indeterminate cell fates. Genes Develop 7:791. Reprinted by permission of the author and Cold Spring Harbor Laboratory Press.)

Medford et al. (1992) identified other Arabidopsis mutants that were interpreted to have problems in maintaining the SAM. One called *forever young* (*fey*) had few true leaves and a flattened SAM. SAM cells in homozygous *fey* mutants were more vacuolated and disorganized than their wild type counterparts. Medford et al. interpreted *fey* to be a mutant that exhausts the cell proliferation capacity of the SAM and, thus, was not self-sustaining. *FEY* has been cloned, however, and it encodes a protein similar to oxidoreductases, such as protochlorophyllide reductase (Callos et al. 1994). The expression of *FEY* is not limited to the SAM. In fact, it is expressed in mature roots, stems, and leaves. From this it is unlikely that FEY is a pattern formation mutant that is involved in the homeostasis of the zonal structure of the SAM; rather, it appears that *fey* is a defect in the function of cells in the central zone, and that this defect impacts the operation of the SAM as the cells radiate out into the peripheral zone.

Mutants that fail to produce or to maintain the SAM in embryos have been found in petunia (*Petunia hybrida*). *No apical meristem* (*nam*) mutants are arrested in seedling development and usually fail to produce first leaves (Souer et al. 1996). The SAM in *nam* seedlings is replaced by

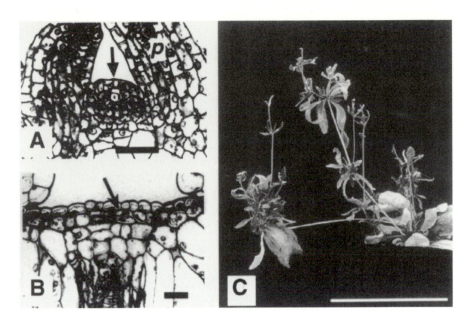

FIGURE 5.7

Comparison of shoot apical meristems (SAMs) in Arabidopsis *wuschel* (*wus*) mutant and wild type. (A) SAM (pointed out by arrow) is domed in 15-day-old wild type seedling. Leaf primordium, p. (B) SAM is flat in 7-day-old *wus* mutant and no leaf primordia are present. Arrow points to periclinal divisions in subepidermal layer that may represent initial stages of secondary meristem formation. (C) During inflorescence development, *wus* mutants develop in a "stop-and-go" mode in which bunches of leaves are produced at the base of inflorescence stems. This is thought to be caused by depletion of the inflorescence SAMs and the occasional outgrowth of inflorescence stalks from secondary meristems. Plant is about 3 months old. Bar = 5 cm. (From Laux, T., et al. 1996. The *WUSCHEL* gene is required for shoot and floral meristem integrity in Arabidopsis. Development 122:89. Reprinted by permission of the Company of Biologists Ltd.)

a group of large vacuolated cells. Some fraction of the seedlings do develop a shoot; however, the flowers on these "escaped" shoots have various abnormalities, including a doubling of primordia in the second whorl of the flower. The *nam* gene has been cloned and is not similar to other genes with known function. The gene is expressed in a ring at the boundaries of the meristems and primordia and is thought to have a role in delimiting the boundaries of the SAM (Souer et al. 1996). In this regard, it is interesting that a mutation called *cup-shaped cotyledon2* (*cuc2*) has been found in an orthologous gene in Arabidopsis (Aida et al. 1997). The mutant was actually recovered as a double mutant, *cuc1* and *cuc2*. Embryonic SAM development is blocked in the double mutant, and the cotyledons fuse to form a funnel, cuplike structure somewhat reminiscent of the effect seen in embryos treated with auxin transport inhibitors

(Fig. 2.10), except that the cuplike structure represents two fused cotyledons in the *cuc1 cuc2* mutants. Adventitious shoots could be induced in the double mutants in tissue culture, and these shoots formed fairly normal stems and leaves, but had abnormal flowers. Stamens were fused with each other, as were sepals, and petals failed to grow properly or were lost. The authors speculated that *CUC1* and *CUC2* are required for SAM development as well as for keeping certain newly forming organs within a whorl from fusing with each other. The fact that a double mutant was required to block SAM development in Arabidopsis while only a single mutation in *nam* produced the same phenotype was interpreted to mean that *cuc1* and *cuc2* have redundant gene functions in Arabidopsis.

Mutants That Overproduce
Proliferative Cells in the Meristem

During the growth of the shoot there is a continual flux of cells from the peripheral zone to the flanks of the apex where organs are formed (morphogenic zone). Cells are recruited for organ formation and replenished by the division of initials in the peripheral zone. The maintenance of the cell populations in these zones is a remarkable feat, but it is essential to the operation of the SAMs. Arabidopsis mutants have been found with defects in maintaining the balance of cell populations in these zones. One mutant, *clavata1* (*clv1*), produces a much larger SAM during vegetative and reproductive growth (Fig. 5.8C and D) (Clark et al. 1993). Weak alleles produce larger meristems with more organs in various whorls of the flower, whereas intermediate alleles cause fasciation and strong alleles result in massive overproliferation of the meristem. Root development is not obviously affected in *clv1*, which suggests that the defect is not some general defect in cell division control.

Another *clavata* mutant has been identified: a recessive mutant called *clavata3* (*clv3*) (Clark et al. 1995). *Clv3* mutants are similar to *clv1*, except that some severe alleles of *clv3* have been found in which the inflorescence meristem grows 1,000-fold larger in volume than wild type. *CLV1* and *CLV3* appear to act on the same pathway because double *clv1 clv3* mutants do not show exaggerated or novel phenotypes. Furthermore, it was found that the semidominance of *clv1* was enhanced in plants doubly heterozygous for certain *clv1* and *clv3* alleles. This, too, was taken as evidence that *CLV1* and *CLV3* interact.

CLV1 has been cloned and found to encode a putative receptor protein kinase (Clark et al. 1997). The interesting feature about the predicted *CLV1* gene product is that it is similar to some receptor kinases involved in plant defense responses (Bent 1996). These kinases have an extracellular

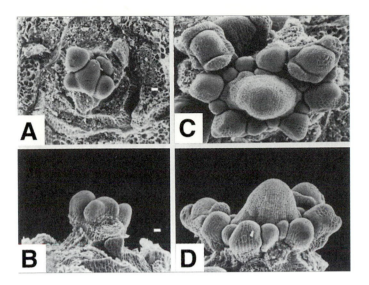

FIGURE 5.8

Comparison of shoot apical meristems (SAMs) in Arabidopsis wild type and *clavata1*-4 (*clv1*-4) mutant. SAMs of 13-day-old wild type (A, B) and *clv1*-4 (C, D) seedlings. (A, C) face or top-down views, (B, D) side views. Note larger central dome and more organ primordia in *clv1*-4. Meristems are visualized by scanning electron microscopy, and more mature organ primordia have been trimmed away. Bar = 10 μM. (From Clark, S. E., Running, M. P., and Meyerowitz, E. M. 1993. *CLAVATA1*, a regulator of meristem and flower development in Arabidopsis. Development 119:401. Reprinted by permission of the Company of Biologists Ltd.)

receptor domain, a transmembrane domain, and an intracellular kinase domain. The extracellular receptor has leucine-rich repeats that are thought to mediate protein–protein interactions between the receptor and its ligand (presumably the pathogen or a product from the pathogen). Future findings on *CLV1* and *CLV3* should be very exciting, particularly the discovery of the ligand for *CLV1*.

Clark et al. (1996) argued that the *CLV* genes might promote cell differentiation at the boundary of the peripheral zone (Fig. 5.9). *CLV1*, however, is expressed in the subepidermal layers in the central and peripheral zones of the Arabidopsis SAM (Clark et al. 1997). If *CLV1* promotes cell differentiation, then the gene must predispose cells in the peripheral zone to differentiate as they approach the outer boundary. In doing so, *CLV1* may act through *WUS* because *wus*-1 is completely epistatic to *clv1*–4 (Laux et al. 1996). Recall that *WUS* normally restrains cell differentiation; therefore, *CLV1* may negatively regulate *WUS*. Clark et al. (1996) proposed that *STM* and *CLV* have opposing effects and that *STM* stimulates the proliferation of apical initials (or slows their differentia-

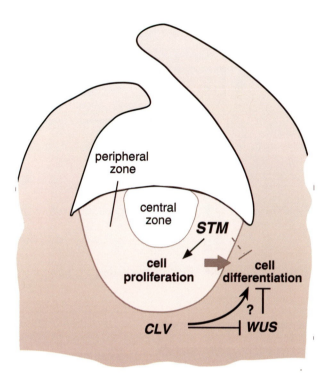

FIGURE 5.9

Model for the interaction of *CLV1* and *STM* genes in maintaining the balance and flow between proliferating and differentiating cell populations in the SAM. Under normal conditions cell proliferation keeps pace with cell differentiation, maintaining the meristem while producing new lateral organs. *STM* is expressed throughout the meristem dome, but not at sites of primordia formation. It is thought that *STM* expression is required for maintaining cell proliferation in the SAM. Mutations in *STM* result in failure to maintain the SAM. *STM* may also act to delay cell differentiation. *CLV1* is expressed across the center of the meristem, and it appears to promote cell differentiation because undifferentiated cells accumulate in *clv1* mutants. The mutant *wuschel* (*wus*) is epistatic to *clv1* and may act downstream from *clv1*. *WUS* may slow the pace of cell differentiation, and in *wus* mutants rapid cell differentiation may deplete the population of proliferating cells. (Based on Clark, S. E., et al. 1996. The *CLAVATA* and *SHOOT MERISTEMLESS* loci competitively regulate meristem activity in Arabidopsis. Development 122:1567–1575; Clark, S. E., Williams, R. K., and Meyerowitz, E. M. 1997. The *CLAVATA1* gene encodes a putative receptor kinase that controls shoot and floral meristem size in Arabidopsis. Cell 89:575–585.)

tion). Cast in this role, the proper balance of *STM* and *CLV1* and/or *CLV3* activity would be required for normal meristem growth and development. Indeed, *STM* and *CLV* appear to have counteracting effects because *clv1* can suppress the effects of *stm-1*. Double *stm-1/stm-1 clv-1/clv-1* mutants do, in fact, produce SAMs that were often larger or smaller than wild type SAMs. The single *stm-1* mutant never forms flowers; however, some of the double mutants produce flowers. Although *clv1* is largely a recessive mutation, *stm-1* can be partially rescued by *clv1* heterozygotes. That is, *stm-1/stm-1 clv-1/+* mutants have a partially restored phenotype; therefore, *clv1* is partially dominant in its ability to suppress *stm-1* (Clark et al. 1996). In a dominant fashion, *stm-1* likewise can partially suppress *clv1* mutations. This was demonstrated by examining the effects of *stm-1* on *clv1* heterozygotes. Flowers in wild type plants have two fused carpels, whereas *clv1* heterozygotes often have three or four carpels. It was shown that the effects of *clv1* increased the number of carpels in populations segregating for *stm-1*. The development and function of the SAM, therefore, is very sensitive to the dose of *STM* and *CLV* genes, and probably to the levels of the gene products.

The remarkable conclusion about the *stm clv* double mutants, however, is that *STM* is not really needed for meristem formation in the absence of *CLV* (Weigel and Clark 1996). This means that *STM* and *CLV* are truly regulators involved in balancing the proliferative and differentiation functions of the meristem, but are not really required for the development of the meristem. The genes are certainly needed for meristem homeostasis because in the absence of their function, the meristem becomes too large or too small. Nonetheless, neither gene is absolutely required for the formation of the shoot meristem; only for balancing its activities.

CELL DETERMINATION IN THE SAM

A basic question in understanding the operation of any generative organ, such as an embryo or meristem, is whether the SAM is a mosaic of determined cells, imprinted with the image of a future shoot, or a self-regulating structure with a plasticity that allows itself to internally reorganize (Kerstetter and Hake 1997). The issue is a classic one, but the problem is relevant in understanding the timing and mechanism of shoot determination. While the question can be addressed in many different ways, microsurgery, which has been used traditionally by embryologists, has been a valuable approach.

Pilkington (1929) addressed the issue by asking whether the SAM in broad bean (*Vicia faba*) or lupine (*Lupinus alba*) could regenerate following bisection with a vertical incision. Each half reorganized into a new meristem with normal symmetry. Bisection did not divide the existing center, which subsequently regenerated the missing half. New centers instead arose at the flanks of the original center. This simple experiment alone is testimony to the unusual plasticity and largely undetermined nature of the SAM.

In other microsurgical experiments, Sussex (1952) tried to find out how far one can subdivide the SAM and still obtain normal regeneration. He resected all but about one twentieth at the flank of the SAM from *Solanum tuberosum* (potato), and found that the remaining vestige regenerated normally. Thus, even cells at the flanks of the meristem can develop a new center. In general, these experiments demonstrate that the meristem is not a mosaic of predetermined cells. The SAM, instead, appears to be like a hologram in which the entire image is integrated over its surface. A hologram can be cut into smaller and smaller pieces without losing the whole image. The regeneration of the SAM indicates that even small subsections of the SAM have the ability to reorganize internally, regenerating a functional SAM.

The resection experiments reveal an interesting point. When the SAM reorganizes, it forms only one center. It has been argued that centers have an integration mechanism that operates through nutritional competition or hormonal action, preventing the development of additional centers. This has been reinforced by experiments that showed when the apical center was punctured, new apical centers formed on the flanking meristem. The numbers of new centers that formed in such experiments varied. In *Solanum* (potato), Sussex (1964) found that one new apical center was formed, while Loiseau (1959) found that in *Impatiens* two to four new apices were usually formed.

The construction of a fate map for the SAM in *Zea mays* (maize) was described in Chapter 2. Similar fate maps for the embryonic SAM in Arabidopsis have been developed, such as one by Irish and Sussex (1992) (Fig. 5.10). They irradiated Arabidopsis dry seed with X-rays and scored white or yellow sectors in which mutations in pigmentation genes had occurred. Sections through the SAM in embryos of dry seeds showed that the SAM was made up of a little more than one hundred cells in three layers. They found that most leaves had sectors covering only a fraction of the leaf, which meant that there was more than one progenitor cell for each leaf even at the dry seed stage. A cell fate map, similar to the one in maize (Chap. 2), was constructed for the embryonic SAM in Arabidopsis (i.e., the map was probabilistic in nature and

FIGURE 5.10

Fate map of the embryonic shoot apical meristem (SAM) in Arabidopsis. Map was developed by X-irradiating dry seed, looking for nonpigmented sectors and estimating the apparent cell number (ACN) for each leaf. The sectoring information was overlaid on a "face on" view of the SAM, and the cells in the layer in view account for about 60 out of approximately 110 cells in the SAM. The location of cells is representational and not actual. (Redrawn from Irish, V. F., and Sussex, I. M. 1992. A fate map of the Arabidopsis embryonic shoot apical meristem. Development 115:745–753.)

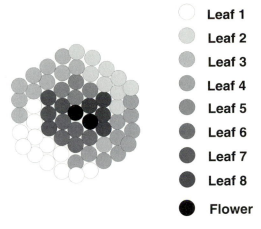

Leaf 1
Leaf 2
Leaf 3
Leaf 4
Leaf 5
Leaf 6
Leaf 7
Leaf 8
Flower

did not describe the fate of each cell). As in maize, a greater portion of the SAM at this stage was devoted to the production of lower leaves (i.e., the ACN was greater for lower leaves than for upper leaves) (Fig. 5.10). Like maize, most of the cells in the embryonic SAM were the progenitors of lower node leaves, and fewer cells were destined to become structures higher on the plant. Several sectors extended from vegetative rosette leaves into the inflorescence, which indicated that the inflorescence or flowers were not separate developmental compartments at the dry seed stage.

SAM fate maps need to be reconciled with the zonal structure of the meristem described earlier. The cells populating the periphery of the embryonic SAM will be among the first to be recruited to form the first leaves. Although the fate of these cells may not be determined, they occupy a place in the SAM where leaf formation will occur. Cells in the peripheral zone are proliferating cells that will repopulate the cells recruited for organ formation. Organs that form later in development do not appear as separate developmental compartments in the embryo, which simply reflects the fact that the cells making up these organs are represented by undifferentiated stem cells whose fate is not yet cast.

SAM MOLECULAR BIOLOGY

There is considerable interest in identifying genes specifically expressed in the vegetative SAM. Because the SAM is so small and difficult to isolate, it has been a challenge to find meristem-specific RNAs simply because it is difficult to obtain enough meristem-specific tissue. Medford et al. (1992) recognized that cauliflower heads were an ample source of SAMs and used them to isolate meristem mRNAs. (Cauliflower heads are formed because of a defect in genes that promote the floral transition. The heads are composed of inflorescences that repeatedly branch, but do not flower, forming a head literally covered with meristems.) A cDNA library was generated and differentially screened to obtain meristem-specific cDNAs. A number of cDNAs were identified by computer database searches.

The SAM is a site of active cell proliferation, and genes involved in the regulation of the cell cycle are expressed in the SAM. In general, the genes involved in regulating the cell cycle in other eucaryotes have also been found in plants. Cell cycle progress is controlled by cell division protein kinases (CDPKs) that are activated by proteins called *cyclins* and held in check by negative regulators. In yeast, cyclins transiently appear at specific cell cycle stages, and their synthesis and proteolysis drives the cell cycle (for example, see King et al. 1996). Cyclins have been cloned from a number of plants, including soybean. The pattern of expression in the SAM of a B-type cyclin, represented by *cyc5Gm*, has been examined by *in situ* hybridization (Kouchi et al. 1995). The cell cycle genes were actively expressed in a punctate pattern in the peripheral zone and rib meristem and rarely in the central mother cells (Fig. 5.11). The punctate pattern is indicative of the fact that cells in the meristem are at different cell cycle stages, and the gene is expressed only during a short interval of the cell cycle (G2-M). The expression of an S-phase marker, histone H4, was also visualized by *in situ* hybridization. The cells highlighted by the two probes were not overlapping.

Another elegant way of identifying genes that are expressed in a small structure such as a SAM is to use gene trap technology. A **gene trap** is a mobile DNA element that contains a promoterless reporter gene that can insert itself at random locations in the genome. If the gene trap inserts within a gene that is expressed in a tissue-specific manner, then the reporter gene adopts the pattern of expression of the disrupted gene. (An **exon trap** is a form of a gene trap that expresses a reporter gene when it inserts into an intron of a plant gene. The construct has splice acceptor sites that fuse the reporter gene to an upstream exon of the plant gene into which the element has inserted.) Gene trap technology has been used to find genes in an Arabidopsis collection that are expressed in various organs or tissues (Springer et al. 1995). One insertion occurred in a gene called *PROLIFERA* (*PRL*), which is expressed in various dividing cells in Arabidopsis including the SAM (Fig. 5.12). *PRL* encodes a gene product that is similar to the *MCM2-3-5* family of genes that are required for DNA replication. Unlike cyclins, *PRL* is not expressed in a punctate or cell cycle stage-specific manner, but it is expressed throughout the SAM.

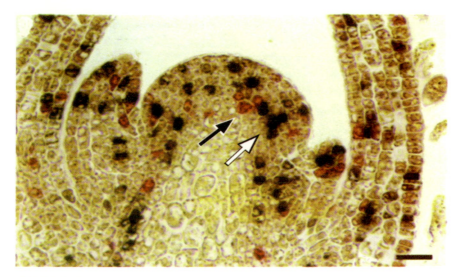

FIGURE 5.11

Pattern of expression of genes involved in cell cycle regulation in an axillary bud of soybean (*Glycine max*). Expression pattern is visualized by *in situ* hybridization. Expression of *cyc5Gm,* encoding a B-like cyclin (black arrow, red staining) and histone H4, a gene expressed during S-phase (white arrow, dark staining). Cyclin gene is expressed during G2-S and histone H4 is expressed during S-phase. Staining patterns for the two genes do not overlap. Note punctate pattern of gene expression and that expression is limited to initials in peripheral zone and not to central mother cells in the central zone. Cyclin gene probe (red stain) is labeled with digoxigenin and reacted with antidigoxigenin antibody conjugated with alkaline phosphatase. Conjugate is detected by an alkaline phosphatase reaction with Fast Red TR. Histone H4 gene probe (dark stain) is fluorescein labeled and reacted with antifluorescein antibody conjugated with alkaline phosphatase. Conjugate is detected by an alkaline phosphatase reaction with nitro blue tetrazolium. Median longitudinal sections are viewed in the light microscope. Bar = 25 μm. (From Kouchi, H., Sekine, M., and Hata, S. 1995. Distinct classes of mitotic cyclins are differentially expressed in the soybean shoot apex during the cell cycle. Plant Cell 7:1149. Reprinted by permission of the American Society of Plant Physiologists.)

PHYLLOTAXY

The arrangement of organs, such as leaves around the axis of the shoot, is called **phyllotaxy.** Phyllotaxy is established in the SAM by the pattern in which leaf primordia arise in the peripheral zone. Nothing is known in molecular terms about how leaf primordia sites are determined; however, a field theory has been used to describe the process in general terms. There are three basic phyllotactic patterns: whorled with a circle of two or more leaves at a node (including decussate arrangements with two opposite leaves per node, alternating at right angles from node to node); distichous with one leaf per node, but arranged in

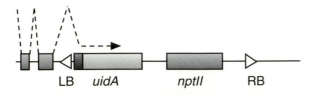

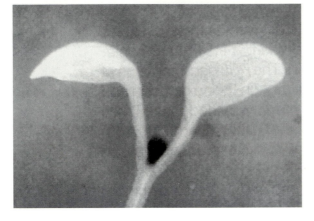

FIGURE 5.12

Gene trap construct expressed in Arabidopsis SAM. (Above) Gene trap contains a promoterless *uidA* (GUS) gene linked upstream to intron acceptor sites. If the element inserts into an intron of an active gene, then GUS will be expressed as a gene fusion with the upstream exons of the disrupted gene. Construct is located near left border, LB, of a T-DNA element that also contains an independent marker, *nptII*, which confers resistance to the antibiotic kanamycin. (Below) Gene trap insertion into the *PROLIFERA* (*PRL*) gene in Arabidopsis. GUS reporter gene in the gene trap construct is expressed in the SAM and in other dividing tissue (not shown) of Arabidopsis seedling. GUS expression is visualized by staining seedling with GUS (β-glucuronidase) substrate, X-gluc. (Reprinted with permission from Springer, P. S., et al. 1995. Gene trap tagging of *PROLIFERA*, an essential MCM2-3-5-like gene in Arabidopsis. Science 268:877. Copyright 1995 American Association for the Advancement of Science.)

two longitudinal rows; or spiraled with leaves arranged in a helical pattern (Esau 1960). The discussion here will focus on spiral phyllotaxy, which is the arrangement pattern of leaves in Arabidopsis. In spiral phyllotaxy, the arrangement is so well ordered that it can be described in mathematical terms. One property of a spiral is its direction or handedness. Handedness is fairly obvious, but it can be determined by draw-

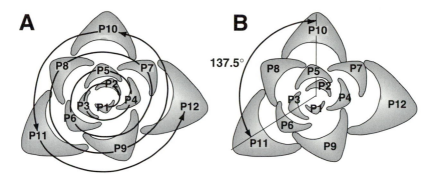

FIGURE 5.13

Characteristics of spiral leaf phyllotaxy patterns in Arabidopsis. Face-on view of shoot apical meristem (SAM) in which successive leaf primordia are numbered starting from the most recently emerged primordium. (A) Generative spiral formed by drawing a line through successive primordia. Handedness of the pattern can be determined from the generative spiral. (B) Divergence angle is the radial angle between successive primordia. In plants with spiral phyllotaxy, the divergence angle approximates 137.5 degrees.

ing a line through successive primordia numbered according to their order of appearance (Fig. 5.13A). The resulting spiral is referred to as a generative spiral. In Arabidopsis, the handedness of the spiral is maintained through vegetative development, but it can be clockwise or counterclockwise with equal frequency (Callos and Medford 1994). This means that some random event must determine the direction of the spiral at an early stage; however, once it is set, it does not change.

Pitch or the period is another characteristic property of a spiral, and pitch or period differs in various species with spiral phyllotaxy. Regardless of pitch, the angle between successive leaf primordia – the divergence angle – is the same at about 137 degrees (Fig. 5.13B). The properties of spirals with different pitch can be described by drawing "winding lines" connecting leaf primordia that differ by primordia numbers that constitute a Fibonacci series, a mathematical series in which each number is the sum of the preceding two. Spirals that differ in pitch are designated by the Fibonacci series represented in these spirals, such as 2 + 3, 3 + 5, and so on. The pitch and the pattern can be derived by measuring plastochron ratios (PR), the ratio of the distance between the center of the apex and any two successive leaf primordia. From the PR one can calculate a phyllotaxis index (PI) = 0.38 − 2.39 loglog (PR) (Richards 1951). For example, Arabidopsis has a PR of 1.20 and a PI of 3.02, which corresponds to a 3 + 5 pattern (Callos and Medford 1994).

Describing phyllotaxy in geometrical terms is important in understanding how the pattern is formed because it has been possible to develop algorithms to predict the positioning of new primordia from the pattern. In considering the formation of the pattern, one might at first think that a spiral pattern is generated on the static surface of the SAM. During growth, however, the surface of the SAM expands from its center (at an exponential rate as a function of the distance from its center), and the pattern develops dynamically by adding units in common relation to each other on an expanding surface. The situation would be somewhat like pulling a rubber sheet from four corners and, while doing so, making marks around the center of the sheet that have some constant relationship to each other. As long as the marks are offset, they will form spirals. To duplicate spiral phyllotaxy in plants, the marks should be offset from each other by 137 degrees, the divergence angle in spiral phyllotaxy. The divergence angle is fixed and can be accounted for if it is assumed that (1) primordia develop as far away as possible from previous primordia, and that (2) the distance between a new primordium and older primordia is an inverse ratio of their ages in plastochrons (Richards and Schwabe 1969).

A novel way of altering the pattern of leaf primordia in the SAM was introduced in Chapter 2. Fleming et al. (1997) demonstrated that individual leaf primordia initials could be coaxed into early emergence, changing the usual order of appearance of leaf primordia. In doing so, the handedness of phyllotaxy could be changed experimentally. Expansin, a protein that enhances cell wall extensibility, was bound to small plastic beads, and individual beads were strategically placed in the apex of tomato (*Lycopersicon esculentum*) next to the site, I2, where a leaf primordium was predicted to emerge (Fig. 5.14A). The local application of expansin induced the unscheduled outgrowth of a leaflike primordium at the I2 position (Fig. 5.14B). This outgrowth influenced the pattern of subsequent leaf primordia and switched the handedness of leaf phyllotaxy for the affected stem (Fig. 5.14C). Thus, the pattern in which leaf primordia appear and the handedness of phyllotaxy are determined by a dynamic process, not by a static plan.

The positioning of new primordia may be mediated by inhibitory factors or forces that prevent the emergence of new leaf primordia in the locale of recently emerged ones. Microsurgical studies support that view. In experiments with lupine (*Lupinus alba*), the newest leaf primordium was isolated from others by surgical incision (Snow and Snow 1931) (Fig. 5.15). The first primordium to arise after the operation appeared in the normal position, but the next was shifted toward the isolated primordium. These experiments demonstrated that the site of new leaf primordium formation is determined one to two plastochrons prior to its

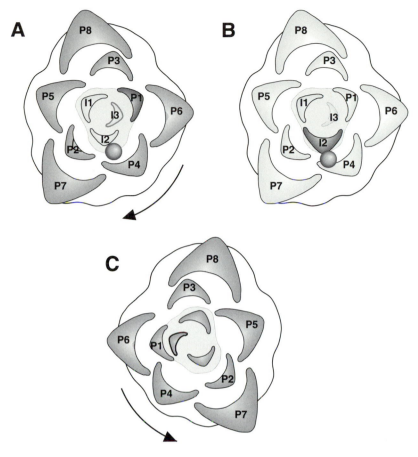

FIGURE 5.14

Experimental reversal of phyllotaxis in Arabidopsis. (A) A small plastic bead is coated with "expansin," which is a protein factor that enhances cell wall extensibility, and placed next to I2, which is the presumptive site of leaf primordium emergence in the shoot apex of a stem with a left-hand spiral. The local application of expansion leads to a bulge 5 days later that develops into a leaflike structure. The precocious emergence of an organ from I2 changes the pattern and order of appearance of leaf primordia. (C) The changed pattern reverses the phyllotaxy to a right-hand spiral, shown here at some stage much later when the new pattern has been re-established. (Based on Fleming, A. J., et al. 1997. Induction of leaf primordia by the cell wall protein expansin. Science 276:1415–1418.)

appearance at the shoot apex. Furthermore, isolation of the primordium allowed the second primordium to arise closer to existing primordia. At the time, the experiments were interpreted to reflect the availability of new space for primordia emergence; however, the experiments were later regarded as evidence for the production of gradients of factors from existing primordia that inhibit the nearby production of new primordia.

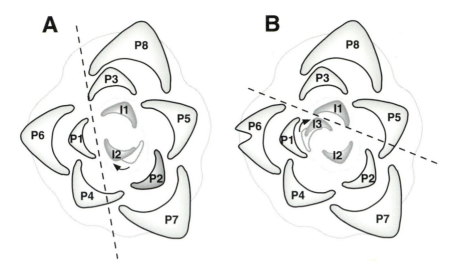

FIGURE 5.15

Using microsurgery to shift the site of appearance of a new leaf primordia in the shoot apical meristem (SAM) of lupine (*Lupinus*). Tangential incisions were made in the SAM to isolate primordia initials from newly emerged primordia or other primordia initials. Primordia are numbered successively starting from the most recently emerged primordia (P1, P2, etc.). Primordia initials are presumptive primordia and are numbered regressively, starting from the initial that is next due to emerge (I1, I2, etc.). (A) Primordia initials are isolated from P1 by incision. Position of I1 does not shift. I2 shifts in position toward P1. (B) Primordia initials are isolated from I1 by incision. Position of I2 does not shift. I3 shifts in position toward I1 (Based on Snow, M., and Snow, R. 1931. Experiments on phyllotaxis. I. The effect of isolating a primordium. Phil. Trans. Roy. Soc. London Ser. B 221:1–43. Adapted from Steeves, T. A., and Sussex, I. M. 1989. Patterns in Plant Development, 2d ed., p. 115. Cambridge, UK: Cambridge University Press. By permission of the publisher.)

It has been argued that auxin may be the inhibitory substance and that local auxin gradients may contribute to the positioning mechanism. The idea is derived from the observation that auxin transport inhibitors, such as 2,3,5-tri-iodobenzoic acid (TIBA) affect the site of leaf primordium emergence in the evening primrose (*Epilobium*) (Schwabe 1971). In the presence of TIBA, the inhibitory effects of older primordia appear to have been abrogated because leaf primordia form a collar around the apex. It is interesting that the new primordium arises as a collar because it suggests that the positioning rules may be determined by more than one gradient or field. It is possible that auxin gradients determine the circumferential location of individual leaf primordia sites, but the radial location may be determined by other factors unaffected by the auxin transport inhibitors.

ELONGATION OF INTERNODES
AND GIBBERELLIC ACID

Plants grow in height by addition of nodes (phytomers) and/or by elongation of internodes. Stature is an important agronomic quality, and the genetics of plant height in crop plants has been extensively investigated. More than fifty mutants that influence plant height have been described in maize (MacMillan and Phinney 1987). The phenotypes of a number of *dwarf* mutants, such as d_1, d_2, can be rescued by treatment with gibberellic acid (GA). Such mutants have been found to accumulate very little bioactive GA and are defective in GA biosynthesis. These mutants have been important in confirming the steps in the pathway of GA biosynthesis.

Other dwarf mutants in maize, such as *D8–1*, cannot be rescued by GA (see Swain and Olszewski 1996). D8–1 is GA insensitive and is thought to be a GA response mutant. In Arabidopsis, a *GA insensitivity* (*gai*) mutant has been identified that has all the characteristics of a GA mutant but is not rescued by GA (Peng and Harberd 1993). The mutant is dwarfed, has increased numbers of axillary shoots, has narrower, darker green leaves, and does not produce fertile flowers in continuous light. The GA insensitive mutants in both maize and Arabidopsis are dominant or semi-dominant and are thought to be gain-of-function mutations. It is not clear, therefore, that they actually are GA response pathway mutants. If they are, then it has been suggested that the mutants could be dominant-negative mutants that might encode a "poison subunit" that interferes with the function of a multisubunit component on the GA response pathway (Swain and Olszewski 1996). Nonetheless, suppressors of *gai* have been selected by screening for mutants that bolt and produce fertile flowers (Carol et al. 1995; Wilson and Somerville 1995). Some of the suppressors that have been found are intragenic suppressors at the *gai* locus, whereas others are mutations at another locus. It has been argued that the suppressor mutations may be in genes that encode components on the GA response pathway because they reduce the dependency of plant growth on GA.

Other GA-related mutants that affect plant stature are mutants that are thought to have increased GA function (GA constitutive response mutants). In an effort to find increased GA response mutants in *Arabidopsis,* Jacobsen and Olszewski (1993) isolated mutants that germinate in the presence of the GA biosynthesis inhibitor, paclobutrazol. The mutants obtained by this screen were alleles at a locus called *spindly* (*spy*). In the absence of the inhibitor the mutants resembled wild type plants that are repeatedly sprayed with GA (Fig. 5.16B). Further indica-

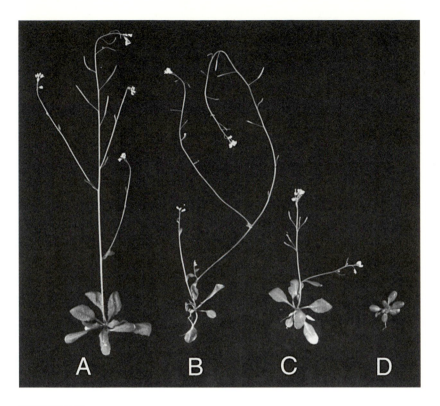

FIGURE 5.16

Phenotype of the *spindly* (*spy*) mutant in Arabidopsis. (A) Wild type plant, (B) *spy*-1 mutant. *Spy* mutants were selected for resistance to paclobutrazol, a gibberellic acid, GA, synthesis inhibitor. The mutants overcome a deficiency of GA brought about by the presence of the inhibitor. In the absence of the inhibitor, as shown here, *spy*-1 has elongated internodes and appears like a wild type plant that has been sprayed repeatedly with GA. The mutant *gibberellic acid1* (*ga1*) is defective in GA biosynthesis. (C) A double *spy*-1 *ga1*-2 mutant has an intermediate phenotype between spy-1 and (D) *ga1*-2. Photographed 4 weeks after germination. (Kindly provided by Neil E. Olszewski.)

tion that the phenotype of the *spy* mutant may, indeed, involve GA relates to the offsetting effect of a mutation in GA biosynthesis, *ga1*, on *spy* (Fig. 5.16C). The phenotype of the double *spy*-1 *ga1*-2 mutant is intermediate between the two single mutants (Fig. 5.16 B and D), which suggests that the *spy* phenotype might be an overresponse to normal GA levels. *SPY* has been cloned and was found to encode a protein with tetratricopeptide repeats (Jacobsen et al. 1996). Related proteins mediate protein–protein interactions and are found in all eucaryotes. Nonetheless, the role of *SPY* in producing a constitutive GA response is not understood. Other mutants with increased GA response have been identified in pea (*la* and *cry^s*), tomato (*pro*), and barley (*sln*). In addition to elongated internodes, these mutants have other characteristics to suggest that they have enhanced GA response. For example, *sln* mutants produce high levels of α-amylase in germinating seeds, and the mutants are male sterile, all of which are characteristics predicted for an overactive GA response (Swain and Olszewski 1996).

GA stimulates the elongation of internodes in stem segments from plants such as oat (*Avena*). The hormone appears to act in a number of ways that bear upon cell elongation. First, GA causes "loosening" of the cell wall (Cosgrove 1993). The cell wall bears considerable stress from the turgor pressure of the cell. Loosening allows stress relaxation of the wall, uptake of water, and expansion of the cell. GA also stimulates synthesis of new cell wall polysaccharides (Montague 1995). Production of new cell wall components is one of the first biosynthetic processes that one can detect after adding GA. In addition, GA alters microtubule and microfibril orientation (Duckett and Lloyd 1994). The direction of cell expansion depends on the orientation of cellulose microfibrils in the cell wall, and transverse orientation of microfibrils encourages cell elongation. The direction in which cellulose microfibrils are laid down in the cell wall is in turn correlated with the orientation of cortical microtubules in the cytoplasm. Duckett and Lloyd (1994) studied the reorientation of microtubules in dwarf (*le*) mutants of pea (*Pisum sativa*) that have defects in GA biosynthesis. They observed a reorientation of microtubules in epidermal cells of the mutant after treating internode segments with GA$_3$. Microtubules were reoriented transversely and correlated in time with the increase in GA-stimulated elongation rate. The stimulus for the reorientation is not known; however, the reorientation correlated with a change in α-tubulin isotypes, and it was speculated that posttranslational modifications of tubulin might be involved in the response.

PHASE CHANGES IN VEGETATIVE GROWTH

There are two very distinctive phases of shoot development: vegetative and reproductive growth. The transition from vegetative to reproductive growth (which will be discussed in Chap. 7) reflects differences in developmental programs carried out in the shoot meristems. During vegetative growth, organs such as leaves, buds, and side branches are produced by shoot meristems, whereas organs of the inflorescence or flower are elaborated during reproductive development. In order to understand the transitions in growth phase one must know what is happening in the SAM.

Although the change from vegetative growth to flowering (reproductive growth) can be quite a dramatic transition in the life of a plant,
other morphological and physiological changes occur during the vegetative phase itself. During their vegetative development, plants undergo a
transition from juvenile to adult forms during which they acquire
reproductive competence. The adult vegetative phase is characterized
by its competence to produce reproductive structures, while the juvenile form is incompetent to do so. Certain woody plants have heteroblastic patterns of development in that juvenile and adult phases are
more distinct and their transitions are more obvious. For example, in
English ivy (*Hedera helix*), many characteristics of the plant change during the juvenile-to-adult transition, including the morphology and
arrangement of leaves (phyllotaxy) and the growth habit of the plant.
In addition, in many herbaceous plants, the size and shape of leaves
change during the course of vegetative development. For example, in
Arabidopsis the first true leaves are similar to cotyledons with long petioles and a rather roundish shape (Fig. 5.17A). Later leaves of the rosette
look more like cauline leaves with short petioles and a more ovate
shape (Tsukaya 1995). The changing size and shape of leaves is a form
of heteroblastic development called **heterophylly.** Heterophylly was
thought to be a major source of evolutionary variation in leaf design. It
was argued nearly a century ago that changes in leaf shape during
development of a plant might reflect the variety of leaf forms in the taxonomic group to which the plant belongs (Goebel 1900) – a form of
"ontogeny recapitulates phyllogeny."

Poethig promoted the idea that the vegetative development can be
further delineated into phases and that the phases are genetically programmed (Poethig 1990; Lawson and Poethig 1995). The number and
characteristics of the phases depend on the plant, and there are not
always sharp transitions during the phase changes from juvenile to
adult forms. Phase transitions are more subtle in maize (*Zea mays*) and
involve changes in leaf morphology and the production of epidermal
hairs, elongation of internodes, reduction in adventitious roots and secondary compounds such as anthocyanin, and epicuticular wax.

In maize, the division of vegetative growth into juvenile and adult
phases is supported by the discovery of mutations that affect the phase
transitions (Dudley and Poethig 1991). Three semidominant gain-of-
function mutations called *Teopod* (*Tp1*, *Tp2*, and *Tp3*) affect this transition
and are thought to extend the juvenile vegetative phase (Fig. 5.18A–C).
Dudley and Poethig (1991) argued that the juvenile phase in the *Tp*
mutants overlaps the adult phase, which results in vegetative growth
that is largely intermediate between juvenile and adult. Tassels and ears

FIGURE 5.17

Leaf heteroblasty or heterophylly in Arabidopsis. Leaves from (A) wild type, (B) *angustifolia,* and (C) *rotundifolia* mutants. From left to right are leaves from progressively higher positions on the shoot: two cotyledons, eight rosette leaves, and three cauline or flower stalk leaves. Leaves were collected when fully expanded. Mutants are discussed in Chapter 6. (From Tsuge, T., Tsukaya, H., and Uchimiya, H. 1996. Two independent and polarized processes of cell elongation regulate leaf blade expansion in *Arabidopsis thaliana* (L.) Heynh. Development 122:1589. Reprinted by permission of the Company of Biologists Ltd.)

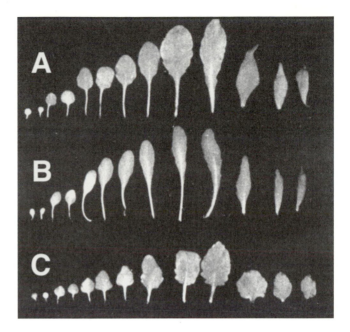

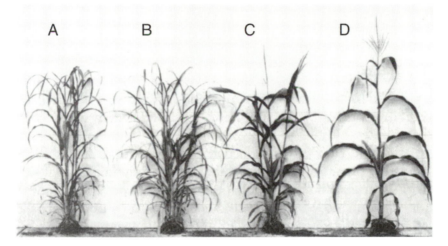

FIGURE 5.18

Teopod mutants in maize have an abundance of slender juvenile leaves. (A) *Tp1/+*, (B) *Tp2/+*, (C) *Tp3/+*, and (D) wild type plants, all in the same inbred background. *Teopod* mutations are dominant and are thought to extend the juvenile phase of vegetative growth. (Reprinted with permission from Poethig, R. S. 1990. Phase change and the regulation of shoot morphogenesis. Science 250:928. Copyright 1990 American Association for the Advancement of Science.)

formed on the normally adult vegetative stem appear to be intermediate in character (i.e., composed of both foliage leaves and flowers).

A question that has been posed about phase transitions is whether the phases are coupled (i.e., whether the onset of one phase is triggered by the end of the succeeding phase). Using the *Tp* mutants, Bassiri et al. (1992) asked if by extending the vegetative phase one would affect the onset of the reproductive phase. The vegetative phase in *Tp2/+* plants is longer (21–26 plastochrons) compared with wild type (18–19 plastochrons). The onset of short day (SD) photosensitivity is considered to mark the beginning of the reproductive phase because this is the time when plants acquire reproductive competence. They found that the onset of SD photosensitivity (~15–18 plastochrons) was nearly the same in both wild type and *Tp2* mutant, which suggests that the beginning of the reproductive phase was independent of the length of the vegetative phase.

Tp1 and *Tp2* are thought to act through diffusible factors because they are not cell autonomous in genetic mosaics. The nature of the diffusible substances is not known; however, Evans and Poethig (1995) found that the hormone GA had profound effects on the juvenile-to-adult phase transitions. Any one of a number of mutants that reduce GA accumulation, particularly *dwarf* mutants, delays the adult phase transition. Through the analysis of double mutants, *dwarf* appears to have synergistic effects with *Tp1* or *Tp2*. In some of the double mutants, the juvenile phase is extended for so long that even leaves in the tassels have juvenile characters (epicuticular wax). Other phase indicators, however, were less affected in the double mutants. This suggests that the *Tp* and GA response pathways are not the same, yet they interact. If the pathways were the same, then one might expect *dwarf* to be epistatic to *TP,* rather than synergistic with it. This picture is consistent with the finding that application of gibberellic acid (GA3) to *Tp1* and *Tp2* can only partially rescue the mutant phenotype. The regulation of phase change therefore appears to be complex with a number of different inputs controlling juvenile-to-adult phase transitions. GA plays an important, although not exclusive, role in promoting the transition.

Although *Tp* genes appear to control phase transitions, the genes are thought to work through other genes that determine the phase characters. Intermediates in this process are genes such as *Glossy15* (*Gl15*), which affect phase transitions in epidermal cells only. Loss-of-function mutations in *Gl15* shorten the time of expression of juvenile phase characters, such as epicutular wax production, and accelerate the onset of adult characters (Moose and Sisco 1994). *GL15* acts in a cell-autonomous manner downstream from *Tp1* and *Tp2*. The interpretation of *GL15* action is that it conditions juvenile traits in epidermal cells in response to more global programs of juvenility operating throughout the vegetative shoot.

REFERENCES

Aida, M., Ishida, T., Fukaki, H., Fujisawa, H., and Tasaka, M. 1997. Genes involved in organ separation in Arabidopsis: An analysis of the *cup-shaped cotyledon* mutant. Plant Cell 9: 841–857.

Barton, M. K. and Poethig, R. S. 1993. Formation of the shoot apical meristem in *Arabidopsis thaliana:* An analysis of development in the wild type and in the shoot meristemless mutant. Development 119: 823–831.

Bassiri, A., Irish, E. E., and Poethig, R. S. 1992. Heterochronic effects of *teopod 2* on the growth and photosensitivity of the maize shoot. Plant Cell 4: 497–504.

Bent, A. F. 1996. Plant disease resistance genes: Function meets structure. Plant Cell 8: 1757–1771.

Callos, J. D., Dirado, M., Xu, B., Behringer, F. J., Link, B. M., and Medford, J. I. 1994. The *forever young* gene encodes an oxidoreductase required for proper development of the Arabidopsis vegetative shoot apex. Plant J. 6: 835–847.

Callos, J. D. and Medford, J. I. 1994. Organ positions and pattern formation in the shoot apex. Plant J. 6: 1–7.

Carol, P., Peng, J., and Harberd, N. P. 1995. Isolation and preliminary characterization of *gas1*-1, a mutation causing partial suppression of the phenotype conferred by the gibberellin-insensitive (gai) mutation in *Arabidopsis thaliana* (L.) Heyhn. Planta 197: 414–417.

Christianson, M. L. 1986. Fate map of the organizing shoot apex in *Gossypium*. Am. J. Bot. 73: 947–958.

Clark, S. E., Jacobsen, S. E., Levin, J. Z., and Meyerowitz, E. M. 1996. The *CLAVATA* and *SHOOT MERISTEMLESS* loci competitively regulate meristem activity in Arabidopsis. Development 122: 1567–1575.

Clark, S. E., Running, M. P., and Meyerowitz, E. M. 1993. *CLAVATA1*, a regulator of meristem and flower development in Arabidopsis. Development 119: 397–418.

Clark, S. E., Running, M. P., and Meyerowitz, E. M. 1995. *CLAVATA3* is a specific regulator of shoot and floral meristem development affecting the same processes as *CLAVATA1*. Development 121: 2057–2067.

Clark, S. E., Williams, R. K., and Meyerowitz, E. M. 1997. The *CLAVATA1* gene encodes a putative receptor kinase that controls shoot and floral meristem size in Arabidopsis. Cell 89: 575–585.

Cosgrove, D. 1993. How do plant cell walls extend? Plant Physiol. 102: 1–6.

Dermen, H. 1945. The mechanism of colchicine-induced cytohistological changes in cranberry. Am. J. Bot. 32: 387–394.

Duckett, C. M. and Lloyd, C. W. 1994. Gibberellic acid-induced microtubule reorientation in dwarf peas is accompanied by rapid modification of an alpha-tubulin isotype. Plant J. 5: 363–372.

Dudley, M. and Poethig, R. S. 1991. The effect of a heterochronic mutation *teopod2* on the cell lineage of the maize shoot. Development 111: 733–740.

Endrizzi, K., Moussian, B., Haecker, A., Levin, J. Z., and Laux, T. 1996. The *SHOOT MERISTEMLESS* gene is required for maintenance of undifferentiated cells in Arabidopsis shoot and floral meristems and acts at a different regulatory level than the meristem genes *WUSCHEL* and *ZWILLE*. Plant J. 10: 967–979.

Esau, K. 1960. The Anatomy of Seed Plants. New York: John Wiley and Sons.

Evans, M. M. S. and Poethig, R. S. 1995. Gibberellins promote vegetative phase change and reproductive maturity in maize. Plant Physiol. 108: 475–487.

Evans, M. W. and Grover, F. O. 1940. Developmental morphology of the growing point of the shoot and the inflorescence in grasses. J. Agricul. Res. 61: 481–520.

Fleming, A. J., McQueen-Mason, S., Mandel, T., and Kuhlemeier, C. 1997. Induction of leaf primordia by the cell wall protein expansin. Science 276: 1415–1418.

Gilbert, S. F. 1994. Developmental Biology. Sunderland, MA: Sinauer Associates, Inc.

Goebel, K. 1900. Organography of plants. I. General organography. New York: Hafner Publishers.

Irish, V. F. and Sussex, I. M. 1992. A fate map of the Arabidopsis embryonic shoot apical meristem. Development 115: 745–753.

Jackson, D., Veit, B., and Hake, S. 1994. Expression of maize KNOTTED1 related homeobox genes in the shoot apical meristem predicts patterns of morphogenesis in the vegetative shoot. Development 120: 405–413.

Jacobsen, S. E., Binkowski, K. A., and Olszewski, N. E. 1996. SPINDLY, a tetratricopeptide repeat protein involved in gibberellin signal transduction in Arabidopsis. Proc. Natl. Acad. Sci. USA 93: 9292–9296.

Jacobsen, S. E. and Olszewski, N. E. 1993. Mutations at the spindly locus of Arabidopsis alter gibberellin signal transduction. Plant Cell 5: 887–896.

Kerstetter, R. A. and Hake, S. 1997. Shoot meristem formation in vegetative development. Plant Cell 9: 1001–1010.

King, R. W., Deshaies, R. J., Peters, J. M., and Kirschner, M. W. 1996. How proteolysis drives the cell cycle. Science 274: 1652–1659.

Kouchi, H., Sekine, M., and Hata, S. 1995. Distinct classes of mitotic cyclins are differentially expressed in the soybean shoot apex during the cell cycle. Plant Cell 7: 1143–1155.

Laux, T., Mayer, K. F. X., Berger, J., and Jürgens, G. 1996. The WUSCHEL gene is required for shoot and floral meristem integrity in Arabidopsis. Development 122: 87–96.

Lawson, E. J. R. and Poethig, R. S. 1995. Shoot development in plants: Time for a change. Trends Genet. 11: 263–268.

Loiseau, J. E. 1959. Observation et experimentation sur la phyllotaxie et le fonctionnement du sommet vegetatif chez quelques Balsaminacees. Ann. Sci. Nat. Bot. Ser. 20: 1–214.

Long, J. A., Moan, E. I., Medford, J. I., and Barton, M. K. 1996. A member of the KNOTTED class of homeodomain proteins encoded by the STM gene of Arabidopsis. Nature 379: 66–69.

Lu, P., Porat, R., Nadeau, J. A., and O'Neill, S. D. 1996. Identification of a meristem L1 layer-specific gene in Arabidopsis that is expressed during embryonic pattern formation and defines a new class of homeobox genes. Plant Cell 8: 2155–2168.

MacMillan, J. and Phinney, H. O. 1987. Biochemical genetics and the regulation of stem elongation by gibberellins. In Physiology of cell expansion during plant growth, ed. Cosgrove, D. J., and Knievel, D. P., 156–171. Rockville, MD: American Society of Plant Physiology.

Medford, J. I. 1992. Vegetative apical meristems. Plant Cell 4: 1029–1039.

Medford, J. I., Behringer, F. J., Callos, J. D., and Feldmann, K. A. 1992. Normal and abnormal development in the Arabidopsis vegetative shoot apex. Plant Cell 4: 631–643.

Montague, M. J. 1995. Hormonal and gravitropic specificity in the regulation of growth and cell wall synthesis in pulvini and internodes from shoots of Avena sativa L. (oat). Plant Physiol. 107: 553–564.

Moose, S. P. and Sisco, P. H. 1994. Glossy15 controls the epidermal juvenile-to-adult phase transition in maize. Plant Cell 6: 1343–1355.

Peng, J. and Harberd, N. P. 1993. Derivative alleles of the Arabidopsis *gibberellin-insensitive* (*gai*) mutation confer a wild type phenotype. Plant Cell 5: 351–360.

Pilkington, M. 1929. The regeneration of the stem apex. New Phytol. 28: 37–53.

Poethig, R. S. 1990. Phase change and the regulation of shoot morphogenesis. Science 250: 923–930.

Richards, F. J. 1951. Phyllotaxis: Its quantitative expression and relation to growth in the apex. Phil. Trans. Roy. Soc. Lond. Ser. B 235: 509–564.

Richards, F. J. and Schwabe, W. W. 1969. Phyllotaxis: A problem of growth and form. In Plant Physiology: A treatise, ed. Steward, F. C., 79–116. New York: Academic Press.

Satina, S., Blakeslee, A. F., and Avery, A. G. 1940. Demonstration of the three germ layers in the shoot apex of *Datura* by means of induced polyploidy in periclinal chimeras. Am. J. Bot. 27: 895–905.

Schwabe, W. W. 1971. Chemical modification of phyllotaxis and its implications. Symp. Soc. Exp. Biol. 25: 301–322.

Sharman, B. B. 1942. Developmental anatomy of the shoot of *Zea mays* L. Ann. Bot. 6: 245–284.

Sinha, N. R., Williams, R. E., and Hake, S. 1993. Overexpression of the maize homeobox gene, *KNOTTED-1,* causes a switch from determinate to indeterminate cell fates. Genes & Devel. 7: 787–795.

Smith, L., Greene, B., Veit, B., and Hake, S. 1992. A dominant mutation in the maize homeobox gene, *knotted-1,* causes its ectopic expression in leaf cells with altered fates. Development 116: 21–30.

Snow, M. and Snow, R. 1931. Experiments on phyllotaxis. I. The effect of isolating a primordium. Phil. Trans. Roy. Soc. London Ser. B 221: 1–43.

Souer, E., Van Houwelingen, A., Kloos, D., Mol, J., and Koes, R. 1996. The *no apical meristem* gene of petunia is required for pattern formation in embryos and flowers and is expressed at meristem and primordia boundaries. Cell 85: 159–170.

Springer, P. S., McCombie, W. R., Sundaresan, V., and Martienssen, R. A. 1995. Gene trap tagging of *PROLIFERA,* an essential MCM2-3-5-like gene in Arabidopsis. Science 268: 877–880.

Steeves, T. A. and Sussex, I. M. 1989. Patterns in Plant Development. Cambridge: Cambridge University Press.

Sussex, I. M. 1952. Regeneration of the potato shoot apex. Nature 170: 755–757.

Sussex, I. M. 1964. The permanence of meristems: Developmental organizers or reactors to exogenous stimuli. Brookhaven Symp. 16: 1–12.

Swain, S. M. and Olszewski, N. E. 1996. Genetic analysis of gibberellin signal transduction. Plant Physiol. 112: 11–17.

Tsuge, T., Tsukaya, H., and Uchimiya, H. 1996. Two independent and polarized processes of cell elongation regulate leaf blade expansion in *Arabidopsis thaliana* (L.) Heynh. Development 122: 1589–1600.

Tsukaya, H. 1995. Developmental genetics of leaf morphogenesis in dicotyledonous plants. J. Plant Res. 108: 407–416.

Vaughn, J. G. 1952. Structure of the angiosperm apex. Nature 171: 751–752.

Weigel, D. and Clark, S. E. 1996. Sizing up the floral meristem. Plant Physiol. 112: 5–10.

Wilson, R. N. and Somerville, C. R. 1995. Phenotypic suppression of the *gibberellin insensitive mutant* (*gai*) of Arabidopsis. Plant Physiol. 108: 495–502.

6

Leaf Development

One only has to look down on the shoot apex of a growing plant to appreciate why leaf formation is of great interest to plant developmental biologists (Fig. 6.1). In an actively growing shoot apex, leaves can be found at many stages of development. When older leaves are peeled

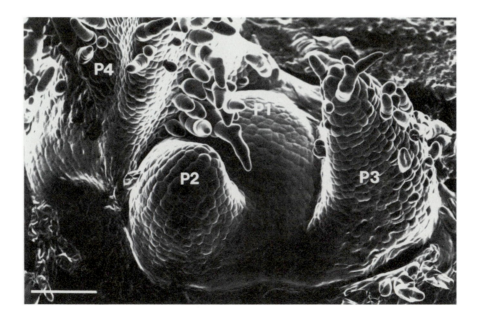

FIGURE 6.1

Shoot apex of tobacco (*Nicotiana tabacum*). Four leaf primordia are visible, P1–P4. Trichomes begin to appear on the apex of P3, and the lamina or leaf blade starts to expand in P4. Shoot apex visualized by scanning electron microscopy. Bar = 100 μm. (From Poethig, R. S., and Sussex, I. M. The developmental morphology and growth dynamics of the tobacco leaf *Nicotiana tabacum* cultivar xanthi-nc. Planta 165:161, Fig. 1, by permission of the author and Springer-Verlag GmbH & Co. KG. Copyright © 1985 Springer-Verlag.)

back from the shoot apical meristem (SAM), younger and younger leaves can be seen. Leaf primordia are formed sequentially during shoot growth, which gives rise to the graded series of developing leaf primordia and leaves at the shoot apex. Because leaf primordia develop sequentially rather than synchronously, they must be able to control much of their own development.

Simple leaves are determinate structures composed of three parts: a leaf blade or lamina, a petiole or stalk, and a leaf base that may surround the stem (Poethig 1997). Leaves, unlike stems, are dorsiventral, that is, the upper (dorsal) surface of the leaf differs from lower (ventral) surface. (New leaves do not develop in an extended position with one surface facing up and the other facing down; rather, they are wrapped closely around the axis of the shoot or stem. Because of that the dorsal surface of the developing leaf, which is adjacent to the stem axis, is called the adaxial surface, and the ventral surface is called the abaxial surface.) The midrib contributes to the dorsiventrality of the leaf in that the midrib protrudes from the ventral or abaxial surface of the leaf.

A dicot leaf primordium first emerges as a small ridge or leaf buttress on the shoot apex. As it elongates, the primordium extends laterally and, in some species, the base can nearly encircle the stem. The primordium elongates forming the apical–basal axis, and the leaf blade or lamina expands laterally from the early midrib. The development of dorsiventrality is an early, committing step in leaf development, and it appears at the time of primordia emergence. Dorsiventrality distinguishes the early development of leaves from the development of a stem or a branch that is radially symmetrical along its axis.

LEAF DETERMINATION

When do primordia become determined to develop into leaves? Steeves and Sussex (1957) addressed this issue by explanting leaf primordia at various stages of development. In such experiments, **determination** is defined as the stage when an explanted leaf primordia will develop into a leaf in tissue culture. When explanted prior to that stage, primordia produce shootlike structures without dorsiventrality. When leaf primordia were explanted at various stages of development in the fern, *Osmunda cinnamomea*, P8–P10 formed leaves at high frequency, while P1–P5 failed to do so (Steeves and Sussex 1957). (As described in the previous chapter, leaf primordia are numbered in order of their appearance with the youngest being P1. Hence, P8–P10 were the eighth to tenth youngest primordia on a shoot apex when they were explanted.)

Thus, leaf determination occurs between P5 and P8 in *Osmunda*. Similar experiments have been conducted in tobacco (*Nicotiana tabacum*) and sunflower (*Helianthus*), and leaf determination occurs at an earlier stage, P2.

Hicks and Steeves (1969) asked whether leaf determination depends upon the influence of more central cells in the apical meristem in *Osmunda*. The answer here appears to be *yes*. When P1 was isolated from the apex by making an incision and inserting a thin mica sheet, shoots arose more frequently than did leaves. Thus, tissue continuity and, possibly, the movement of diffusible substances from the apical center to the incipient leaf primordium are required for the decision about whether development follows the course of determinate growth (leaf development) or indeterminate growth (shoot development).

Some important points emerge from the microsurgical studies. First, leaf determination occurs in these systems after primordia emerge. There is a brief period, therefore, when a young primordia is not determined to form a leaf, and other routes of development may be followed depending on environment and developmental cues. Many structures or organs of the shoot, such as floral organs, are modified leaves (see Chap. 8), and through evolution some of these structures may have arisen by the imposition of other developmental programs on uncommitted leaf primordia. Second, leaf primordia become determined soon after emergence and, in doing so, acquire the capacity to develop independently. It is remarkable that a leaf primordia less than 1 mm in length, such as a leaf primordia from sunflower, can give rise to a normal leaf in culture (Steeves et al. 1957) (Fig. 6.2). Leaf primordia must have mechanisms to guide their own development at very early stages. Third, leaf determination does not appear to be a one-step process. There are decisions to be made whether growth will be determinate or indeterminate and further decisions about whether growth will have dorsiventral symmetry or be radial. There may also be progressive decisions that involve different regions of the prospective leaf at different stages of primordia development. For example, Sachs (1969) found that the leaf structures derived from the margins of the leaf primordia in pea (*Pisum sativa*) could be regenerated if the margins were microsurgically removed when the leaf primordia were less than 30 μm in length. Removal of margins later in development resulted in loss of leaf parts, tendrils, leaflets, and/or stipules. (Peas have compound leaves composed of a various leaf parts, which will be discussed in a later section on compound leaves.)

FIGURE 6.2

Sunflower (*Helianthus annus*) leaf grown to maturity in sterile culture from an explanted leaf primordium (P8). Growth of a normal leaf in culture demonstrates that sunflower leaf primordia are determined to form leaves at an early stage. (Reprinted with permission from Steeves, T. A., Gabriel, H. P., and Steeves, M. W. 1957. Growth in sterile culture of excised leaves of flowering plants. Science 126:350. Copyright 1957 American Association for the Advancement of Science.)

ORIGINS OF LEAF PRIMORDIA IN DICOTS

As discussed in Chapter 5, leaf primordia arise at predicted positions in the shoot apex, and a position is established about one to two plastochrons before emergence. One can therefore examine presumptive primordial sites and determine what events occur prior to emergence. Emergence of primordia is accompanied by changes in both the plane and rate of division of subepidermal cells. In pea, changes in the rate of division at the presumptive leaf primordium site are also evident one to two plastochrons prior to leaf emergence (see Chapter 5 for the definition of plastochron). Changes in the plane of division appear later about one-half plastochron prior to emergence (Lyndon 1983). Reorientation of cell wall cellulose microfibrils foretell the site of emergence of a leaf primordium. Microfibrils in the L1 layers of periwinkle (*Vinca major*) encircle the shoot apex, presumably constraining growth along the shoot axis (Jesuthasan and Green 1989). Leaf primordium emergence is preceded by a local reorientation of microfibrils that produces a cytoskeletal array which surrounds the incipient leaf primordium. The reorientation of microfibrils is thought to shift the polarity of growth associated with primordium emergence.

How large is the founding cell population for a leaf primordium before it emerges? Is a leaf derived a single cell in the SAM? We actually

FIGURE 6.3

Periclinal chimeras in geranium (*Pelargonium*). In these chimeras, L2 or L3 layers are either green (chlorophyll containing) or white (albino). (L1 is mostly nonchlorophyllous and does not contribute to the pigmentation pattern.) (A) Green-over-white chimera, which means that the L2 layer is green and the L3 layer is white. The periphery of the leaves, which is made up almost exclusively of L2 cells, is green, whereas the central area, composed of L2 and L3, is a pale green. (B) White-over-green chimera, which means that L2 is white and L3 is green. The periphery of the leaves in these chimeras is almost pure white because it is made up exclusively of albino L2 cells. The central area in green-over-white chimeras is smaller than in white-over-green chimeras and is thought to result from an unknown compensatory mechanism that maximizes green leaf surface area. (From Tilney-Bassett, R. A. E. 1963. Genetic and plastid physiology in *Pelargonium*. Heredity 18:485–504.)

know that leaves cannot be derived from a single cell because leaves are composed of cells from all three cell layers, L1–L3. All three layers, therefore, must contribute to the formation of a new leaf. L1 is generally a nonchlorophyllous layer of epidermal cells that covers the surface of the leaf. The bulk of a dicot leaf is made up of L2 (Tilney-Bassett 1963). L2 gives rise to the palisade and spongy mesophyll cells at the center of the leaf and the spongy mesophyll cells at the periphery. Much of the vasculature of the midrib and the inner spongy mesophyll layers in the middle of the leaf arise from L3. The relative contribution of L2 and L3 can be seen in periclinal (layer) chimeras in which either L2 or L3 have defects in plastid development and are white or nonchlorophyllous layers. The leaves of green-over-white chimeras in geranium (*Pelargonium*) have a pale central region where cells in the L3 layers are located (Fig. 6.3) (Tilney-Bassett 1963). The central region is pale green because it is composed of green L2 cells overlaying white L3 cells. On the other hand, leaves of white-over-green chimeras have green centers and white margins. The periphery of these chimeric leaves is made up exclusively of nonpigmented L2 cells and is almost pure white. (The leaf pigmentation patterns in the two chimeras are not an exact complement because the green sectors are larger than expected. An unknown compensation mechanism that allows for preferential development of the photosynthetic layer apparently operates in such chimeric plants.)

The size of the founding cell population for a dicot leaf was determined by Poethig and Sussex (1985), who used cell lineage analysis in tobacco (*Nicotiana tabacum*), as described in Chapter 2. Clonal sectors in the tobacco leaf were identified following X-ray irradiation of buds heterozygous for genes conferring albinism (*a1* and *a2*). The resulting sectors were composed of subepidermal cells, mostly palisade and mesophyll cells (L2 cells), that had lost either *a1* or *a2* and were darker green, or cells that had lost the corresponding wild type alleles and were yellow. (Sectors in the epidermal layer were not scored because most cells in the epidermal layer do not contain chloroplasts).

The number of leaf founder cells (apparent cell number, ACN) was determined at different developmental stages. When irradiated at 3 days prior to leaf initiation, the sectors occupied only a fraction of the leaf, which indicated that there are many founder cells for a leaf even before the primordium appears on the shoot apex (Fig. 6.4A). They estimated from the average length of sectors along the axis of the leaf that the presumptive leaf primordium at 3 days before emergence must be about three cell tiers high. From the average width of the sectors, they estimated that each tier must be about twelve to thirteen cells wide. The total number of cells in the shoot apex giving rise to the leaf 3 days prior to leaf primordia emergence must therefore be about thirty-five to forty cells per layer, or a total of 100–200 cells.

Despite the fact that more than 100 cells are destined to become a leaf primordium at 3 days prior to emergence, their fate is not determined until close to the time of primordium emergence. As described by the microsurgery experiments of Snow and Snow in Chapter 5, the position of the new primordia can be shifted one to two plastochrons prior to their appearance. To rationalize this, one might think of the leaf primordium "anlage" (cells in the meristem destined to become a primordium) as a region defined by a small "field" of a diffusible growth-promoting substance. The field would be ethereal in the sense that it can be moved by perturbing the shoot apex, and yet it cuts across all three cell layers. On the other hand, the location of the prospective primordium could be determined by biophysical pressures that could change depending on other distortions or compression forces in the area.

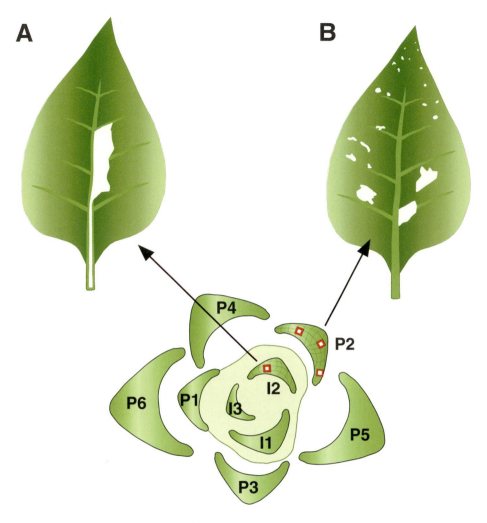

FIGURE 6.4

Sector analysis in developing tobacco (*Nicotiana tabacum*) leaves. Drawing shows a shoot apex (below) that was X-irradiated; marked cells that give rise to unpigmented sectors are outlined in red. Leaf primordia that have emerged are designated P1, P2 and so forth. Primordia initials that have not emerged from the shoot apex are indicated as I1, I2, and so on. (A) Nonpigmented sector in leaf that results from irradiation of primordia initial (I2) 1–2 plastochrons before primordia emergence. Apparent cell number (ACN) was determined by estimating the lateral and longitudinal dimensions of sectors from many sectored leaves. (B) Nonpigmented sectors produced in leaf by X-irradiation of emerged leaf primordium (P2) were smaller and indicated growth by intercalary cell divisions. Sectors were larger at the base of the leaf than at the tip because cell division frequencies declined from the tip to the base of the leaf during maturation. (Based on Poethig, R. S., and Sussex, I. M. 1985a. The cellular parameters of leaf development in tobacco *Nicotiana tabacum:* a clonal analysis. Planta 165:170–184.)

DICOT LEAF DEVELOPMENT

In dicots, leaf development involves growth along the apical–basal axis of the leaf (growth along the midrib in pinnately veined leaves) and growth of the leaf blade or lamina. In tobacco, the leaf primordium develops into a structure about 0.5 mm long before the lamina begins to expand (Poethig and Sussex 1985b). Are the various parts of the leaf, such as the lamina, midrib, and petiole, derived from different cells in the leaf primordium? Other observations in the studies described earlier by Poethig and Sussex (1985) relate to the origin of various leaf parts. The authors observed three general boundaries for sectors generated by irradiation prior to leaf initiation. The sectors ran (1) from the middle of the midrib to the leaf margin, (2) from the internode through the midrib to the leaf margin, and (3) from the internode through the midrib terminating before leaf margin (the latter is shown in Fig. 6.4A). They concluded that the major parts of the leaf (i.e., lamina, midrib, petiole) were derived from the same initials because many clones encompassed all three structures. Because a number of the sectors involved only the midrib and petiole, the midrib and its position appeared to be fixed very early in development. Since midrib sectors that also included the lamina often did not do so in the basal portion of the leaf (as in Fig. 6.4A), it could be concluded that the midrib formed first (and was composed of the largest sectors) followed by the formation of the lamina. That is consistent with the morphological observations on tobacco leaf development described earlier.

Dicot leaves, in particular, vary enormously in shape and size from species to species. How is the shape of the lamina in a dicot leaf determined? Despite the difference in leaf shapes at maturity, leaf primordia and young leaves before lamina expansion look quite similar in different species. The events that shape the dicot leaf, therefore, happen largely after the leaf primordia stage and during the expansion of the lamina or leaf blade. Poethig and Sussex (1985) tested a theory that leaves grow from meristematic cells at their tip or at their margin, and that stereotypic patterns of divisions in these cells determine leaf shape. They studied the development of tobacco leaves, which are simple and ovate. One might expect that if such leaves grow by proliferation of cells at the tip and margins, the highest rates of cell division would occur at the tip. Poethig and Sussex found that was not the case; rather, higher rates of cell division were found toward the base of the leaf than at the tip.

Poethig and Sussex (1985) used sector analysis to describe the growth patterns of the lamina in tobacco leaves. To do so, leaf sectors were analyzed as described earlier, except the sectors were generated at various times after leaf initiation. As expected, sectors were smaller when leaves were irradiated later in development (Fig. 6.4B, where P2 is shown as an example). If leaves grew from their tips or margins, then one might expect that most sectors would arise from those sites. Instead, the sectors were scattered and somewhat irregular, however, which suggests that the lamina expands by intercalary growth, not by patterned growth at the leaf tip or margins. Submarginal sectors were found, but they were generally oriented parallel to the leaf margin. Growth at the margins, therefore, did not appear to contribute significantly to the prevailing growth direction of the leaf.

The leaf sectors observed in the Poethig and Sussex studies (1985) indicated that cell divisions were not oriented entirely at random (Fig. 6.4B). In general, the sectors angled obliquely from the midrib toward the margins, particularly at the leaf tip. In the early stages of leaf development, it was observed that most cell divisions were similarly oriented, creating an outward expansion of the leaf. Small clones of cells were also classified, whether they were elongate (indicating oriented divisions) or isodiametric (indicating random divisions or divisions in alternating perpendicular planes). It is interesting that the larger clones were elongate while the smaller clones were isodiametric. The interpretation was that the earlier divisions (represented by the larger clones) were oriented and later divisions (represented by the smaller clones) tended to alternate in planes. These observations indicated that although growth of the lamina is intercalary, cell divisions are modestly oriented and regulated in frequency. As might be predicted from the frequencies of cell divisions over the surface of the leaf, sectors were larger toward the base of the leaf than they were at the tip, especially when the leaves were irradiated later in development (Fig. 6.4B), thereby indicating that cell divisions decline at the tip of the leaf earlier than at the base during development. Thus, the cell division patterns in tobacco leaves appear to have a modest relationship, if any, to leaf shape.

Tobacco leaves in studies of Poethig and Sussex had smooth margins; however, understanding how leaves with teeth or lobes are formed presents even greater challenges. Little is known about the determinants involved; however, new ways to generate lobed leaf patterns have provided some insight into the process. Recall from the Chapter 5 that the expression of the maize *knotted1* gene (encoding a homeodomain transcription factor) in tobacco results in adventitious meristem development on leaves. Members of the *knotted1* gene family (*knox* genes) that are expressed in shoots are normally active in meristematic tissue and are silenced in developing leaves (Lincoln et al.

FIGURE 6.5

Formation of leaf lobes through the action of *knotted1*-like gene (*KNAT1*) in Arabidopsis. Transgenic Arabidopsis plants bearing the 35S:*KNAT1* transgene showed different degrees of lobe formation. Leaves shown here from left to right were derived from untransformed plants (left) and mild, moderate, and severely lobed plants. Note that the leaves are smaller in plants that are more deeply lobed. (From Chuck, G., Lincoln, C., and Hake, S. 1996. *KNAT1* induces lobed leaves with ectopic meristems when overexpressed in Arabidopsis. Plant Cell 8:1279. Reprinted by permission of the American Society of Plant Physiologists.)

1994). It is interesting that ectopic expression in transgenic Arabidopsis of *KNAT1*, a *knotted1* ortholog from Arabidopsis, transforms simple leaves into lobed leaves (Fig. 6.5) (Chuck et al. 1996). Leaves on normal Arabidopsis plants, particularly ones formed later, are serrated, and lobes appear at the positions of serrations in the 35S:*KNAT1* plants. Primary leaf veins branching from the midrib terminate at serrations in normal leaves, whereas in transgenic 35S:*KNAT1* plants these veins are the midveins of the lobes and terminate at their ends. More highly lobed leaves were smaller in size and had deep sinuses with stipules at the base of the sinus. In some transgenic lines, ectopic inflorescence meristems formed at the base of the sinuses on the adaxial surfaces of the leaf. It appeared, therefore, that the lobes resulted from an inhibition of lamina expansion between the lobes. The base of the sinus had features in common with the base of the leaf, including the presence of stipules (normally found at the base of Arabidopsis leaves) and, in some cases, the formation of meristems. Leaves produced in these transgenic plants have unique properties caused by the unregulated expression of *KNAT1;* however, the effects of the gene may provide clues as to how lobed-shaped leaves might have evolved and how they are formed normally. It is clear that the spatial regulation of genes, such as those in the *knotted1* family that exercise control over cell proliferation, must be highly influential in the generation of leaf shape.

GENETICS OF LEAF DEVELOPMENT

Dicots

Interesting leaf mutants have been described in a variety of dicots, including Arabidopsis (Telfer and Poethig 1994). There are clearly many important questions that could be asked about leaf development through mutants, such as: What steps in leaf development are genetically definable? Are there master regulatory genes that control leaf development? Is leaf development a pattern formation process similar in principle to the processes by which flowers or embryos form?

As described earlier in the chapter, the development of dorsiventrality is a committing step in leaf development and is one of the first steps that distinguishes the development of a leaf from a shoot or branch. Mutants in *Antirrhinum* (snapdragon) called *phantastica* (*phan*) affect the development of dorsiventrality. Waites and Hudson (1995) argued that *phan* mutants are defective in the expression of dorsalizing functions. Snapdragon leaves are asymmetrical with respect to dorsal–ventral surfaces (Fig. 6.6A). The lamina arises from the dorsal midrib, and the bulk of the midrib is on the ventral side of the leaf. In addition, the leaf lamina and the vasculature show dorsal–ventral asymmetry. Cell layers of the lamina appear in order from the dorsal to the ventral surface as follows: the dorsal epidermis, palisade mesophyll, spongy mesophyll, and ventral epidermis. The midrib vasculature also has dorsal–ventral asymmetry. For example, xylem bundles are located dorsal to the phloem elements. In addition, the dorsal and ventral epidermis that surrounds the midrib can be distinguished by differences in the size of cells and the presence of hairs on the dorsal surface.

Phan mutants showed no dorsiventrality in leaves at and above the fifth node. The upper leaves were needlelike and symmetrical and were composed of ventral tissue types, xylem, phloem, parenchyma, and ventral epidermis (Fig. 6.6B). The lower leaves and cotyledon were broader than wild type and had unusual patches of ventral tissue on the dorsal epidermis. The patches were more frequent in the proximal part

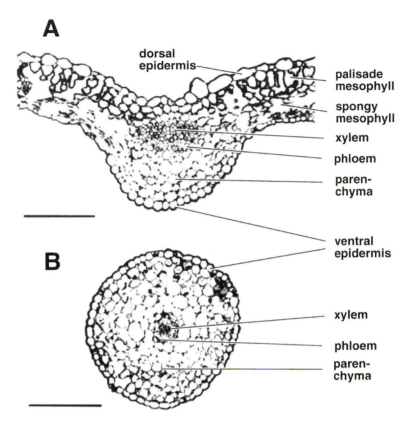

FIGURE 6.6

Comparison of leaves from Arabidopsis wild type and *phantastica* (*phan*) mutant. (A) Wild type leaf exhibits pronounced dorsal–ventral asymmetry with respect to the order of cell types from the dorsal to ventral surface of the leaf. (B) Needlelike *phan* leaf shows only ventral cell types arranged with radial symmetry around a central vascular bundle. Transverse leaf sections stained with toluidine blue and viewed by light microscopy. Bar = 250 µm. (From Waites, R., and Hudson, A. 1995. Phantastica: A gene required for dorsoventrality of leaves in Antirrhinum majus. Development 121:2143–2154. Reprinted by permission of the Company of Biologists Ltd.)

of the leaf, and it was proposed that these patches may be made up of cells that are clonally related and initiated later in development. The authors noted that the boundary between the ventral patches and the dorsal tissue were like leaf margins, where dorsal and ventral tissue normally meet. Ridges formed at the boundaries and only a spongy mesophyll layer separated the epidermal layers at the ridge.

Waites and Hudson (1995) argued that *phan* mutants were defective in a dorsalizing function, not just in their ability to produce the lamina. The argument is based largely on the observation that upper leaves of mutant plants were composed almost exclusively of ventral cells. In

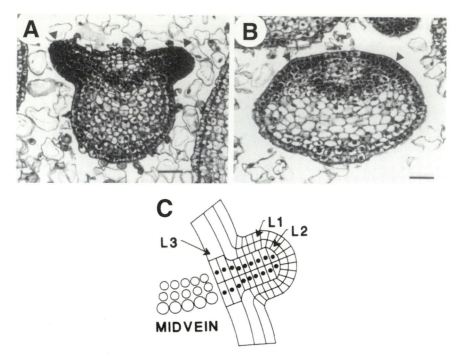

FIGURE 6.7

Leaf blade initiation in (A) wild type and (B) *lam-*1 mutant of *Nicotiana sylvestris.*
Median transverse sections of leaf primordia at the P3 stage. Arrowheads indicate sites presumptive of blade initiation. Note dorsal–ventral asymmetry in *lam-*
1 leaf, particularly with respect to the location of the developing vasculature.
Defect in *lam-*1 mutant is thought to arise from disorder in the normal orderly
divisions in L3 during the outgrowth of the lamina (C). Defects in L3 divisions
are thought to block anticlinal divisions in L1 and L2 and subsequent development of the lamina. Sections were stained with hemalum and safranin and
viewed by light microscopy. Bar = 50 μm. (From McHale, N. A. 1993. *LAM-1*
and *FAT* genes control development of the leaf blade in *Nicotiana sylvestris.* Plant
Cell 5:1030 and 1033. Reprinted by permission of the American Society of Plant
Physiologists.)

addition, they contended that upper leaves have a greater dependence
on the dorsalizing function and that the lower leaves are intermediate,
showing patches of ventral tissues on the dorsal surface.

Bladeless leaf mutations in tobacco (*Nicotiana sylvestris*), such as *lam-1,*
are defective in the initiation of the lamina (McHale 1993). Unlike
phan, lam-1 develops leaf dorsiventrality; however, leaf blades do not
form because the meristematic regions that flank the midrib on the
adaxial surface of the leaf and which give rise to the lamina do not
develop (Fig. 6.7B). Leaves in *lam-1* are nearly rod-shaped and lengthen
normally, which shows that elongation of the midrib and expansion of
the lamina are genetically separable processes. In cross-sections of the

vestigial *lam-1* leaves, it appears that the outpouching of cells from the adaxial side of the midrib, which normally gives rise to the lamina, aborts in the early stages of lamina formation (Fig. 6.7B). During lamina initiation in normal plants, L1 and L2 cells divide by anticlinal division and wrap around the middle mesophyll core, which is made up of L3 cells (Fig. 6.7C). In *lam-1*, L3 cells of the middle mesophyll core become vacuolated and divide in random planes, while L1 and L2 cells divide normally. McHale suggested that defects in the normal pattern of anticlinal divisions in L3 limit further development of the lamina. Constraints imposed by L3 are somewhat surprising given that L3 contributes very little to the lamina, particularly at the margin. This obstruction clearly could be tested by producing genetic chimeras in which *lam-1* was only expressed in the L3 layer.

Another tobacco leaf development mutant described by McHale (1993) is the *fat* mutant with thickened leaves. Leaves of *fat* plants have up to four layers of middle mesophyll cells that arise from abnormal periclinal divisions in L2 or L3 cells. Periclinal divisions occur in clusters of three to six mesophyll cells, and the displaced cells resume normal anticlinal division intercalating between existing layers of cells. Together the *lam-1* and *fat* mutants demonstrate that alterations in the planes of cell division affect the expansion and the thickness of the lamina.

Other leaf mutations have effects on lamina expansion. Much of the growth of the lamina is by cell expansion, not by cell division. Cells may increase as much as two- to forty-fold in size during the unfolding of the lamina (Dale 1988). Two mutants in Arabidopsis suggest that cell expansion along the axis (leaf length) and across the axis of the leaf (leaf width) may be controlled by different mechanisms. Tsuge et al. (1996) identified *angustifolia* (*an*), a mutant in which the leaves are narrower than wild type, and a group of *rotundifolia* (*rot1–rot 3*) mutants in which the leaves are shorter, but rounder, than wild type (see Fig. 5.17B and C). In describing the mutants, they compared the fifth true leaf in the mutants with wild type. The number of cells in the palisade layer of *rot3* leaves was similar to wild type, but the cells were about 10 percent shorter along the leaf axis. On the other hand, palisade layer cells in *an* mutants did not expand in width as much as wild type, particularly at the earliest stages of leaf expansion; rather, they tended to increase in thickness. Double *an rot3* mutants showed both reductions in leaf width and length, indicating that the effects were independent and additive. Thus, *an* and *rot* mutants are thought to have defects in factors that restrain growth preferentially along one axis or the other. The genes responsible for the mutations might simply encode cytoskeletal elements or cell wall components that preferentially exert restraining forces in one direction or another.

Monocots

Monocot leaves appear to be simpler than dicot leaves because they usually have striate or parallel venation. Grass leaves, such as maize (*Zea mays*) leaves, do not have the blade/petiole/base structure of dicot leaves. Maize leaves are instead differentiated along their length into a blade and a sheath that wraps around the stem. At the junction between the blade and sheath, a specialized structure is formed that consists of auricles and a ligule (Fig. 6.8A). Auricles allow the leaf blades to bend down and to form a seal around the stem. The ligule is a fringe around the edge of sheath that is formed on the adaxial surface of the leaf.

In maize, leaf primordia are founded by a ring of about 250 cells surrounding the shoot (Sylvester et al. 1990; Freeling 1992). Early in development, cell divisions occur throughout the primordium; the leaf then differentiates into the blade and sheath. Further growth of the leaf results from transverse divisions that occur near the base of the leaf, giving rise to parallel files of cells. Growth from the base of leaf means that the developmental history of the leaf can be traced from the tip of the leaf to the base. Hence, later developmental events in the older part of the leaf occur toward the tip, while earlier events occur toward the base.

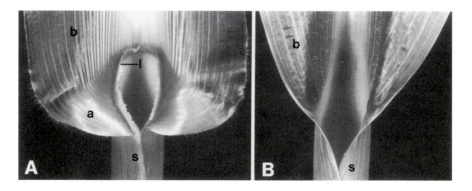

FIGURE 6.8

Comparison of leaves of wild type and *liguleless-1* (*lg1*) mutant in maize (*Zea mays*). Leaves have been removed from plants and the adaxial surface of a leaf blade in the ligular region is shown. (A) In wild type leaf, the blade, b, and sheath, s, are connected by wedge-shaped auricles, a. The ligule is the fringe on the adaxial surface where sheath and auricles meet. (B) Both auricles and ligule are missing in *lg1* mutant. (From Becraft, P. W., et al. 1990. The *liguleless-1* gene acts tissue specifically in maize leaf development. Develop. Biol. 141:221. Reprinted by permission of the author and Academic Press.)

The pattern of elements across the width of the maize leaf, such as the arrangement of vascular elements, is set early in leaf development. For example, Sharman (1942) demonstrated that the pattern of lateral veins is laid down by the end of the primordial stage of leaf development. The establishment of the leaf pattern early in leaf primordia development is evident from the effects of two mutations in maize: *narrow sheath*-1 and -2 (*ns1* and *ns2*) (Scanlon et al. 1996). Mutations in both *ns* genes produce narrow leaf sheaths and blades, particularly in the older or more basal leaves. These leaves are narrow because they appear to be missing leaf margins. The typical sawtooth hairs at the edge of the maize leaf are missing as well as a marginal structure that spans the upper and lower surfaces of the leaf. The defect in the *ns* mutants can be traced back to the development of the leaf primordium where the edges of P3 (the third youngest primordium) fail to develop and enclose the shoot apex (Fig 6.9D). In normal plants, P3 wraps entirely around the shoot apex (Fig 6.9C), but the shoot apex in *ns* mutants is not entirely enclosed, opening a window in the shoot apex. The authors determined whether or not expression of *knotted1*-like homeodomain proteins, which are detected by immunolocalization, were normally downregulated in the mutant. In nonmutant plants, the expression pattern of *knotted1*-like homeodomain proteins forms a ring around P0 that demarcates boundaries of the leaf primordia in the vegetative meristem. In the mutant, this ring is reduced to a crescent of cells that does not completely encircle the primordium.

Scanlon et al. (1996) argued that the absence of leaf margins in the *ns* mutants was due to the fact that during the formation of the leaf primordia, cells at the margin of the primordia were never recruited for leaf development. As a result, margins were missing from the mature leaf. There are other ramifications of the *ns* mutation because the leaf primordia specifies the whole phytomer, not just the leaf. In fact, the marginal side of each internode was shorter, forming sinuses and curved stems. In any case, the authors argued that developmental domains are set up in the leaf primordium and that certain domains (the margins) were missing in the *ns* mutant. This defect in the primordia is one for which the developing leaf cannot compensate and results in a leaf without margins.

Quite a different view about maize leaf development is illustrated by the *tangled-1* mutant in maize (introduced in Chap. 1). The *tangled-1* mutant has been used as an example to demonstrate that the shape of a leaf is not determined by set patterns of cell division, but by other forces (Smith et al. 1996). Recall that longitudinal divisions in the *tangled* mutant were frequently misaligned, producing files of cells in the leaf that were

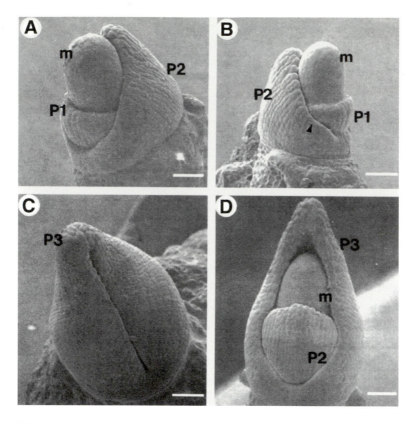

FIGURE 6.9

Leaf primordia in wild type and *narrow sheath* (*ns*) leaf mutant in maize (*Zea mays*). Note that the margins of the leaf primordium at the P2 stage in *ns* mutant (B) do not entirely encircle the shoot apex as they do in wild type (A). At the P3 stage, the leaf primordium in wild type (C) completely encloses the shoot apex, whereas a window is left open in the *ns* mutant (D). Shoot apices viewed by scanning electron microscopy. Bar = 35 μm. (From Scanlon, M. J., Schneeberger, R. G., and Freeling, M. 1996. The maize mutant *narrow sheath* fails to establish leaf margin identity in a meristematic domain. Development 122:1683. Reprinted by permission of the Company of Biologists Ltd.)

tangled (Fig. 6.10D). Cells in maize leaves, such as the epidermal cells, normally lie in parallel files. Despite disorder in the alignment of epidermal cells and vascular elements, the leaves on the mutant plants were normal in shape, although they were smaller in size (Fig. 6.10B). This mutant therefore provides a dramatic example showing that the planes of cell division are not the final arbiters of leaf and organ shape.

The *tangled-1* mutant seems to demonstrate that there are regulative aspects of leaf development in maize. The example of the *ns* mutant cited earlier, however, indicates that other aspects of leaf development are not regulative. The leaf cannot make up for elements that are lost in

FIGURE 6.10

Comparison of leave surfaces in the wild type and *tangled-1* (*tan-1*) mutant of maize (*Zea mays*). (A) Adaxial surface of wild type leaf is glossy and smooth. (C) Similar leaf surface on *tan-1* mutant is rough. Detail of surface of leaf primordia in (B) wild type and (D) *tan-1* mutant. Leaf primordia were 0.5–1.5 cm long; and microphotograph is oriented with long axis of leaf aligned vertically. Note uniform alignment of cell in wild type and tangled array in mutant. Black arrowheads indicate recent transverse, t, or longitudinal, l, divisions. White arrows in (D) indicate recent divisions with misoriented planes. Bar = 30 μm. (From Smith, L. G., Hake, S., and Sylvester, A. W. 1996. The *tangled-1* mutation alters cell division orientations throughout maize leaf development without altering leaf shape. Development 122:485. Reprinted by permission of the Company of Biologists Ltd.)

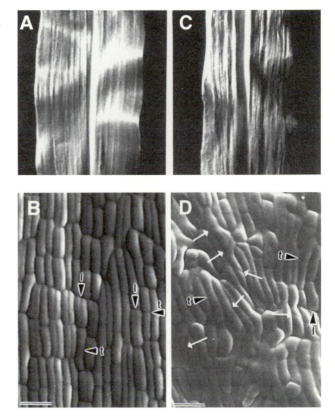

the formation of the leaf primordium. The conclusion concerning *tangled-1* might also seem at odds with those reached in the analysis of the *lam-1* and *fat* mutants in tobacco. Defects in anticlinal divisions in L3 in *lam-1* resulted in leaves with unexpanded lamina, and added periclinal divisions in *fat* mutants gave rise to thicker leaves. In the tobacco mutants, it was the frequency or pattern of division (whether anticlinal or periclinal) that was of issue. In *tangled-1* mutants in maize, however, the relative frequency of transverse and longitudinal divisions are about the same in the mutant and wild type; it is just that the longitudinal divisions are misaligned (Smith et al. 1996). Perhaps this is the deciding point in distinguishing between the effects of the mutants, or it may be that there are fundamental differences between leaf development in monocots and dicots.

Another interesting maize mutant that affects leaf form is *liguleless-1* (*lg1*), which is a mutant that interferes with the development of the auricles and ligules. In wild type leaves, the ligule is initiated by a special series of anticlinal divisions in the epidermis that gives rise to a band of cells, which then produces the ligule by periclinal divisions (Fig. 6.8A). The epidermis usually grows by anticlinal divisions; therefore, the formation of the ligule is a departure from normal growth. Auricles are wedge-shaped structures formed next to ligules and are derived from epidermis and more internal tissues.

Liguleless-1 (*lg1*), which is a recessive mutation, fails to produce auricles and ligules, and *lg1* leaves, which have little distinction between the blade and sheath, tend to clasp the stalk (Fig. 6.8B). The mutant fails to undergo the anticlinal divisions involved in ligule initiation. Becraft et al. (1990) generated genetic mosaics by X-irradiating heterozygotes and found that the expression of the *lg1* gene was generally cell autonomous in generating the ligule from the epidermis and the auricle in subepidermal tissue. Ligules were generally not formed in *lg1* epidermal tissue over subepidermal sectors in which the wild type gene was expressed except when *lg1* epidermal tissue directly contacted wild type tissue, in which case a rudimentary ligule was formed. The most interesting situation was revealed when a mutant sector passed through the ligule region of an otherwise wild type leaf. In these cases, the ligule on the marginal side of the sector was displaced toward the base of the leaf (Becraft and Freeling 1991). This suggested that some signal emanating from the midrib region of the leaf was involved in coordinating ligule development. It was argued that delayed development of the ligule on the marginal side of a sectored leaf did not result simply from slowing the movement of the signal across the mutant sector. The mutant sector instead blocked the movement of the signal, and the signal was spontaneously reinitiated with some delay on the marginal side of the sector. The argument was based on the observation that the extent of displacement of the ligule was not related to the size of the sector.

Compound Leaves

The leaves discussed so far have largely been simple leaves. Compound leaves have divided blades or leaflets, and the leaflets can be attached to a rachis (pinnate arrangement) or joined together and attached to the distal end of the petiole (palmate arrangement). A leaf can be distinguished from a leaflet in that only the former has an axillary bud at its base. Single gene mutations have been found that affect the compounding of leaves. Tomato leaves are normally compound with leaflets borne on petioles along a central rachis. The dominant *lanceolate* (*La*) mutation in a heterozygous state results in plants with leaves that are not compound and dentate; rather, they are simple and entire (see Mathan and Jenkins 1962). (In a homozygous state, the extreme form of *La* gene is nearly lethal and results in plants that lack cotyledons, foliage leaves, and a shoot apex.) On the other hand, the leaves

FIGURE 6.11

Higher order compounding of leaves by the action of the *knotted1* (*kn1*) gene ortholog in tomato (*Lycopersicon esculentum*). (A) Standard cultivar has a compound leaf with a terminal leaflet, TL, and lateral leaflets, LT, along the central rachis, R. (B) Higher order compounding of leaves in transgenic tomato expressing the 35S:*Kn1* construct. Leaves in transgenic plant are anatomically similar to standard cultivar except that they show higher order of compounding and leaflets are smaller. (From Hareven, D., et al. 1996. The making of a compound leaf: Genetic manipulation of leaf architecture in tomato. Cell 84:736. Reprinted by permission of the author and Cell Press.)

on *Petroselinum* (*Pet*) mutant are very finely divided and compounded to the third order.

It has been found that, again, the action of the constitutively expressed *knotted1* gene (35S:*Kn1*) from maize or the tomato ortholog (similar gene from tomato) enhances the compounding of leaves in transgenic tomato (Hareven et al. 1996) (Fig. 6.11B). It is of note that *knotted1* (*kn1*) expression produces higher order compounding only when the leaves are already compounded. Constitutive *kn1* gene expression will not produce compound leaves in tomatoes such as *La*/+ in which the leaves are not already compounded. The authors suggest that compound leaves result from the action of leaflet meristems that form along the flanks of the leaf primordium. Higher order compounding arises from a reiteration of leaflet meristem formation before lamina expansion. They argue that competition between lamina expansion and leaflet meristem development determines whether the leaf will be simple or compound and the degree to which it will be compounded. In that case, the action of 35S:*Kn1* is thought to tip the balance toward development and activity of the leaflet meristems. Thus, the effect of

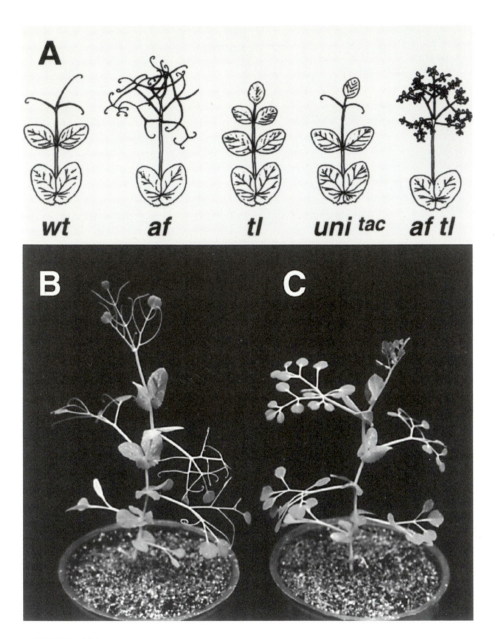

FIGURE 6.12

Leaf mutants in pea (*Pisum sativa*). (A) Wild type leaf is compound and composed of a central rachis, two basal stipules, a pair of leaflets and terminal tendrils. In *afila (af)*, leaflets are replaced by tendrils; in *tendril-less (tl)*, tendrils are replaced by leaflets; and in *unifoliata tendrilled acacia (uni^tac)*, terminal tendrils are replaced by leaflets. A novel "parsley leaf" phenotype is produced in the *af tl* double mutant that has a highly branched rachis and tiny laminate structures at the rachis tips. (B) Triple *af uni^tac apu* mutant shows highly branched rachis with terminal tendrils replaced by leaflets. (C) Quadruple *af uni^tac apu tl* mutant in

(continued)

35S:*Kn1* on leaf compounding is probably similar to its effects on the formation of leaf lobes, knots, and adventitious shoots described earlier in this chapter and in previous chapters. Ectopic expression of the *kn1* gene appears to stimulate meristematic activity and, in the cases described, to disrupt its spatial control. Manipulation of *kn1* gene expression can clearly have interesting developmental consequences.

There are a number of fascinating mutants in pea (*Pisum sativa*) that affect the identification of leaflets in the compound leaf. Compound pea leaves are composed of two large leaflike stipules at the base, one or more pairs of leaflets, and tendrils at the tip (Fig. 6.12). Stipules and tendrils are modified leaflets. Marx (1987) argued that development of the pea leaf is a pattern formation process and that mutants in which one leaf part is substituted for another are homeotic mutants. An example is *af* (*afila*), a recessive mutant in which tendrils substitute for the normal leaflets (Fig. 6.12). Another example is *tl* (*tendril-less*), a recessive mutant in which the tendrils are replaced by leaflets. It could be argued, however, that these mutants might be better classified as developmental rather than homeotic mutants because the pattern of leaflet formation can change during development in pea plants. For example, leaves on older wild type plants may consist entirely of tendrils.

TRICHOME DEVELOPMENT AND PATTERNING

Development of the leaf also involves the formation of specialized epidermal cells, such as trichomes and stomata. In Arabidopsis, trichomes are usually located on the upper or adaxial side of rosette leaves and stomata on the lower side or abaxial side. Arabidopsis leaf trichomes are single cells that develop into branched leaf hairs (Fig. 6.13). In other species, trichomes can be small, unbranched leaf hairs or large, multicellular spikes. In Arabidopsis and tobacco, the first leaf trichomes are

FIGURE 6.12 *(continued)*

which the tendrils on the highly branched rachis are replaced by leaves. The *apulvinic* (*apu*) mutation conditions leaflets to appear on stalks rather than to be sessile (stalkless). (Upper panel from Murfet, I. C., and Reid, J. B. 1993. Developmental mutants. In Peas: Genetics, Molecular Biology and Biotechnology, ed. Casey, R. and Davies, D. R., 165–216. Wallingford UK: CAB International. *In* R. Casey and D. R. Davies, eds., Peas: Genetics, Molecular Biology and Biotechnology. CAB International, Wallingford UK, p. 168. Reprinted by permission of CAB International. Lower panel from the collection of G. A. Marx kindly provided by N. F. Weeden, Geneva Experiment Station, Cornell University.)

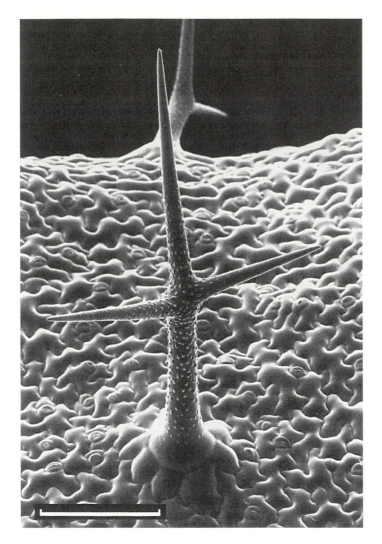

FIGURE 6.13

Three-branched trichome on the surface of an Arabidopsis leaf. Branching occurs in two successive branching events, the primary event, which gives rise to the proximal branch (lower branch on left), and the secondary event, which results in the distal branch (higher branch on right) and the main stem. Visualized by scanning electron microscopy. Bar = 100 μm. (From Linstead, P., Doan, P., and Roberts, K. In Bowman, J. 1994. Arabidopsis: An Atlas of Morphology and Development. Springer-Verlag, New York, p. 43. Reprinted by permission of Springer-Verlag.)

formed progressively from the tip of the leaf down, appearing at the leaf tip soon after the leaf primordia begin to elongate (Fig. 6.1).

Hülskamp et al. (1994) collected and analyzed a group of Arabidopsis mutants with defects in trichome development. The group of 70 trichome mutants represented 21 different genes. They used the mutants to define steps in a pathway of trichome development (Fig. 6.14). It was found through the analysis of double mutants that most of the genes acted on independent pathways. In any case, the mutants demonstrate that trichome formation involves a series of genetically defined steps, such as genome endoreduplication or nuclear enlargement, outgrowth of the trichome cell, primary branching of the trichome, secondary branching, and lignification to form a hardened structure.

Severe alleles of trichome mutants, such as *glabrous1* (*gl1*) or *transparent testa glabrous* (*ttg*), lack trichomes altogether (Fig. 6.14) (Marks 1997). These mutants are epistatic to all other trichome mutants and are defective in the earliest steps of trichome initiation, including nuclear enlargement. Mature trichomes are polyploid cells, and endoreduplication accompanied by nuclear enlargement is one of the first recognizable events in the differentiation process. The *GL1* gene appears to act in a cell autonomous fashion because glabrous (trichomeless) sectors were observed in X-ray–induced sectors derived from *GL1/gl1* heterozygotes (Redei 1967).

GL1 has been cloned, and the sequence predicts a helix–loop–helix transcription factor (Oppenheimer et al. 1991). The role of *GL1* in initiating trichomes suggests that it is a master gene controlling trichome formation. Mutant *gl1* can be rescued by expressing *GL1* as a transgene from a constitutively expressed promoter, the CaMV 35S promoter (35S:*GL1*) (Larkin et al. 1994). This was quite interesting because the CaMV 35S promoter and the native *GL1* promoter have different patterns of expression. The *GL1* promoter is expressed at low levels in the epidermis of leaves and stems and at higher levels in developing trichomes (Larkin et al. 1993). The CaMV 35S promoter is a strong, constitutive promoter, expressed at high levels in almost all cells. The experiment demonstrates that the pattern of *GL1* gene expression does not determine the pattern of trichome formation on the leaf surface. Rescued *gl1* mutants have nearly normal trichome patterns despite the fact that the 35S:*GL1* gene construct is presumably expressed in all epidermal cells.

It is interesting that the *ttg* mutation can be rescued by introducing a similar construct into mutant plants that encodes another helix–loop–helix transcription factor. In this case, the *ttg* mutant can be rescued by the maize *R* gene expressed under the control of the CaMV 35S promoter (35S:*R*) (Lloyd et al. 1992). The maize *R* gene is required for the expression of a number of different epidermal cell functions in maize, including anthocyanin biosynthesis. Expression of the CaMV 35S promoter-*R* gene constructs (35S:*R*) rescues the *ttg* mutation, but not *gl1* mutations. As described earlier, CaMV 35S promoter-*GL1* con-

159

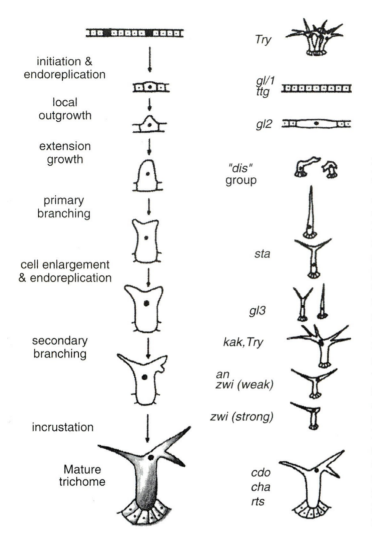

initiation &
endoreplication

local
outgrowth

extension
growth

primary
branching

cell enlargement
& endoreplication

secondary
branching

incrustation

Mature
trichome

Try

gl/1
ttg

gl2

"dis"
group

sta

gl3

kak, Try

an
zwi (weak)

zwi (strong)

cdo
cha
rts

FIGURE 6.14

Trichome development in Arabidopsis (left), and effects of various trichome mutants. Mutants are blocked at different stages of trichome development, and mutant phenotypes may provide some indication of when the normal genes act in development. *Triptychon* (*try*) mutations produce trichome clusters. *Distorted* (*dis*) group is composed of eight mutants with similar phenotypes and *zwichel* (*zwi*) mutants have different phenotypes based on allele strength. (From Hülskamp, M., Misera, S., and Jürgens, G. 1994. Genetic dissection of trichrome cell development in Arabidopsis. Cell 76:558. Reprinted by permission of the author and Cell Press.)

structs (35S:*GL1*) will rescue *gl1* mutations; however, the construct will not rescue the *ttg* mutation (Larkin et al. 1994). Although both *gl1* and *ttg* can be rescued by the action of helix–loop–helix transcription factors, they require the action of different transcription factors that apparently act at different points on the trichome formation pathway.

Trichomes are distributed nonuniformly on the leaf surface, and the pattern suggests either that inhibitory signals establish the spacing between developing trichomes or that cell lineage patterns separate trichomes from each other (Larkin et al. 1997). In general, it is thought that an inhibitory signal might better account for the distribution of trichomes than cell lineage models. Trichomes and neighboring cells do not appear to be clonally related because sector boundaries generated by

35S:*Ac*:GUS expression (described in Chap. 2) pass randomly through trichomes and neighboring cells (Larkin et al. 1996).

The spacing pattern of trichomes is compromised in weak alleles of *gl1* and *ttg* that do not completely suppress trichome production. Trichomes are clustered in these mutants, and it has been argued that this may result from the fact that the genes required for trichome initiation might also be involved in spacing (Larkin et al. 1997). The situation has parallels to the patterning of sensory hairs in Drosophila that are regulated by the *achaete-scute* complex, which, like *TTG* and *GL1*, also encode helix–loop–helix transcription factors. *Achaete-scute* is required for the initiation of sensory hairs, but it also activates the expression of other genes that inactivate its own expression in neighboring cells (Ghysen et al. 1993). The mutation *TRYPTYCHON* (*TRY*) also increases the number of the trichome clusters (Hülskamp et al. 1994). This mutation has been found to act downstream and may weaken the inhibitory effects of *TTG* and *GL1*. An interesting effect on the patterning of trichomes was observed in various combinations of transgenes and mutations when the CaMV 35S promoter-*GL1* construct (35S:*GL1*) was expressed in plants heterozygous for the *ttg*-1 allele (Larkin et al. 1994). Trichomes emerged in clusters in these plants. While there is no simple explanation for this effect, the authors argued that the level of *TTG* inhibitory action might fall below some critical threshold in *ttg*-1 heterozygotes with one copy of 35S:*GL1*, thereby allowing for the appearance of clusters. In any case, trichome development and spacing appear to be related.

Leaf trichomes in Arabidopsis are generally three-branched structures that form from two successive branching events, one which results in a proximal branch and the other which gives rise to a distal branch and the main stem (Fig. 6.13). Folkers et al. (1997) found that two genes *STACHEL* (*STA*) and *ANGUSTIFOLIA* (*AN*) specify the primary and secondary branching events, respectively (Fig. 6.14). (*AN* was discussed earlier in the chapter as a mutation that also affects leaf size.) Developing trichomes undergo the primary branching event in *an* mutants, but distal stems fail to split again. As a result, trichomes with two branches, a proximal branch and a distal stem, form in *an* mutants. Primary branching fails in *sta* mutants and, again, only two-branched trichomes are formed. The branches in *sta* mutants, however, consist of the distal branch and stem. Double *an sta* mutants have only a main stem and no branches at all, which indicates that the genes control branching events on two independent pathways.

Strong mutations in another Arabidopsis gene, *STICHEL* (*STI*) also result in leaf trichomes with no branches (Folkers et al. 1997). In double *an sti* mutants, *sti* is epistatic to *an* which suggests that *STI* acts upstream of *AN* (also, presumably before *STA*). Mutations in *NOECK* (*NOK*) have just the opposite effect. Trichomes with supernumery branches are formed in recessive *nok* mutants which suggests that the function of the normal *NOK* gene is to suppress supernumerary branching. Double mutants between weak *sti* and *nok* alleles generally reduce the effects of both mutations which argues that the genes have opposite functions, one to promote branching and the suppress branching. Double *nok an* and *nok sta* mutations were constructed to determine whether *NOK* suppresses primary or secondary branching. It was found that double *nok an* mutants had an *an* phenotype with mostly two-branched trichomes rather than with multiple branched trichomes (Folkers et al. 1997). It was concluded that *an* is epistatic to *nok* and probably that *NOK* normally suppresses multiple branching in the secondary branching events specified by *AN*. (Recall that recessive mutations in downstream genes are epistatic to recessive mutations in upstream genes if the upstream genes inhibit or suppress the action of the downstream gene.) Double *nok sta* mutants, in constrast to double *nok an* mutants, had multiple branches which suggested that *NOK* does not act on primary branching events specified by *STA*.

In conclusion, branching in Arabidopsis trichrome development requires the action of positive and negative regulators (Folkers et al. 1997). *STI* promotes the action of *STA* and *AN* which specifies primary and secondary branching, respectively. *NOK* acts as a negative regulator which prevents the repetition of the secondary branching events specified by *AN*.

STOMATAL PATTERNING AND SPACING

Stomata are most characteristically found on the underside or abaxial side of leaves and provide passageways for gas exchange. A stoma is composed of guard cells that form a pore which opens and closes in response to various environmental conditions. Stomata are epidermal derivatives and are associated in some species with other epidermal accessory cells called subsidiary cells. Because stomata are involved in gas exchange, their spacing is of vital importance to the operation of a leaf. Protagonists in the argument over cell lineage versus cell position have frequently clashed about issues that involve stomatal spacing in leaves. According to Bünning's (1956) theories, the very regular stomatal spacing in many monocots is a product of cell lineages and stereotypic patterns of cell division, while the less-regular stomatal spacing in dicots is influenced by cell position (Fig. 6.15). In dicots, stomatal ini-

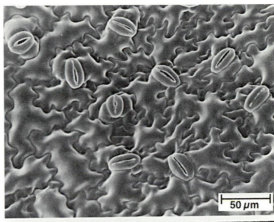

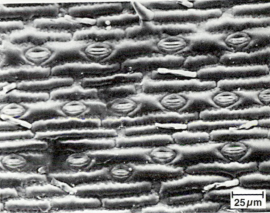

FIGURE 6.15

Stomata on abaxial surface of potato (*Solanum tuberosum*) and maize (*Zea mays*) leaves. Oriented arrangement with regular spacing is characteristic of the stomatal spacing patterns in monocots, particularly grasses. In maize each pair of guard cells is flanked by subsidiary cells. Visualized by scanning electron microscopy. (From Raven, P. H., Evert, R. F., and Eichhorn, S. E. 1986. Biology of Plants. Worth Publishers, New York, 4th ed., p. 428. Reprinted by permission of Michele McCauley Szupica.)

tials appear randomly on the leaf with the constraint that a minimum distance separates any one stoma from another. The spacing pattern suggested that stomata inhibit the formation of other stomata by a mechanism that acts over short distances.

Stomatal spacing in most monocots does, indeed, appear to result from regular patterns of cell division. In monocots, an epidermal cell typically undergoes an asymmetric and polarized division to produce a larger cell toward the base of the leaf and smaller initial toward the leaf tip (Larkin et al. 1997) (Fig. 6.16A). The smaller initial becomes a guard mother cell (GMC) and undergoes a symmetric division to form two guard cells. In so doing, an alternating pattern of stomatal initials and neighboring cells is created within cell files in many monocots (Fig. 6.15, lower panel). In dicots, several asymmetric cell divisions occur in the formation of GMCs. The cell initiating the series of divisions is called a **primary meristemoid.** Primary meristemoids appear to be positioned randomly and undergo a few further asymmetric divisions in which larger

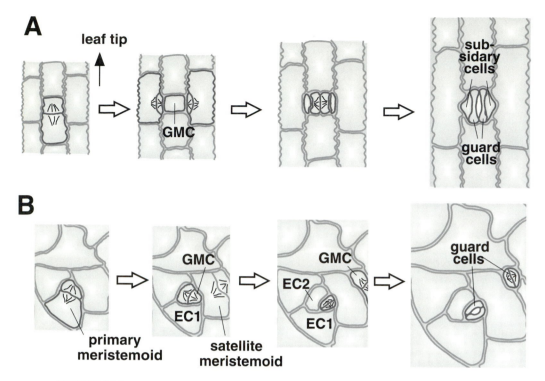

FIGURE 6.16

Development and patterning of stomata. (A) In monocot grasses, such as maize (*Zea mays*), an asymmetric, transverse division of an epidermal cell produces a larger cell toward the leaf base and smaller cell toward the leaf tip which functions as a guard mother cell (GMC). Subsidiary cells are produced from asymmetric, longitudinal divisions of neighboring cells in adjacent cell files. A GMC undergoes a symmetric longitudinal division forming two equal guard cells. Because stomata arise from oriented, asymmetric cell divisions, they are always separated from each other in a cell file by at least a single cell. (B) In a dicot, such as Arabidopsis, an epidermal cell that serves as a primary meristemoid undergoes an unoriented, asymmetric division which gives rise to a larger epidermal cell (EC1) and a smaller meristemoid cell. The primary meristemoid cell undergoes one or two further asymmetric divisions in which the smaller daughter cell serves a meristemoid and gives rise to a GMC. The GMC undergoes a symmetric division forming two guard cells of equal size. In Arabidopsis, stomata are also formed from satellite meristemoids that neighbor existing stomata. Divisions of the satellite meristemoid are also asymmetric, and the first division is oriented so that the smaller daughter cell, which functions as the meristemoid, arises away from the existing stoma. Stomatal patterning in Arabidopsis, therefore, is determined by random placement of primary meristemoids and oriented divisions of satellite meristemoids. (Adapted from Larkin, J.C., et al. 1997. Epidermal cell fate and patterning in leaves. Plant Cell 9:1114.)

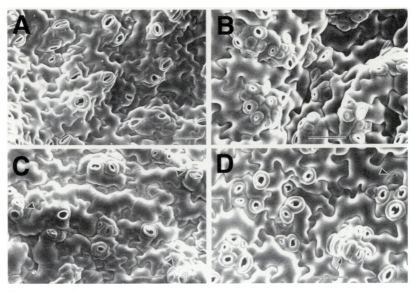

FIGURE 6.17

Pattern of stomata on abaxial surface of cotyledons of wild type and stomatal mutants of Arabidopsis. (A) Wild type. (B) *Too many mouths* (*tmm*) mutant has clustered stomata in various arrangements. Some stomata have incomplete pore development. (C) In *four lips* (*flp*) mutant, stomata are sometimes found in pairs (see arrow heads) in which the stomatal pores are laterally aligned with each other. (D) Double *tmm flp* mutants have a novel phenotype in which stomata are usually found in clusters in which some are laterally aligned. (From Yang, M., and Sack, F. D. 1995. The *too many mouths* and *four lips* mutations affect stomatal production in Arabidopsis. Plant Cell 7:2231. Reprinted by permission of the American Society of Plant Physiologists.)

daughters become epidermal cells (Fig. 6.16). After the last asymmetric division, the smaller meristemoid cell becomes a GMC which gives rise to two equal-size guard cells by a symmetric division. Some dicots, such as Arabidopsis, have another type of meristemoid called a satellite meristemoid. The satellite meristemoid is formed by an asymmetric division of a cell which neighbors a stoma; however, the division of the neighboring cell is polarized such that the smaller daughter cell (the satellite meristemoid) is positioned away from the stoma. The satellite meristemoid can convert to a GMC or undergo additional asymmetric division placing additional neighboring cells between the satellite meristemoid and the stoma. In this way stomata in Arabidopsis are spaced away from each other.

In an Arabidopsis mutant called *too many mouths* (*tmm*), the controls on stomatal spacing have been compromised. Some stomata appear to form arc-shaped clusters (Fig. 6.17B). The clusters are varied in shape, number, state of development of the stomata, and arrangement of stomata with respect to each other. In developing *tmm* cotyledons, meristemoids are often found next to stomata or GMCs. This situation has been interpreted to be one in which satellite meristemoids are overproduced and some of the asymmetric divisions are not polarized so that meristemoids end up contacting existing stomata or each other. Another Arabidopsis mutant called *four lips* (*flp*), is characterized by groups of mostly two adjacent stomata and a few unpaired guard cells (Fig. 6.17C). The groups of stomata have about the same spacing as individual stomata in wild type; therefore, *flp* does not appear to be a spacing pattern mutant. Stomata within a group generally had the same orientation, which suggests that the defect may be in the division of the GMCs. The defect in *flp*, therefore, appears to act later than *tmm*, affecting one or more of the stereotypic divisions in the pathway of stomatal formation. In double *tmm flp* mutants, an additive phenotype was observed (Fig. 6.17D) which suggests that the mutations affect independent processes in the stomatal development pathway.

REFERENCES

Becraft, P. W., Bongard, P. D. K., Sylvester, A. W., Poethig, R. S., and Freeling, M. 1990. The *liguleless-1* gene acts tissue specifically in maize leaf development. Dev. Biol. 141: 220–232.

Becraft, P. W. and Freeling, M. 1991. Sectors of *liguleless-1* tissue interrupt an inductive signal during maize leaf development. Plant Cell 3: 801–807.

Bowman, J. 1994. Arabidopsis: An Atlas of Morphology and Development. New York: Springer-Verlag.

Bünning, E. H. 1956. General processes of differentiation. In The Growth of Leaves, Milthorpe, F. L., ed. London: Butterworths.

Chuck, G., Lincoln, C., and Hake, S. 1996. *KNAT1* induces lobed leaves with ectopic meristems when overexpressed in Arabidopsis. Plant Cell 8: 1277–1289.

Dale, J. E. 1988. The control of leaf expansion. Ann. Rev. Plant Physiol. Plant Mol. Biol. 39: 267–295.

Folkers, U., Berger, J., and Hülskamp, M. 1997. Cell morphogenesis of trichomes in Arabidopsis: Differential control of primary and secondary branching by branch initiation regulators and cell growth. Development 124: 3779–3786.

Freeling, M. 1992. A conceptual framework for maize leaf development. Dev. Biol. 153: 44–58.

Ghysen, A., Dambly-Chaudiere, C., Jan, L. Y., and Jan, Y. 1993. Cell interactions and gene interactions in peripheral neurogenesis. Genes Devel. 7: 723–733.

Hareven, D., Gutfinger, T., Parnis, A., Eshed, Y., and Lifschitz, E. 1996. The making of a compound leaf: Genetic manipulation of leaf architecture in tomato. Cell 84: 735–744.

Hicks, G. S. and Steeves, T. A. 1969. *In vitro* morphogenesis in Osmunda cinnamomea. The role of the shoot apex in early development. Can. J. Botany 47: 575–80.

Hülskamp, M., Misera, S., and Jürgens, G. 1994. Genetic dissection of trichrome cell development in Arabidopsis. Cell 76: 555–566.

Jesuthasan, S. and Green, P. B. 1989. On the mechanism of decussatte phyllotaxis: Biophysical studies on the tunica layer of *Vinca major*. Am. J. Bot. 76: 1152–1166.

Larkin, J. C., Marks, M. D., Nadeau, J., and Sack, F. 1997. Epidermal cell fate and patterning in leaves. Plant Cell 9: 1109–1120.

Larkin, J. C., Oppenheimer, D. G., Lloyd, A. M., Paparozzi, E. T., and Marks, M. D. 1994. Roles of the *GLABROUS1* and *TRANSPARENT TESTA GLABRA* genes in Arabidopsis trichome development. Plant Cell 6: 1065–1076.

Larkin, J. C., Oppenheimer, D. G., Pollock, S., and Marks, M. D. 1993. Arabidopsis *GLABROUS1* gene requires downstream sequences for function. Plant Cell 5: 1739–1748.

Larkin, J. C., Young, N., Prigge, M., and Marks, M. D. 1996. The control of trichome spacing and number in *Arabidopsis*. Development 122: 997–1005.

Lincoln, C., Long, J., Yamaguchi, J., Serikawa, K., and Hake, S. 1994. A *knotted1*-like homeobox gene in Arabidopsis is expressed in the vegetative meristem and dramatically alters leaf morphology when overexpressed in transgenic plants. Plant Cell 6: 1859–1876.

Lloyd, A. M., Walbot, V., and Davis, R. W. 1992. Arabidopsis and Nicotiana anthocyanin production activated by maize regulators *R* and *C1*. Science 258: 1773–1775.

Lyndon, R. F. 1983. The mechanism of leaf initiation. In The Growth and Functioning of Leaves, ed. Dale, J. E. and Milthorpe, F. L., 3–24. Cambridge: Cambridge University Press.

Marks, M. D. 1997. Ann. Rev. Plant Physiol. Plant Mol. Biol. 1997 48: 137–163.

Marx, G. A. 1987. A suite of mutants that modify pattern formation in pea leaves. Plant Mol. Biol. Rep. 5: 311–335.

Mathan, D. S. and Jenkins, J. A. 1962. A morphogenetic study of *lanceolate,* a leaf-shape mutant in the tomato. Am. J. Bot. 49: 504–514.

McHale, N. A. 1993. *LAM-1* and *FAT* genes control development of the leaf blade in *Nicotiana sylvestris*. Plant Cell 5: 1029–1038.

Murfet, I. C. and Reid, J. B. 1993. Developmental mutants. In Peas: Genetics, Molecular Biology and Biotechnology, ed. Casey, R. and Davies, D. R., pp. 165–216. Wallingford, U.K.: CAB International.

Oppenheimer, D. G., Herman, P. L., Sivakumaran, S., Esch, J., and Marks, M. D. 1991. A *myb* gene required for leaf trichome differentiation in Arabidopsis is expressed in stipules. Cell 67: 483–493.

Poethig, R. S. 1997. Leaf morphogenesis in flowering plants. Plant Cell 9: 1077–1087.

Poethig, R. S. and Sussex, I. M. 1985a. The cellular parameters of leaf development in tobacco *Nicotiana tabacum:* A clonal analysis. Planta 165:170–184.

Poethig, R. S. and Sussex, I. M. 1985b. The developmental morphology and growth dynamics of the tobacco leaf *Nicotiana tabacum* cultivar xanthi-nc. Planta 165: 158–169.

Raven, P. H., Evert, R. F., and Eichhorn, S. E. 1986. Biology of Plants. New York: Worth Publishers.

Redei, G. P. 1967. Genetic estimate of cellular autarky. Experientia 23: 584.

Sachs, T. 1969. Regeneration experiments on the determination of the form of leaves. Israel J. Bot. 18: 21–30.

Scanlon, M. J., Schneeberger, R. G., and Freeling, M. 1996. The maize mutant *narrow sheath* fails to establish leaf margin identity in a meristematic domain. Develop. 122: 1683–1691.

Sharman, B. B. 1942. Developmental anatomy of the shoot of *Zea mays* L. Ann. Bot. 6: 245–284.

Smith, L. G., Hake, S., and Sylvester, A. W. 1996. The *tangled-1* mutation alters cell division orientations throughout maize leaf development without altering leaf shape. Develop. 122: 481–489.

Steeves, T. A., Gabriel, H. P., and Steeves, M. W. 1957. Growth in sterile culture of excised leaves of flowering plants. Science 126: 350–351.

Steeves, T. A. and Sussex, I. M. 1957. Studies on the development of excised leaves in sterile culture. Am. J. Bot. 44: 665–673.

Sylvester, A. W., Cande, W. Z., and Freeling, M. 1990. Division and differentiation during normal and *liguleless-1* maize leaf development. Development 110: 985–1000.

Telfer, A. and Poethig, R. S. 1994. Leaf development in Arabidopsis. In Arabidopsis, ed. Meyerowitz, E. M. and Somerville, C. R., 379–401. Cold Spring Harbor, NY: Cold Spring Harbor Press.

Tilney-Bassett, R.A.E. 1963. Genetic and plastid physiology in *Pelargonium*. Heredity 18: 485–504.

Tsuge, T., Tsukaya, H., and Uchimiya, H. 1996. Two independent and polarized processes of cell elongation regulate leaf blade expansion in *Arabidopsis thaliana* (L.) Heynh. Development 122: 1589–1600.

Waites, R. and Hudson, A. 1995. Phantastica: A gene required for dorsoventrality of leaves in Antirrhinum majus. Development 121: 2143–2154.

Yang, M. and Sack, F. D. 1995. The *too many mouths* and *four lips* mutations affect stomatal production in Arabidopsis. Plant Cell 7: 2227–2239.

7

Transition to Flowering

The transition to flowering can be a remarkable change in the life of a plant. In many species, the transition marks the end of vegetative growth and the beginning of reproductive development. In other species, such as in many perennials, reproductive development occurs in certain regions of the plant, but vegetative growth of the plant continues. The transition occurs in shoot meristems, which are reprogrammed to make inflorescence or floral organs rather than vegetative organs upon receiving appropriate environmental or development signals. From a developmental perspective, therefore, the floral transition is as much about reprogramming the shoot meristems as it is about the actual production of inflorescences or flowers. In Arabidopsis, the inflorescence meristem differs somewhat in anatomy from the vegetative meristem. The inflorescence meristem is more dome-shaped, and the rib meristem (also called the *file meristem*) is larger and more active in supporting growth of longer internodes of the flower stalk (Vaughan 1955). It is not known, however, whether the anatomical changes are the cause or the result of changes in growth status of the meristem.

The floral transition marks the beginning of reproductive development, and in many plants, such as those with single flowers, it also signifies the end of indeterminate growth. In plants such as Arabidopsis, which bear flowers in clusters on an inflorescence (flower stalk), there are two transitions: one to form inflorescences and the other to produce flowers. The two transitions are distinct processes that can be distinguished genetically. Different types of inflorescences are formed in determinate and indeterminate species. In determinate species, the inflorescence meristem forms a terminal flower that ends any further inflorescence growth. In indeterminate species, flowers are formed on lateral branches or inflorescences (coinflorescences), and not from terminal buds (Bradley et al. 1997).

Understanding what controls the transition to flowering is a fundamental concern in developmental biology, but it is also of practical interest in floriculture and agriculture. There is an enormous literature on the control of flowering, but a unified physiological model has not emerged (McDaniel 1996). This chapter, therefore, focuses on the control of flowering in a just a few species. Where the process has been extensively studied, such as in Arabidopsis, one appreciates the complexity of the problem as well as the power that can be brought to bear through genetic analysis.

TRANSITION TO FLOWERING IN PEA

The floral transition is most easily studied in plants where flowering can be induced by temperature or photoperiod (Bernier 1988). In such species, environmental signals are perceived in the vegetative parts of the plant and transmitted to the meristem. For example, in pea (*Pisum sativa*), day length changes are sensed in leaves, and the flowering signal is transmitted to the shoot apex.

One of the novel features about the studies in pea is that the tools of grafting and genetics can be combined to examine both the source and target of the flowering signal. Flowering in most pea cultivars is induced by photoperiod transitions from short to long days. Murfet et al (1985) described a number of mutants that affect photoperiod control of flowering (Table 7.1) and determined their site of action by grafting the apical bud of a mutant (the scion) onto another mutant or wild type plant (the stock). In this way, they could determine whether the mutant was expressed in the shoot apex or in the leaves (or other vegetative parts of the plant). It was found that recessive mutations in the *VEG1* (vegetative) gene totally block flowering. *VEG1* is expressed in the shoot apex

Table 7.1 *Genes That Control Time to Flowering in Pea*

Gene	Mode of Action	Site of Action
VEG1	Required for flower initiation	Shoot apex
LF	Specifies minimum flowering node, affects sensitivity to flowering signal	Shoot apex
GI	Promotes flowering, promotes floral stimulus production	Leaves
SN	Delays flowering, promotes photoperiod response and floral inhibitor production	Leaves and cotyledons
DNE	Delays flowering, promotes photoperiod response and floral inhibitor production	Leaves and cotyledons
PPD	Delays flowering, promotes photoperiod response and floral inhibitor production	Leaves and cotyledons
HR	Enhances effect of *SN*, *DNE*, and *PPD*	Leaves

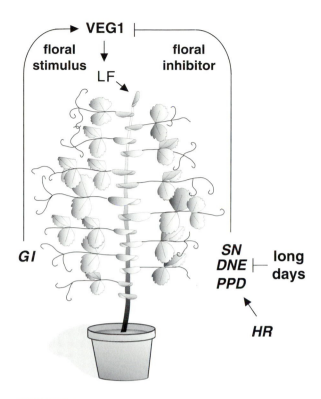

FIGURE 7.1

Genes that affect time to flowering in pea (*Pisum sativa*). Site of action of genes determined by grafting shoot apices (scions) from various mutants onto different stocks. Genes such as *VEG1* and *LY act in the shoot apex, whereas* GI, SN, DNE, PPD, and *HR* act in leaves. *GI* promotes early flowering and the production of floral stimulus. On the other hand, *SN, DNE,* and *PPD* inhibits flowering under short day conditions and are thought to promote the production of a floral inhibition. Long days are thought to inhibit the action of *SN, DNE,* and *PPD. HR* further enhances the photoperiod effect. (Activation is indicated by an arrow, ↓, whereas inactivation is represented by ⊥.) (Based on Murfet, I. C. 1989. Flowering genes in *Pisum*. In Lord, E., and Bernier, G., eds., Plant Reproduction: From Floral Induction to Pollination, pp. 10–18. Rockville MD: American Society of Plant Physiologists; and Beveridge, C. A., and Murfet, I. C. 1996. The *gigas* mutant in pea is deficient in the floral stimulus. Physiol. Plant. 96:637–645.)

as demonstrated by grafting the shoot apex from *veg1* onto a nonmutant plant (Fig. 7.1). In such grafts, the *veg1* scion fails to flower under any day length condition. These mutations are epistatic to all other flower mutations that also act at the shoot apex. Since *VEG1* is required under all conditions, *VEG1* might not be involved in the perception of the flowering signal but may be required for flower initiation. The *LF* (late flowering) gene appears to delay the onset of the reproductive phase because various recessive *lf* alleles flower early. The effects of *lf* alleles are not graft transmissible, therefore, *LF* is thought to act in the shoot apex. It has been proposed that *LF* sets the meristem's level of sensitivity to the ratio of inducer/inhibitor and that loss-of-function mutants require more inducer action (Murfet et al. 1985).

SN (sterile nodes), *DNE* (day neutral), and *PPD* (photoperiod response) act in the photoperiod induction pathway in leaves and cotyledons (Fig. 7.1) (see Murfet and Reid 1993). These genes are thought to operate at different points in a pathway that controls the production of a graft-transmissible inhibitor. Recessive alleles in any one of the genes knock out the function of the inhibitor, promoting flowering under short-day conditions; therefore, if a day-neutral scion (such as *sn, DNE, PPD*) is grafted to a wild type stock (*SN, DNE, PPD*), the graft requires long days to flower. In wild type plants, it has been argued that long days suppress the activity of *SN, DNE,* and *PPD* and promote flowering (Murfet et al. 1985). Another gene, *HR* (high response), appears to promote inhibitor production by *SN, DNE,* and *PPD* plants under short day conditions. Dominant alleles of *HR* make the long day requirement brought about by *SN, DNE,* and *PPD* nearly obligate, while recessive *hr* alleles relaxed the requirements for long days. Grafting experiments indicate that floral inhibitor production is high in grafts from *SN, DNE,* and *PPD* plants grown under short-day conditions, but that inhibitor levels rapidly decline with age in short-day-grown *hr* plants (Weller et al. 1997).

The *GI* (gigas) gene appears to promote flowering because recessive *gi* mutations result in late flowering (Table 7.1). Unvernalized *gi* plants maintained under short-day conditions grow extremely tall without flowering (Beveridge and Murfet 1996). (Vernalization or cold treatment can bypass photoperiod requirements for flowering, and *gi* mutants are very responsive to vernalization.) Late flowering in *gi* mutants might be due to enhanced production of a floral inhibitor in leaves, reduced production of floral stimulus in leaves, or reduced ability to respond to a floral stimulus in the shoot apex (Beveridge and Murfet 1996). Enhanced production of the inhibitor in leaves was tested by grafting scions from an early-flowering day-neutral line onto stocks of the *gi* mutant. It was found that flowering was no more delayed in grafts to *gi* stocks than it was in grafts to wild type stocks. It does not appear, therefore, that *gi* mutants produce high levels of a floral inhibitor. The opposite grafts were made to test whether *gi* mutants have a reduced response to a floral stimulus; that is, *gi* mutant scions were grafted onto early-flowering, day-neutral stocks. The stocks were effective in promoting early flowering in both grafts, thereby indicating that the *gi* mutant was as responsive to the floral stimulus as its nonmutant parental counterpart. Finally, it was tested whether the *gi* mutant was defective in the production of a graft-transmissible floral stimulus. To do so, reciprocal grafts were made between the epicotyls (shoots) of a *gi* mutant and nonmutant parental line seedlings. (Lateral branches were trimmed from the stocks in most cases to avoid the contribution in the analysis of leaf tissue from the stocks.) Under long-day conditions where *gi* mutants do not usually flower, *gi* mutant scions on nonmutant stocks flowered like the parental lines. Under short-day conditions, flowering in parental lines was delayed, and the floral stimulus was thought to be weaker. Under these conditions, only grafts of *gi* mutant scions on nonmutant parental line stocks bearing lateral branches were able to flower. This was interpreted to mean that leaf tissue in the nonmutant parental line stocks provided the floral stimulus. Thus, *GI* is thought to control or be involved in the production of a floral stimulus in leaf tissue, which is able to cross graft junctions and induce flowering in the shoot (Fig. 7.1).

Thus, flowering in pea is thought to be regulated by graft-transmissible activators that are constitutively produced and by graft-transmissible inhibitors that are regulated by environmental conditions such a photoperiod (Weller et al. 1997). The interaction between activators and inhibitors is not necessarily thought to be direct. Because of other effects of the flowering genes, it has been hypothesized that the inhibitors might influence the direction of nutrient flow in the plant and may delay flowering by diverting the flow of floral stimulus away from the apical region. Identifying the graft-transmissible floral activator (sometimes called "florigen") has been a lifelong quest of a number of investigators in this field. Such a substance remains elusive and has not been found, so far.

TRANSITION FROM VEGETATIVE TO INFLORESCENCE DEVELOPMENT IN ARABIDOPSIS

Flowers in Arabidopsis are borne on an inflorescence, so flowering involves two transitions – development of the inflorescence and development of the flower. The first transition, formation of the inflorescence (or bolting), is influenced by environmental signals, such as long days and/or cold temperature. The second transition from inflorescence to flower development requires no other known environmental signals.

The transition from vegetative to inflorescence development in Arabidopsis has been largely studied by analyzing mutants that are either late or early flowering (Weigel 1995). Arabidopsis is a facultative long-day plant, which means that long-day photoperiod conditions promote early flowering, but are not absolutely required for flowering (Koornneef 1997). Time to flowering, however, in Arabidopsis is relatively invariant under constant photoperiod and temperature conditions, and both early and late flowering mutants have been identified. It is interesting that mutants have not been found that never flower at all, which suggests that there is considerable genetic redundancy in the control of the flowering process.

The interplay between environmental and genetic factors has been assembled into a model for the control of flowering time in Arabidopsis (Martinez-Zapater et al. 1994). The model is based on the central role of a floral repressor that blocks the normal transition from vegetative to inflorescence development (Fig. 7.2). A candidate for the floral repressor is the gene responsible for the mutant *embryonic flower1* (*emf1*) described by Sung et al. (1992) and Yang et al. (1995) (Fig. 7.3). Loss-of-function *emf1* mutants bypass vegetative growth (rosette leaf formation) altogether and produce inflorescences without forming rosette leaves. *EMF1* is required, therefore, for vegetative development (formation of the rosette). In the absence of *EMF1* function, Arabidopsis undergoes a rapid transition from seedling to flowering plant. Thus, by promoting vegetative development, the normal gene represses the formation of the inflorescence. *EMF1* is thought to play a central role in the control of the time to flowering because recessive *emf1* mutations are epistatic to any late-flowering mutations and act early in Arabidopsis seedling development. The *emf* phenotype is independent of photoperiod or vernalization; therefore, the mutation disengages the regulation of the flowering process from other environmental signals.

Other genes that affect time to flowering are thought to reinforce or reduce the effects of *EMF1*. These genes have been assembled into three pathways (Fig. 7.2) (Koornneef 1997). The constitutive repression pathway further enhances the repressive effects of *EMF1*. Loss-of-function mutations in this pathway, such as *terminal flower 1* (*tfl1*) and *early flowering 1* and *2* (*elf1* and *elf2*), relieve the floral repression effects of *EMF1* and promote early flowering. These genes and their gene products are not involved in environmental responses to flowering and, therefore, these genes are thought to act constitutively on flowering. (*Tfl1* will be described further in the next section.)

Genes in the other two pathways counteract the floral repression effects of *EMF1* (Fig. 7.2). Loss-of-function mutants in these pathways

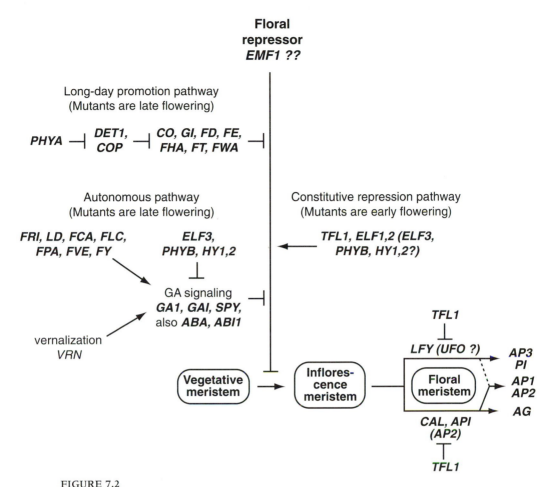

FIGURE 7.2

Model for control of flowering in Arabidopsis. The transition from vegetative to inflorescence meristem development is controlled by a floral repressor. Other genes that affect time to flowering (late flowering or early flowering) are thought to modify, directly or indirectly, the action of the floral repressor. (Activation is indicated by an arrow, ↓, whereas inactivation is represented by ⊥.) Late-flowering mutations occur in genes in two pathways: the long-day promotion and autonomous pathways. Early-flowering mutations occur in genes in the constitutive repression pathway. The transition from inflorescence to floral meristem is not acted upon by environmental factors in Arabidopsis; rather, it is controlled by other flowering genes. (Based in part on Martinez-Zapater, J. M., et al. 1994. The transition to flowering in Arabidopsis. In Meyerowitz, E. M., and Somerville, C. R., eds., Arabidopsis, pp. 403–433. Cold Spring Harbor, NY: Cold Spring Harbor Press.)

are late flowering because they relax the counteracting effects on *EMF1* and allow *EMF1* to repress the floral transition. Mutants in the long-day promotion pathway are delayed in flowering under inductive (long-day) conditions but not under noninductive (short-day) conditions;

FIGURE 7.3

Embryonic flower1 (*emf1*) mutant of Arabidopsis. Mutant produces an inflorescence at the seedling stage. Twenty-day old plant has cotyledons and does not produce rosette leaves, but it does form an inflorescence stalk bearing cauline or flower stalk leaves (arrows) and floral buds. Bar = 0.48 μm. (Reprinted with permission from Sung, Z. R., et al. 1992. *EMF* an Arabidopsis gene required for vegetative shoot development. Science 258:1645. Copyright © 1992 American Association for the Advancement of Science.)

therefore, the genes or gene products on this pathway are thought to be involved in responses to environmental (photoperiod) conditions. Mutants in the "autonomous pathway" are delayed in flowering under noninductive (short-day) conditions; therefore, the genes or gene products in this pathway are not thought to be involved in the responses to environmental conditions. Mutations in upstream genes in the autonomous pathway can be rescued by vernalization (cold treatment) which suggests that vernalization bypasses or provides a function similar to these genes.

A late-flowering mutant called *luminidependens* (*ld*) bears a mutation in a gene on the autonomous pathway (Lee et al. 1994) (Fig. 7.2). The mutant *ld* is late flowering even under long-day conditions; however, the effect of the mutant can be overcome by vernalization. *LD* has been cloned by T-DNA tagging and appears to encode a transcription factor. The gene product has a bipartite nuclear localization signal (for targeting to the nucleus) and a glutamine-rich C-terminus, which is characteristic of a number of transcription factors. *LD* RNA accumulation is not regulated by daylength, which is consistent with it being as a gene on the autonomous pathway.

Another gene member of the autonomous pathway *FLOWERING CA*

(*FCA*) has also been cloned (Macknight et al. 1997). *FCA* encodes a protein with RNA-binding and protein-interaction domains. The RNA-binding domains are similar to those encoded by a Drosophila gene that posttranscriptionally regulates important developmental pathways in the fly, such as sex determination. The *FCA* gene in Arabidopsis is large (8.1 kb), and the RNA transcripts produced from the gene are alternatively spliced. The most abundant spliced form (transcript β) encodes a truncated form of the predicted protein; however, the β form is not thought to be involved in time to flowering because production of the β transcript from a cDNA transgene does not restore early flowering in the late-flowering *fca-1* mutant. Only one of the spliced RNAs (γ transcript) encodes a full-length protein which is thought to be functional. The γ transcript represents a third of the total *FCA* transcripts. Alternative splicing, therefore, may control the regulation of the *FCA* gene because only one of the minor transcripts is functional in flowering.

Gibberellic acid (GA) signaling is also thought to be involved in the autonomous pathway (Fig 7.2). Chandler and Dean (1994) found that the effects of some of the late-flowering mutants could be overcome by application of GA, which suggests that GA acts downstream from late flowering mutations. GA is required for flowering under noninductive (short-day) conditions because mutants that fail to produce significant amounts of GA, such as *ga1*, do not flower under SD conditions (Wilson et al. 1992). The effect on flowering is similar in *gibberellic acid insensitive (gai)* mutants. Both *ga1* and *gai* mutants flower under continuous light or long days, and it is not clear whether they do so because the effects of long days and GA are redundant or that the GA threshold is lower under long-day conditions. Wilson et al. (1992) favored the former interpretation because they used an allele that is not thought to be leaky in their analysis of the *ga1* mutants. Although GA mutants do not have an effect under long-day conditions, *spindly (spy)* mutants that appear to be constitutive GA responders (see Fig. 5.16) tend to flower early even under long-day conditions (Jacobsen and Olszewski 1993). This may be due to the fact that the effects of *spy* are quite severe.

A late-flowering mutant called *constans (CO)* on the long-day promotion pathway has been cloned (Putterill et al. 1995). *CO* also appears to be a transcription factor and is a member of the zinc-finger class of DNA-binding proteins. It is interesting that unlike *LD*, *CO* expression is regulated by photoperiod. *CO* is expressed in long-day light regimes, which is consistent with the observation that *CO* has a phenotype in long days, but not short days.

In conclusion, the model described in Figure 7.2 for the control of flowering time in Arabidopsis (Martinez-Zapater et al. 1994) is similar to the model for pea in that both have stimulatory and inhibitory functions; however, the Arabidopsis model differs operationally. Late-flowering mutants in the Arabidopsis model are not thought to be defective in the production of a floral stimulus, but in the inactivation of the central floral repressor. Thus, the concept of a floral stimulus has not yet dominated the thinking of those who study floral induction in Arabidopsis.

TRANSITION FROM INFLORESCENCE TO FLORAL DEVELOPMENT IN ARABIDOPSIS

Flowers are produced from floral meristems on inflorescences (flower stalks) in Arabidopsis and snapdragon (*Antirrhinum*). The transition from inflorescence to floral meristem involves changes in the identity, phyllotaxy and determinacy of the meristem. Much work in Arabidopsis has been directed toward understanding the changes in the identity of the meristem during the floral transition. The roles of the inflorescence and floral meristems are quite different. Inflorescence meristems generate lateral inflorescences (paraclades or coinflorescences) or flowers, and floral meristems produce flower organs. The transition from inflorescence meristem to floral meristem requires the action of a number of genes, including *APETALA1* (*AP1*), *CAULIFLOWER* (*CAL*), and *LEAFY* (*LFY*) (Fig. 7.2). As will be discussed in Chapter 8, *AP1* functions in the floral transition and in the specification of flower organs. In *ap1* mutants, petals are not usually formed, and leaflike organs appear in place of sepals. At the base of these leaflike organs, axillary floral meristems arise that produce flowers. On its own, the *cal-1* mutant has almost no detectable phenotype. *Cal-1*, however, intensifies the effects of *ap1* in *cal-1 ap1* double mutants, which indicates that the function of *CAL* is largely redundant to *AP1* (Irish and Sussex 1990). In *cal-1 ap1* double mutants, meristems that would have produced flowers in single *ap1* mutants act as inflorescence meristems. Thus, in double *cal-1 ap1* mutants the formation of floral meristems is blocked, and indeterminate inflorescence meristems of increasingly higher order are produced. As a result, flower stalks in *cal-1 ap1* double mutants become densely packed masses of meristems, like a cauliflower head (Fig. 7.4).

AP1, CAL, and *LFY* have been cloned and all encode proteins predicted to be transcription factors. LFY has an acidic domain similar to other transcription factors (Weigel et al. 1992). *AP1* and *CAL* are MADS-box transcription factors, which are members of a family of transcription factors that include the flower homeotic genes (Mandel et al. 1992; Kempin et al. 1995). (Both MADS-box transcription factors and flower homeotic

FIGURE 7.4

Inflorescence from a double *cauliflower*-1 *apetala1*-1 (*cal*-1 *ap1*-1) mutant in Arabidopsis. Production of many higher-ordered inflorescence meristems forms dense clusters like that of a cauliflower head. Inflorescence is visualized by scanning electron microscopy. Bar = 100 µm. (From Bowman, J. 1994. Arabidopsis: An Atlas of Morphology and Development. Springer-Verlag, New York, p. 205. Reprinted by permission of Springer-Verlag.)

genes will be discussed in Chap. 8.) *CAL* and *AP1* have been used as probes to detect their own expression in Arabidopsis, and *CAL* has been used to detect orthologs (orthologous genes) in other plants. Both *CAL* and *AP1* expression were detected in young floral primordia in Arabidopsis consistent with their role in the formation of floral meristems. Because the double *cal-1 ap1* mutants in Arabidopsis appear like cauliflower heads, Kempin et al. (1995) examined the *CAL* gene ortholog in the garden variety cauliflower (*Brassica oleracea* var. *botrytis*). They found that the *CAL* gene in cauliflower was defective when compared with the orthologous gene in a flowering variety of *Brassica oleracea*. The authors speculated that a mutation in the *CAL* gene in *Brassica oleracea* may have given rise to the common cauliflower! (If the relationship between *CAL* and *AP1* is the same in cauliflower as it is in Arabidopsis, then the corresponding *AP1* gene may also be nonfunctional in cauliflower.)

Bowman et al. (1993) proposed that *CAL* and *AP1* control the sustained expression of *AP1* itself and *LFY*. They reached this conclusion when they could not find significant *AP1* and *LFY* expression in *cal-1 ap1* double mutants. Gustafson-Brown et al. (1994) also concluded that *CAL and AP1* might stabilize *AP1* expression because they found that *AP1* expression is reduced in older primordia.

Leafy (*lfy*) mutants are partially defective in the transition from inflorescence to flower meristem (Weigel et al. 1992). Inactivation of *LFY* causes inflorescences to develop in place of flowers, particularly in the

FIGURE 7.5

Terminal flower1 (*tfl1*-1) mutant in Arabidopsis. Lateral branch that terminates in a solitary flower is seen on the left of the main flower stalk. The *tfl1*-1 mutant flowers early, and flowers are produced soon after the inflorescence is formed. (From Shannon, S., and Meeks-Wagner, D. R. 1991. A mutation in the *Arabidopsis TFL1* gene affects inflorescence meristem development. Plant Cell 3:879. Reprinted by permission of the American Society of Plant Physiologists.)

early flowers. The flowers that do form are abnormal with a spiral phyllotaxy characteristic of inflorescences rather than flowers. In addition, the outermost organs are leaflike, and additional flowers form within mutant flowers. *LFY* is expressed in early floral primordia before the flower homeotic genes are expressed. *LFY* and *AP1* enhance each other's function, but they do not enhance each other's transcription. That was demonstrated by the fact that there was a greater loss of function when weak or intermediate *lfy* and *ap1* alleles were combined in double mutants, rather than when they acted alone in single mutants.

Ap1 mutants in combination with *lfy* block the transition from inflorescence to flower (Huala and Sussex 1992). *AP1* and *LFY* are largely redundant (Fig. 7.2), but they have somewhat different functions in flower development (Shannon and Meeks-Wagner 1993). *AP1* identifies the pedicel of the inflorescence as a floral stem, and *LFY* suppresses bract formation (Bowman et al. 1993). *AP1* RNA accumulates in floral meristems, but not in inflorescence meristems (Gustafson Brown et al. 1994). The expression of *AP1* in inflorescences is under negative regulation by another gene called *terminal flower* (*TFL1*) (Fig. 7.2). In *tfl1* mutants, *AP1* and *LFY* RNAs accumulate to higher levels in inflorescence meristems (Bowman et al. 1993).

Tfl1 was identified as an early flowering mutant by Shannon and Meeks-Wagner (1991). Wild type plants bear branched inflorescences with flowers. In contrast, *tfl1* produces a flower stalk with some lateral branches, but often with a single terminal flower (Fig. 7.5). Floral development at the terminal meristem stops further growth of the inflorescence. Shannon and Meeks-Wagner argued that wild type *TFL1* counteracts the expression of *LFY* and *AP1/AP2*, which are genes that promote the floral transition. In *tfl1* mutants, therefore, *LFY* and *AP1/AP2* expression are not suppressed during inflorescence development, and the mutant plant flowers precociously and usually forms a terminal flower on the inflorescence.

TFL1 in Arabidopsis has been cloned by virtue of its homology (similarity in function) with a cloned gene in *Antirrhinum* called *CENTRORADIALIS* (*CEN*), which will be discussed later in this chapter (Bradley et al. 1997). When *TFL1* was used as a hybridization probe to detect its own expression, it was found that *TFL1* RNA transcripts accumulated below the dome in floral meristems and throughout the stem of the inflorescence. The sequence of *TFL1* reveals little about its function. It was found to be similar to phosphatidylethanolamine-binding proteins, which can be associated with membrane protein complexes (Bradley et al. 1997).

CONTROL OF FLOWERING THROUGH THE ACTION OF TRANSGENES

To ask whether *LFY* or *AP1* expression alone is sufficient to induce flowering, the genes were introduced into transgenic Arabidopsis plants under the control of the CaMV 35S promoter, which is a promoter that is constitutively expressed in many tissues, including the flower (Mandel and Yanofsky 1995; Weigel and Nilsson 1995). Expression of the 35S:*LFY* transgene in wild type Arabidopsis accelerated the transition to floral development (Fig. 7.6) (Weigel and Nilsson 1995). Constitutive *LFY* expression appeared to have a greater effect on branches than on the primary shoot. Meristems on lateral inflorescences (paraclades or coinflorescences) were transformed to floral meristems soon after they were formed, which resulted in short, unbranched lateral inflorescences with single flowers. It was found that primary inflorescences were usually terminated by single flowers; however, the primary shoot did not flower immediately after bolting. Weigel and Nilsson (1995) argued that the primary inflorescence was not competent to respond to the expression of 35S:*LFY* immediately after bolting and must acquire that competence. In any case, the action of the 35S:*LFY* transgene was consistent with a gain-of-function phenotype for a gene that converts the shoot into flowers.

Blázquez et al. (1997) examined the expression of *LFY* during vegetative growth and during the transition to flowering. *LYF* transcription, as assessed by the activity of *LFY*:GUS constructs, was upregulated in

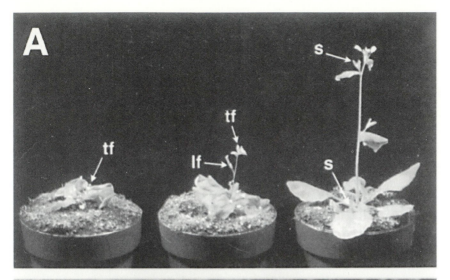

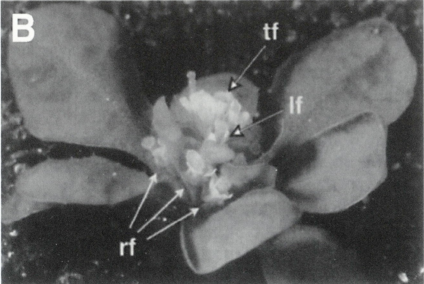

FIGURE 7.6

Transgenic Arabidopsis plants in which the *LEAFY* (*LFY*) gene is constitutively expressed, driven by the CaMV 35S promoter (35S:*LFY*). LFY expression rapidly converts all shoot meristems into floral buds. (A) Vegetative and inflorescence development in the two transgenic 35S:*LFY* plants on the left has been prematurely terminated. On the right is a nontransgenic control in which the main inflorescence has elongated and additional shoots, s, and shoot branches are formed before flowers are initiated at the apex. (B) Closer view of transgenic 35S:*LFY* plant. Inflorescence is short with a terminal flower, tf, and a few lateral flowers, lf. Transgenic plant has also formed solitary flowers (rf) in the axils of rosette leaves. Plants are about 4 weeks old and grown in long days. (From Weigel, D., and Nilsson, O. 1995. A developmental switch sufficient for flower initiation in diverse plants. Nature 377:496. Reprinted by permission of the author and Nature.)

plants grown under long days, conditions that induce flowering, as described earlier in this chapter. *LFY* is expressed in shoot apices under inductive conditions, largely in leaf primordia. *LFY* expression is much reduced under noninductive, short day conditions; however, *LFY* expression was enhanced by treating seedlings with GA which enhances flowering under short conditions (as described earlier in the chapter.) Hempel et al. (1997) found that they could decouple flowering and LFY expression by manipulating the spectral quality of the light source used in photoinduction. They observed that extension of short days by continous red or far-red light induced *LFY* expression. Only the red light conditions, however, induced flowering. The *LFY* gene, therefore, appears to be a rather direct target of the photoperiod mechanisms that control flowering in Arabidopsis. Nonetheless, there must be additional parallel controls that regulate the overall process.

To study the interaction of *LFY* with other floral induction genes, the 35S:*LFY* transgene was introduced by crossing it into other floral mutants. It was found that *ap1*-1 suppressed the effect of 35S:*LFY* on lateral inflorescences (paraclades). This indicates that *AP1* may act downstream from *LFY* in these inflorescences. Mandel and Yanofsky (1995) constructed a 35S:*AP1* transgene and introduced the construct into transgenic Arabidopsis plants. They found that the effects of the 35S:*AP1* transgene were similar to the 35S:*LFY* transgene in that primary and lateral inflorescences were prematurely converted to single flowers. Together, the two studies on the 35S:*LFY* and 35S:*AP1* transgenes indicate that although the genes are redundant in their function, either one is sufficient to convert shoots to flowers in transgenic plants. As described earlier, the principal effect of 35S:*LFY* on lateral inflorescences is suppressed in an *ap1*-1 mutant background. The expression of 35S:*AP1*, however, is not suppressed in a *lfy*-6 background. This is again interpreted to mean that *AP1* may act downstream from *LFY* and that *LFY* may activate *AP1*. (The effects of *Ap1* and *Lfy* are shown in parallel in Fig. 7.2 to indicate the redundancy in their functions.)

As described earlier, flowering can be induced and *LFY* upregulated in Arabidopsis by long days, and it was asked whether the constitutive expression of *LFY* and *AP1* could bypass the photoperiod requirement. It was found that the constitutive expression of either gene accelerated the floral transition in plants grown under short-day conditions. For example, when wild type plants are grown under short days, they flower about 10 weeks after germination. 35S:*AP1* plants grown under short days, however, flowered after 3 weeks. Weigel and Nilsson (1995) determined whether the constitutive expression of *LFY* affected the onset of the adult vegetative phase as well as the floral transition. In Arabidopsis, the transition from juvenile to adult vegetative phases is marked by the appearance of trichomes on the abaxial (ventral) surface of rosette leaves. They found that 35S:*LFY* had no effect on the vegetative transition. It was argued, therefore, that the effect of constitutive *LFY* expression is to truncate the adult vegetative phase, without affecting the juvenile to adult transition, and to accelerate the floral transition.

In conclusion, *LFY* appears to play an important role in mediating the controls between inflorescence and floral development (Fig. 7.2) (Blázquez et al. 1997). *LFY,* itself, appears to be a rather direct target of the photoperiod induction system and, in turn, *LFY* activates the expression of organ identification genes (to be discussed in Chap. 8). Although *LFY* is a central player in the control of flowering, the flowering mechanism appears to integrate signals from other players as well.

TRANSITION FROM INFLORESCENCE TO FLORAL DEVELOPMENT IN *ANTIRRHINUM*

Recall that the transition from inflorescence to floral meristem also involves changes in the phyllotaxy and determinacy. Special attention has been given to these changes in the floral transition in *Antirrhinum* (snapdragons). The transition from inflorescence to flower development in *Antirrhinum* involves the action of genes orthologous to those in Arabidopsis (Table 7.2). The transition from inflorescence to flower development is accompanied by a change from spiral to whorled phyllotaxy in that flowers with organs in whorls are borne in the axils of bracts that spiral up the inflorescence. Floral induction (meristem identity) genes, *floricaula* (*flo*) and *squamosa* (*squa*), control this transition. *Flo* and *squa* mutants produce inflorescence-like shoots instead of flowers and have an indeterminate growth pattern.

Carpenter et al. (1995) examined the effects of *flo* and *squa* on phyllotaxy in the transition from inflorescence to flower development. In wild type plants, the floral meristem first emerges as an eye-shaped structure and then expands along the medial–lateral axis to form a loaf

Table 7.2 *Genes That Control Inflorescence Meristem Identity and Determinacy in Arabidopsis and* Antirrhinum

Function	Arabidopsis	Antirrhinum
Meristem identity	*APETALA 1 (AP1)*	*SQUAMOSA (SQA)*
	LEAFY (LY)	*FLORICAULA (FLO)*
	CAULIFLOWER (CAL)	ND
Determinacy	*TERMINAL FLOWER1 (TFL1)*	*CENTRORADIALIS (CEN)*

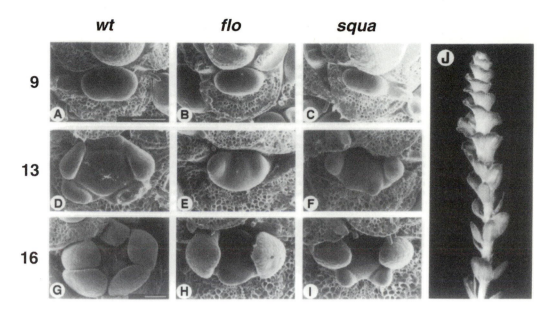

FIGURE 7.7

Effects of *floricaula* (*flo*) and *squamosa* (*sqa*) mutants on the transition to whorled phyllotaxy in flower development in *Antirrhinum*. Floral meristems in (A,D,G) wild type, (B,E, H) *flo*, (C,F, I) *squa*. Developmental stage indicated on left by the node number at which the floral meristem is found. At node 9, wild type floral meristem becomes loaf shaped. At node 13, wild type meristem acquires a pentagonal shape and shows development of whorled sepal arrangement, while *flo* and *sqa* meristems largely retain loaflike shape with bracts at ends. By node 16, petal primordia are visible in wild type and additional bract formation in spiral phyllotaxy are observed particularly in *squa*. Inflorescence meristem is visualized by scanning electron microscopy. Bar in A and G = 100 μm. (J) Spiral of bracts in *flo* inflorescence. Bracts toward tip fuse in this mutant creating a continuous spiral. (From Carpenter, R., et al. 1995. Control of flower development and phyllotaxy by meristem identity genes in *Antirrhinum*. Plant Cell 7:2004 and 2005. Reprinted by permission of the American Society of Plant Physiologists.)

(Fig. 7.7A). Next, the meristem expands more along the dorsal–ventral axis to form a pentagon structure, from the sides of which five petal primordia emerge with whorled phyllotaxy (Fig. 7.7D and G). In the *flo* mutant, development is similar to wild type until the loaf stage, but the meristem then fails to form the pentagon. Instead, bract primordia form at the ends of the loaf structure (Figure 7.7H). This meristem gives rise to an indeterminate inflorescence with a spiral array of bracts. The effects of *squa* are somewhat intermediate between *flo* and wild type. The meristem on the *flo* mutant forms two ventral bract primordia in addition to the lateral primordia; however, the ventral meristem only gives rise to threadlike structures (Fig. 7.7I).

Although *flo* and *squa* mutants have a profound effect on the shape

FIGURE 7.8

Effects of *centroradialis* (*cen*) mutant on inflorescence determinacy in snapdragon (*Antirrhinum majus*). Wild type inflorescence is indeterminate (left) in contrast to the *cen* inflorescence (right) which is determinant, has a terminal flower, and has numerous flowers on inflorescence branches. (From Bradley, D., et al. 1996. Control of inflorescence architecture in *Antirrhinum*. Nature 379:791. Reprinted by permission of the author and from Nature.)

of the meristem, it is not clear whether that is the cause or the effect of a change in phyllotaxy. Carpenter et al. (1995) argued that the defects in the *flo* and *squa* mutants might simply be a change in the synchrony of primordia emergence events. In wild type, the floral organ primordia appear synchronously in a pentagonal array. In the *squa* mutant, the appearance of the ventral bract primordium is delayed with respect to the lateral bracts. In the *flo* mutants, the ventral bracts never appear. *In situ* hybridization analysis shows that the *flo* and *squa* genes in wild type plants are expressed in axillary floral buds early in development. *Flo* mutants are epistatic to *squa,* and neither *flo* nor *squa* genes are expressed in axillary meristems in *flo* mutants. Hence, the expression of *flo* and *squa* genes is thought to be necessary for the shape changes in the axillary bud meristem that give rise to the whorled array of floral organs. The *FLO* gene is also thought to influence the partitioning of primordia into discrete structures. Older inflorescences in *flo* mutants have regions along the inflorescence stalk where bracts are fused, creating a continuous double spiral of bracts (Fig. 7.7J). It is thought that the bract primordia fail to individualize during development in the mutant, and that another function of the *FLO* gene is to promote the partitioning of bract primordia.

The *centroradialis* (*cen*) mutant in *Antirrhinum* affects determinacy of the inflorescence meristem. Recall that the gene is thought to be the ortholog of the *TFL1* gene in Arabidopsis (Bradley et al. 1996). The inflorescence meristem in *Antirrhinum* is normally indeterminate (capable of indefinite growth). Determinate flower development occurs in axillary bud meristems along the length of the inflorescence. The *cen* mutation leads to the formation of a terminal flower that ends growth of the inflorescence (Fig. 7.8). The effects of *cen* are almost completely opposite to *flo*. In *flo* mutants, flower development is suppressed, and flowers are replaced by lateral inflorescences (coinflorescences). Because of that the inflorescences in *flo* mutants are highly branched. In *cen* mutants, the shoot is abruptly terminated by the formation of a terminal flower.

At an early stage in development, floral meristems can be distinguished morphologically from inflorescence meristems by their production of flower organ primordia. In the *cen* mutant, the terminal meristem undergoes a transition to flowering and starts to form flower organ primordia after about ten axillary floral meristems are produced. (In wild type plants, flower organs are not produced by the apical inflorescence meristem.) In addition, it was found using *in situ* hybridization that *FLO* was expressed in the terminal meristem. In wild type plants, *FLO* is expressed only in the young axillary meristems where flowers are formed, but not in the SAM. Thus, it appears that *CEN* normally prevents *FLO* expression in the shoot apex. *CEN* was cloned by Tam element transposon tagging and used as a probe to determine the sites of expression (Bradley et al. 1996). It is interesting that the patterns of *CEN* and *FLO* expression did not overlap. *CEN* was expressed in a subapical region of the terminal meristem, below the region where FLO was ectopically expressed in *cen* mutants. Thus, the argument was made that *CEN* acts non–cell autonomously to limit the expression of *FLO*. (It was also argued that the SAM in the *cen* mutant did not undergo the transition to flowering until after about ten axillary flower buds were formed because the SAM was recruited for flower development at a later stage in development than the axillary meristems.)

FLORAL TRANSITION IN NONINDUCIBLE SYSTEMS

The floral transition in noninducible plants, such as day-neutral tobacco (*Nicotiana tabacum*), is regulated by developmental signals and less so by environmental factors. Nonetheless, flowering can be studied in day-neutral tobacco because the process is genetically controlled and quite reproducible under constant conditions. Any particular *N. tabacum* genotype produces a constant number of nodes before forming a terminal flower; however, the number of nodes produced varies from one genotype to another. At present, the control of flowering in tobacco is better understood operationally than mechanistically, and much of our knowledge about the transition to flowering in these plants is still at a phenomenological level.

The sort of questions that have been asked about flowering in noninducible plants include: How do the plants know when to flower? How do they count nodes? When does the meristem become deter-

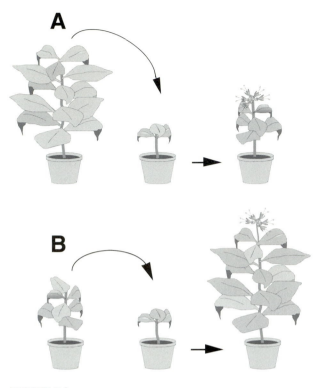

FIGURE 7.9

Floral determination in day-neutral tobacco (*Nicotiana tabacum*). Plants were decapitated at various times during vegetative development (after having produced a certain number of nodes), and shoots were rerooted. (A) Shoots with a critical number of nodes were "determined" to flower. They produced a few leaves and then flowered. (B) Shoots with less than a critical number of nodes were not yet determined to flower. They recapitulated most of vegetative development, produced a critical number nodes, and then flowered. (Based on McDaniel, C. N., Hartnett, L. K., and Sangrey, K. A. 1996. Regulation of node number in day-neutral *Nicotiana tabacum:* A factor in plant size. Plant J. 9:55–61.)

mined to flower? (Determination occurs when the fate of a developing structure is fixed. A developing structure is said to be determined when its fate does not change in a new environment. See Chap. 6.) Singer and McDaniel (1986) showed that tobacco plants (*N. tabacum* Wisconsin-38 plants grown under specified conditions) produced terminal flowers at node 41 (counting nodes starting from the base). To find out when the apex became determined to flower, plants were decapitated and the shoots rerooted (Fig. 7.9). Shoots from plants that reached 37 nodes at the time of decapitation flowered according to schedule prior to decapitation (i.e., with very little further development after rerooting). Shoots from plants that had not reached 37 nodes at the time of decapitation, recapitulated the course of vegetative development before flowering at 41 nodes. From these experiments Singer and McDaniel (1986) concluded that the shoot apex was determined for flower formation at node (or plastochron) 37 even though there was no visible evidence of the transition at this time. The rerooting experiments also showed that the passage of time was not the developmental signal that triggers flowering. Tobacco plants can be maintained indefinitely in a vegetative stage, if they are continuously decapitated and rerooted prior to the time when they become determined to flower. The signal to flower is also not based on the accumulation of some critical leaf mass. When leaves were progressively removed during development, leaving only a few mature ones at any position, the plant flowered with the normal number of nodes (Gebhardt and McDaniel 1991).

These experiments demonstrate that flowering is a predictable event in day-neutral tobacco. It has been proposed that a shoot becomes determined to flower when sufficient "floral stimulus" acts on a meristem that has become competent to respond to the stimulus. The nature of the floral-stimulus substance is not known, but it is thought to exist in a gradient in which the uppermost internodes have the highest floral stimulus activity. The gradient of floral stimulus activity has been demonstrated by regenerating shoots from axillary buds on stem pieces (containing six or more internodes) taken from different positions along the stem (McDaniel and Hartnett 1993). Regenerated shoots from stem pieces taken from the uppermost internodes flowered very quickly, usually producing four or fewer nodes before forming a terminal flower. Stem pieces from basal internodes produced more than four nodes, but substantially less than a whole plant before flowering. These regeneration experiments have been interpreted to mean that higher internodes have more floral stimulus activity than lower internodes (McDaniel 1996). (The same argument could be made for a gradient of a floral inhibitor in which the upper nodes have the lowest inhibitor activity.)

The competence of the shoot meristem to respond to the floral stimulus was determined by grafting tobacco shoot tips derived from plants of different stages in vegetative development onto stocks that had already flowered. (Such stocks were thought to provide a constant amount of floral stimulus.) It was found that the sensitivity to the floral stimulus increased with ontogenic age of the shoot tip (Singer et al. 1992). It was reasoned, therefore, that the number of nodes to flowering is controlled in tobacco plants by the gradient of the floral stimulus and the competency to respond. These parameters weigh differently in different genotypes and contribute to the variation in flowering time (McDaniel 1996).

REFERENCES

Bernier, G. 1988. The control of floral evocation and morphogenesis. Ann. Rev. Plant Physiol. Plant Mol. Biol. 39: 175–219.

Beveridge, C. A. and Murfet, I. C. 1996. The *gigas* mutant in pea is deficient in the floral stimulus. Physiol. Plant. 96: 637–645.

Blázquez, M.A., Soowal, L.N., Lee, I., and Weigel, D. 1997. *LEAFY* expression and flower initiation in Arabidopsis. Development 124: 3835–3844.

Bowman, J. 1994. Arabidopsis: An Atlas of Morphology and Development. New York: Springer-Verlag.

Bowman, J. L., Alvarez, J., Weigel, D., Meyerowitz, E. M., and Smyth, D. R. 1993. Control of flower development in Arabidopsis thaliana by *APETALA1* and interacting genes. Development 119: 721–743.

Bradley, D., Carpenter, R., Copsey, L., Vincent, C., Rothstein, S., and Coen, E. 1996. Control of inflorescence architecture in *Antirrhinum*. Nature 379: 791–797.

Bradley, D., Ratcliffe, O., Vincent, C., Carpenter, R., and Coen, E. 1997. Inflorescence commitment and architecture in Arabidopsis. Science 275: 80–83.

Carpenter, R., Copsey, L., Vincent, C., Doyle, S., Magrath, R., and Coen, E. 1995. Control of flower development and phyllotaxy by meristem identity genes in *Antirrhinum*. Plant Cell 7: 2001–2011.

Chandler, J. and Dean, C. 1994. Factors influencing the vernalization response and flowering time of late flowering mutants of *Arabidopsis thaliana (L.) Heynh*. J. Exp. Botany 45: 1279–1288.

Gebhardt, J. S. and McDaniel, C. N. 1991. Flowering response of day-neutral and short-day cultivars of *Nicotiana tabacum L*. Interactions among roots, genotype, leaf ontogenic position and growth conditions. Planta 185: 513–517.

Gustafson Brown, C., Savidge, B., and Yanofsky, M. F. 1994. Regulation of the Arabidopsis floral homeotic gene *APETALA1*. Cell 76: 131–143.

Hempel, F.D., Weigel, D., Mandel, M.J., Ditta, G., Zambryski, P.C., Feldman, L.J., and Yanofsky, M.F. 1997. Floral determination and expression of floral regulatory genes in Arabidopsis. Development 124: 3845–3853.

Huala, E. and Sussex, I. M. 1992. Leafy interacts with floral homeotic genes to regulate Arabidopsis floral development. Plant Cell 4: 901–913.

Irish, V. F. and Sussex, I. M. 1990. Function of the apetala-1 gene during Arabidopsis floral development. Plant Cell 2: 741–754.

Jacobsen, S. E. and Olszewski, N. E. 1993. Mutations at the spindly locus of Arabidopsis alter gibberellin signal transduction. Plant Cell 5: 887–896.

Kempin, S. A., Savidge, B., and Yanofsky, M. F. 1995. Molecular basis of the cauliflower phenotype in Arabidopsis. Science 267: 522–525.

Koornneef, M. 1997. Timing when to flower. Curr. Biol. 7: R651-R652.

Lee, I., Aukerman, M. J., Gore, S. L., Lohman, K. N., Michaels, S. D., Weaver, L. M., John, M. C., Feldmann, K. A., and Amasino, R. M. 1994. Isolation of *LUMINIDEPENDENS:* A gene involved in the control of flowering time in Arabidopsis. Plant Cell 6: 75–83.

Macknight, R., Bancroft, I., Page, T., Lister, C., Schmidt, R., Love, K., Westphal, L., Murphy, G., Sherson, S., Cobbett, C., and Dean, C. 1997. *FCA*, a gene controlling flowering time in Arabidopsis, encodes a protein containing RNA binding domains. Cell 89:737–745.

Mandel, M. A., Gustafson-Brown, C., Savidge, B., and Yanfosky, M. F. 1992. Molecular characterization of the Arabidopsis floral homeotic gene APETALA1. Nature 360: 273–277.

Mandel, M. A. and Yanofsky, M. F. 1995. A gene triggering flower formation in Arabidopsis. Nature 377: 522–524.

Martinez-Zapater, J. M., Coupland, G., Dean, C., and Koornneef, M. 1994. The transition to flowering in Arabidopsis. In Arabidopsis, ed. Meyerowitz, E. M. and Somerville, C. R., pp. 403–433. Cold Spring Harbor, NY: Cold Spring Harbor Press.

McDaniel, C. N. 1996. Developmental physiology of floral initiation in *Nicotiana tabacum L.* J. Exp. Bot. 47: 465–475.

McDaniel, C. N. and Hartnett, L. K. 1993. Floral stimulus activity in tobacco stem pieces. Planta 189: 577–583.

McDaniel, C. N., Hartnett, L. K., and Sangrey, K. A. 1996. Regulation of node number in day-neutral *Nicotiana tabacum:* A factor in plant size. Plant J. 9: 55–61.

Murfet, I. C. 1985. *Pisum sativum* L. In Handbook of Flowering, ed. Halevy, A. H., pp. 97–126. Boca Raton, FL: CRC Press.

Murfet, I. C. 1989. Flowering genes in *Pisum.* In Plant Reproduction: From Floral Induction to Pollination, ed. Lord, E. and Bernier, G., pp. 10–18. Rockville MD: American Society of Plant Physiologists.

Murfet, I. C. and Reid, J. B. 1993. Developmental mutants. In Peas: Genetics, Molecular Biology and Biotechnology, ed. Casey, R. and Davies, D. R., pp. 165–216. Wallingford, UK: CAB International.

Putterill, J., Robson, F., Lee, K., Simon, R., and Coupland, G. 1995. The *CONSTANS* gene of Arabidopsis promoters flowering and encodes a protein showing similarities to zinc finger transcription factors. Cell 80: 847–857.

Shannon, S. and Meeks-Wagner, D.R. 1993. Genetic interactions that regulate inflorescence development in Arabidopsis. Plant Cell 5: 639–655.

Shannon, S. and Meeks-Wagner, D. R. 1991. A mutation in the *Arabidopsis TFL1* gene affects inflorescence meristem development. Plant Cell 3: 877–892.

Singer, S. R., Hannon, C. H., and Huber, S. C. 1992. Acquisition of competence for floral development in *Nicotiana* buds. Planta 188: 546–550.

Singer, S. R. and McDaniel, C. N. 1986. Floral determination in the terminal and axillary buds of *Nicotiana tabacum L.* Devel. Biol. 118: 587–592.

Sung, Z. R., Belachew, A., Shunong, B., and Bertrand Garcia, R. 1992. *EMF* an Arabidopsis gene required for vegetative shoot development. Science 258: 1645–1647.

Vaughan, J. G. 1955. The morphology and growth of the vegetative and reproductive apices of *Arabidopsis thaliana* (L.) Hyeyh., *Capsella bursa-pastoris* (L.) Medic. and *Anagallis arvensis,* L.J. Linn. Soc. Lond. Bot 55: 279–301.

Weigel, D. 1995. The genetics of flower development: From floral induction to ovule morphogenesis. Ann. Rev. Gen. 29: 19–39.

Weigel, D., Alvarez, J., Smyth, D. R., Yanofsky, M. F., and Meyerowitz, E. M. 1992. LEAFY control floral meristem identity in Arabidopsis. Cell 69: 843–859.

Weigel, D. and Nilsson, O. 1995. A developmental switch sufficient for flower initiation in diverse plants. Nature 377: 495–500.

Weller, J. L., Reid, J. B., Taylor, S. A., and Murfet, I. C. 1997. The genetic control of flowering in pea. Trends Plant Sci. 2: 412–418.

Wilson, R. N., Heckman, J. W., and Somerville, C. R. 1992. Gibberellin is required for flowering in *Arabidopsis thaliana* under short days. Plant Physiol. 100: 403–408.

Yang, C. H., Chen, L. J., and Sung, Z. R. 1995. Genetic regulation of shoot development in Arabidopsis: Role of the *EMF* genes. Devel. Biol. 169: 421–435.

8

Flower Development

The discoveries that have led to our understanding of the genetic basis of flower development have been a triumph in modern plant developmental biology. The description of flower development as a pattern formation process (recall Chap. 1) has provided a theoretical framework to understand the forces that control flower development and make sense out of the vast number of flower mutants in Arabidopsis and snapdragon (*Antirrhinum majus*). Much of this chapter is devoted to a discussion of pattern formation in flower development and its description in molecular terms.

Flower formation is a function of floral meristems. During flowering, the SAM switches from vegetative to inflorescence development, as discussed in Chapter 7, after which inflorescence meristems undergo further transitions to become floral meristems. In the subtropical flowering legume (*Neptunia pubescens*), the meristem grows in size to become more highly mounded, and primordia for flower organs begin to emerge (Fig. 8.1). Floral organ primordia develop in concentric whorls, starting from the outer whorl. Sepal primordia arise first, in the outer whorl, around the periphery of meristem dome. Petal primordia emerge next in positions alternating with the sepals. Stamen primordia appear simultaneously in positions alternating with the petals, and finally, carpel primordia in the innermost whorl are produced. In general, sepals and petals are dorsiventral almost from inception, whereas stamens and carpels are radially symmetrical (during early stages of development).

Flowering may also involve a change from indeterminate to determinate growth. Depending on species, the switch to determinate growth occurs either at the transition from vegetative to inflorescence meristem or at the transition from inflorescence to flower. In some species, inflorescence growth is indeterminate and flowers, flower branches, or lateral inflorescences (coinflorescences) arise from axillary bud meristems on the

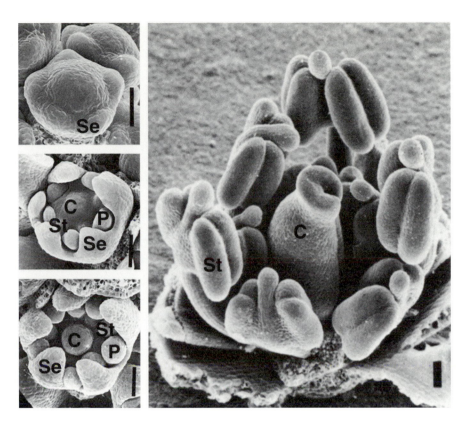

FIGURE 8.1

Development of the floral primordium in the perfect flower of *Neptunia pubescens*. *Neptunia* has pentagonal flowers with whorled phyllotaxy and two whorls of stamens. (Left panel, upper) Floral bud in which five sepal primordia, Se, have begun to emerge around the floral apex. (Left panel, middle, and lower) Five petal primordia, P, have emerged interdigitated between the sepals. Five stamen primordia, St, have been initiated at positions that alternate with the petals, and carpel primordia, C, are emerging. (Right panel) Older flower bud from which petals and sepals have been removed. Stamens have differentiated into anthers and filaments. Carpel is starting to differentiate into stigma and ovary. Depression in stigma is evident. Bar = 50 μm. (From Tucker, S. C. 1988. Heteromorphic flower development in *Neptunia pubescens*, a mimsoid legume. Am. J. Bot. 75:201–224. Reprinted by permission of the author and the American Journal of Botany.)

main inflorescence stem (Bradley et al. 1997). In determinate species, the terminal meristem on the inflorescence forms a flower. The termination of growth correlates with a loss of cells with proliferative potential in the meristem during flower development. Cells with proliferative potential at the center of the floral meristem are converted to or are replaced by cells that differentiate into floral organs. As will be described later, mutations that interfere with the development of carpels, which are the most central floral organs, sometimes confer indeterminate growth. It is interesting, however, that carpel formation and determinate growth are genetically separable. Mutants have been found in Arabidopsis in which carpels are formed, yet flower development is indeterminate in that reiterations of other floral organs are produced (Sieburth et al. 1995). Determinacy in flower development, therefore, is not as simple as the idea that carpel formation depletes proliferative cells in the meristem.

FOUNDING CELL POPULATIONS FOR FLOWERS

It was demonstrated in previous chapters (Chaps. 2 or 5) that founding cell populations for inflorescences could be defined during embryogenesis or during early seedling development, before cells giving rise to individual flowers and their organs had been sorted out. Estimates of the founding cell population for individual flowers in Arabidopsis were made by Bossinger and Smyth (1996) using sector boundary analysis. To generate sectors, they used a construct discussed in Chapter 2 (Fig. 2.1) with an *Ac* transposon inserted between the CaMV 35S promoter and the *uidA* (β-glucuronidase or GUS) reporter gene (35S:*Ac*:GUS). Excision of *Ac* generates sectors that can be visualized by histochemical stains for GUS activity.

They found that large sector boundaries usually divided the flower into halves, and the most frequent boundary occurred along the medial plane (Fig. 8.2). A less frequent boundary crossed the medial plane dividing the flower into quarters. It was concluded that the flower primordium is probably initiated from four cells derived from the longitudinal division of two cells that lay side by side. Each of the two cells gives rise to the bulk of the left half or right half of the flower. This was thought to happen because the descendants of the four primordial cells largely maintain their radial orientation in the formation of the flower, not because of some strict cell lineage considerations. Sector boundaries were frequently displaced from the medial plane, and displacement was thought to result from misoriented or unscheduled divisions that occurred during the early stages in organ formation. There was no effect on the development of the organ in instances where sector boundaries were displaced. This, again, points to the fact that flower development is not determined by cell lineages. The analysis demonstrates that flowers, like leaves, are polyclonal in origin, although the founding cell population for flowers appears to be small in Arabidopsis.

Similar sector boundary analysis was carried out for each floral organ. As expected, this analysis revealed that individual floral organs are specified at a later stage of flower primordium development (when

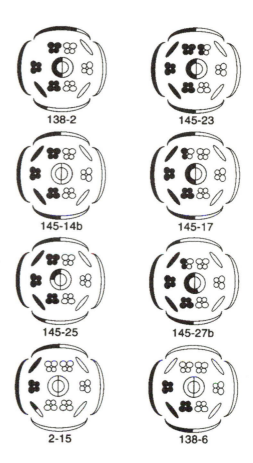

FIGURE 8.2

Sector boundary analysis of Arabidopsis flowers. Sectors were generated using the 35S:*Ac*:GUS construct described in Fig. 2.1B. Flower is represented diagrammatically with floral organs from the outside in: sepals, petals, stamens and carpels. GUS stained sectors are shaded darkly. Eight of sixty flowers used in the analysis are shown. Only one sector was observed in any flower with sectors. The most frequent sector boundary divided the flower in half, down the vertical, medial plane. A less-frequent boundary crossed the horizontal, medial plane dividing the flower into quarters, such as in flower 138-6, except for the lateral sepal on the left which is fully included in the sector. (From Bossinger, G., and Smyth, D. R. 1996. Initiation patterns of flower and floral organ development in *Arabidopsis thaliana*. Development 122:1096. Reprinted by permission of the Company of Biologists Ltd.)

the floral primordium is composed of more cells) rather than when the flower itself is specified. It was found that eight cells in a row founded each sepal and carpel, while four cells founded each stamen and two each petal (Bossinger and Smyth 1996). The number of founding cells for the various organs corresponded closely with the number of cells that can be seen at the site of organ primordium formation.

MOLECULAR GENETICS AND THE ABC MODEL

The use of molecular genetics to analyze flower development was championed by Meyerowitz and co-workers, who studied Arabidopsis, and the groups of Coen and of Saedler, who worked on snapdragons (*Antirrhinum*). These laboratories and others identified the genes required for floral organ formation and defined the interactions between the genes. (Several of their studies are cited in this chapter). The process of flower development was conceptualized along the lines of pattern formation in

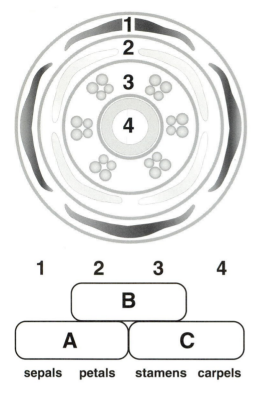

FIGURE 8.3

The Arabidopsis flower is composed of four concentric whorls of floral organs. (Upper) whorl 1, sepals; whorl 2, petals; whorl 3, stamens; whorls 4, carpels. (Lower) Simple representation of the ABC model. Model demonstrates that floral organs are specified by the overlapping domains of class A, B, or C homeotic gene expression in whorls 1–4.

Drosophila, as we discussed in Chapter 1. Recall that the first steps in pattern formation involved establishing a body plan, laying out the positions of organs and their relationship to each other, and then identifying organs in those positions. The flower "plan" is composed of four concentric whorls of organs (Fig. 8.3). Whorls are the equivalents of segments or developmental compartments in animal pattern formation terms, and the organs in each whorl assumed different identities in the normal flower. Sepals are found in whorl 1, petals in whorl 2, stamens in whorl 3, and carpels (or the gynoecium) in whorl 4.

The mutants that led to the idea that flower development was a pattern formation process were the homeotic selector mutants (Bowman et al. 1989). In homeotic mutants, the right organs developed in the wrong place (i.e., the wrong whorl). For example, petals arose in whorls where stamens would normally form, and so forth. Another way of describing the defect in the mutants is that the organs in various flower whorls were misidentified. To explain how the organs in whorls are normally identified and why the organs were misidentified in mutants, a simple but powerful model, now called the **ABC model,** was proposed (see Weigel and Meyerowitz 1994). The ABC model posits three classes of **homeotic genes** that determine the identity of various floral organs: class A, B, and C genes (Table 8.1).

Table 8.1 *Floral Homeotic Genes in Arabidopsis and* Antirrhinum

Class	Arabidopsis	Antirrhinum
Class A	*APETALA 1 (AP1), APETALA 2 (AP2)*	*SQUAMOSA (SQA)*
Class B	*APETALA 3 (AP3), PISTILLATA (PI)*	*DEFICIENS (DEF), GLOBOSA (GLO)*
Class C	*AGAMOUS (AG)*	*PLENA (PLE)*

In each whorl, a different combination of one or more homeotic genes is expressed, and it is the particular combination of functions that determines organ identity in each whorl (Fig. 8.3). Class A genes function in whorls 1 and 2, class B in whorls 3 and 4, and class C in whorls 2 and 3 (Figs. 8.3 and 8.4). Thus, in whorl 1, the production of sepals is determined by the class A function alone; in whorl 2, petals are formed as a result of the action of both class A and B functions; in whorl 3, stamens are identified by a combination of class B and C functions; and in whorl 4, carpels are the product of the class C function acting alone.

An important feature of the operation of homeotic genes in flower development is that the genes are both involved in identifying floral organs and in determining where they will be expressed. Class A and class C compete with each other, and in doing so mutually exclude each other from functioning in the same whorls. The function of class A genes in whorls 1 and 2 prevent class C genes from functioning in the same whorls, and vice versa. As a result, in class A loss-of-function mutants, class C genes are expressed in whorls 1 and 2, as well as in whorls 3 and 4, which leads to the formation of carpels in whorl 1 and stamens in whorl 2 (Bowman et al. 1991). In class C loss-of-function mutants, class A functions invade whorls 3 and 4 resulting in the formation of petals in whorl 3 and sepaloid structures in whorl 4. The territorial competition between homeotic genes of different classes is described as being the **cadasteral activity** of these genes. Floral homeotic genes with cadasteral activity appear to operate much like the gap genes in Drosophila development in defining boundaries between neighboring domains of gap gene expression (Chap. 1). There are two class A and two class B genes in Arabidopsis (Table 8.1). These genes function cooperatively in that the expression of both genes is required for class A or B function. Because of that, loss-of-function mutations in either gene incapacitate the pattern-formation function of both genes. The molecular basis for the cooperativity will be discussed shortly.

Knowing the relationship among the genes in the ABC model, one can predict the phenotypes of various homeotic mutants. The strength of the ABC model is that the phenotypes of mutant plants largely fulfill these predictions. Class A mutants, such as *apetala1* (*ap1*), involve alteration in organ specification in whorls 1 and 2 due to the loss of A func-

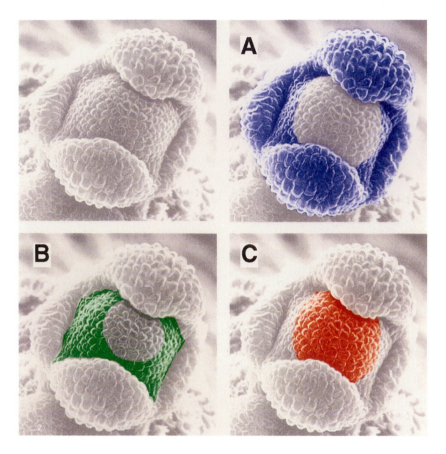

FIGURE 8.4

Pattern of expression of floral homeotic genes in Arabidopsis. Shown is a scanning electromicrophotograph of a stage 4 flower bud and the territory of expression of Class A, B, and C genes, indicated in false color. Class A pattern is typical of *AP1*, but not of *AP2* in which RNA is accumulated in all four whorls. (Based on Weigel, D., and Meyerowitz, E. M. 1994. The ABCs of floral homeotic genes. Cell 78:205. Figure adapted from Bowman, J. 1994. Arabidopsis: An Atlas of Morphology and Development. Springer-Verlag, New York, p. 149. Reprinted by permission of Springer-Verlag.)

tion, and the encroachment of C function into territory normally controlled by class A genes (Fig. 8.5A). In *ap1* mutants, sepals in the first whorl are converted to bractlike structures that have some carpelloid characteristics in extreme cases (Gustafson-Brown et al. 1994). Second whorl organs are usually missing in *ap1* mutants, but stamens can develop in intermediate alleles. Organ specification in whorls 3 and 4 are similar to wild type. Class B mutants, such as *apetela3* (*ap3*) or *pistillata (pi)*, are characterized by altered organ specification in whorls 2 and 3 due to the loss of B function (Fig. 8.5B) (see Bowman et al. 1991). In *ap3* mutants, sepals replace petals in whorls 2, and carpels replace sta-

FIGURE 8.5

Flower phenotypes of floral
homeotic mutants. Wild type;
Class A mutant, *apetala1*;
class B mutant, *apetala3;* class
C, *agamous*-1. Shown below
each picture is the interpretation of phenotype according
to the ABC model. Mutations
are loss-of-function mutations, and if there are two
homeotic genes in a class, as
in class A and B, both are ren-
dered nonfunctional by the
mutation. Class A and class C
genes compete for territories
of expression, and in the
absence of class A function in
whorls1 and 2, as in *apetala1*-
1, class C genes are expressed
in these whorls. Likewise, in
the absence of class C function
in whorls 3 and 4, as in *agamous*-1, class A genes are
expressed in these whorls.
(Weigel and Meyerowitz
1994). Bar = 1 μm. (Upper
panel from Gustafson Brown,
C., Savidge, B., and Yanofsky,
M. F. 1994. Regulation of the
Arabidopsis floral homeotic
gene *APETALA1*. Cell 76:132.
Reprinted by permission of
the author and Cell Press.
Lower panel from Bowman,
J. L., Smyth, D. R., and
Meyerowitz, E. M. 1989.
Genes directing flower development in *Arabidopsis*. Plant
Cell 1:39. Reprinted by permission of the American Soci-
ety of Plant Physiologists.)

mens in whorl 3. (In fact, the organs in whorl 3 range from stamenoid
to carpelloid, depending on the strength of the allele.) *AGAMOUS* (*AG*)
is the only class C gene in Arabidopsis, and class C mutants are characterized by misidentification of reproductive organs in whorls 3 and 4
(Fig. 8.5C). Flowers in *ag* mutants consist of many sepals and petals,
and most alleles have no stamens or carpels. *Ag* flowers are usually

FIGURE 8.6

Phenotype of flower that lacks all functional floral organ identity genes. (A) Wild type Arabidopsis flower. (B) Flower of triple *apetala2*-1 *pistillata*-1 *agamous*-1 (*ap2*-1 *pi*-1 *ag*-1) mutant. All floral organs in triple mutant appear as leaves. (From Weigel, D., and Meyerowitz, E. M. 1994. The ABCs of floral homeotic genes. Cell 78:204. Reprinted by permission of the author and Cell Press.)

indeterminate, and many sepals and petals arise from higher ordered flowers within the primary flowers.

In all the mutants, at least one floral homeotic gene still functions in each whorl. If all the genes required for floral organ identification are inactivated as in class ABC triple mutants, *ap2*-1 *pi*-1 *ag*-1, then leaves (or sepals) are formed in all whorls by default (Fig. 8.6B). Thus, leaves are the organs produced in the ground state, when no floral organs are specified. Botanists have long regarded flowers as modified leaves, and modern genetics has confirmed it as so.

In theory, the ABC model predicts that any type of floral organ can be produced in any whorl at will, given that the right combination of homeotic selector genes are expressed in that whorl. For example, it has been possible to produce any one of the four different floral organs in whorl 1 in Arabidopsis flowers (Krizek and Meyerowitz 1996). The genetic manipulations have involved combining the mutations described earlier with various floral homeotic genes as transgenes.

IDENTIFICATION OF FLORAL HOMEOTIC GENES

The operation of the ABC model can also be described in molecular terms because the floral homeotic genes have been isolated and identified. The first flower homeotic gene to be cloned was the *DEFICIENS A* (*DEFA*) gene in *Antirrhinum* (Sommer et al. 1990). *DEFA* is a class B gene that is an ortholog of *AP3* or *PI* in Arabidopsis. Mutations in *DEFA,* such as *deficiens*[globifera] (defA-1), are recessive and transform male organs into abnormal female organs and petals into sepals. *DEFA* was cloned by differential screening of a cDNA library with flower-specific probes from wild type and mutant flowers. (It's worth pointing out that screening cDNA libraries to find mutant genes is not recommended for a number of reasons. In most mutants, the RNA encoding the defective gene product might still be produced at normal levels. In this case, the mutation was caused by the insertion of a transposon, making it more likely that the mutant would not accumulate RNA from the gene.)

The sequence of *DEFA* predicted that the gene would encode a transcription factor similar to the minichromosome maintenance factor (MCM1) and pheromone receptor transcription factor (PRTF) in yeast, and the serum response transcription factor (SRF) in mammals. These factors contain a characteristic fifty-five to sixty amino acid segment, dubbed the MADS box, which includes a DNA-binding domain that recognizes a DNA sequence with CArG motifs containing a palindromic core 5'-CCTAATTAGG-3' (Schwarz-Sommer et al. 1992). The X-ray crystal structure of a related transcription factor, SRF, has been determined, and the core protein that binds to DNA is a dimer composed of an antiparallel coiled coil of two amphipathic α-helices that are each derived from one of the monomers (Fig. 8.7) (Pellegrini et al. 1995). The helixes lie in the minor groove of DNA in the center of the CArG sequence. The key features of this general class of transcription factors are that they form dimers (homodimers or heterodimers) and that the dimers bind to specific palindromic sequences in the promoters of target genes.

The next floral homeotic gene, *AGAMOUS* (*AG*), was cloned from Arabidopsis by T-DNA tagging (Yanofsky et al. 1990). The *AGAMOUS* gene product is also a MADS box protein, as are all of the other floral homeotic genes with the exception of *AP2.* All the major floral homeotic genes in both Arabidopsis and *Antirrhinum* have been cloned, and all of the homeotic selector genes encode MADS box transcription factors (except *AP2*).

CADASTERAL FUNCTIONS OF HOMEOTIC GENES

Pattern formation in flower development results from the interactions between floral patterning genes and their gene products (Fig. 8.8). There are three general types of interactions: type 1 – interactions between members of the same gene class (intraclass); type 2 – cadasteral func-

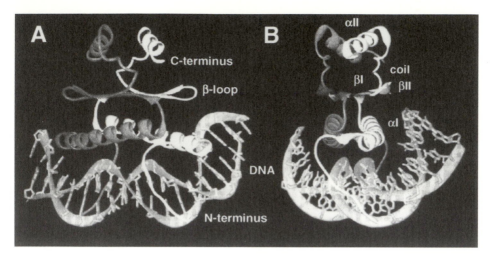

FIGURE 8.7

Structure of the serum response factor core-DNA complex. SRF is a MADS box transcription factor, and the general structure of the DNA-binding core in other MADS box factors is thought to be similar. (A) View of the complex perpendicular to the helical axis of the DNA. SRF is a dimer and the α-helixes from either monomer are antiparallel and lie along the minor groove of the DNA. (B) View along the helical axis of the DNA in which individual monomers can be more easily discerned. Protein-protein interactions in dimerization domain that constitutes most of the core can be seen. SRF recognition element with CArG recognition site with a palindromic core (5′-CCTAATTAGG-3′). (From Pellegrini, L., Tan, S., and Richmond, T. J. 1995. Structure of serum response factor core bound to DNA. Nature 376:494. Reprinted by permission of the author and Nature).

tions or competitive interactions between members of different classes (interclass) expressed in different parts of the flower; and type 3 – combinatorial interactions between members of different classes (interclass) expressed in the same regions of the flower. The intraclass interactions among the class A and B genes are cooperative or positive, whereas the interclass interactions that result from the cadasteral activities of class A and C genes are competitive or negative.

Class A and C genes mutually exclude each other's expression in the same whorl by their competitive interactions or cadasteral activity. The cadasteral functions of the class C gene, *AG*, were further demonstrated by Mizukami and Ma (1992) who ectopically expressed *AG* in transgenic plants through the action of the constitutive CaMV 35S promoter (35S:*AG*). *AG* is normally expressed only in whorls 3 and 4, but it was expected that AG would be expressed in all whorls of the flower (as well as all vegetative tissues) in transgenic plants. It was found that 35S:*AG* expression produced a phenocopy of *ap2* mutants in which class C functions are expressed in all four whorls (Fig. 8.9B). Flowers of

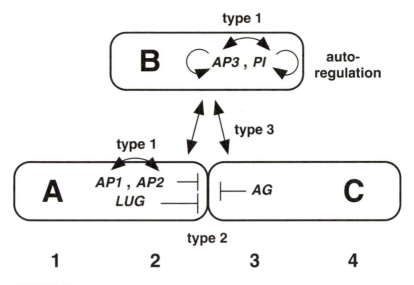

FIGURE 8.8

Interactions among class A, B, and C floral homeotic genes. *LEUNIG* (*LUG*) is also shown. It is a gene that has cadasteral function, but lacks organ specification function. Type 1 interactions between members of the same class are cooperative or positive. Type 2 interactions between two different classes, class A and C, are competitive or negative. Type 3 interactions are combinatorial and represent the interaction between different classes of homeotic genes expressed in whorls 2 and 3. Other gene controls include the autoregulation of the class B genes.

FIGURE 8.9

Phenotype of flowers from transgenic Arabidopsis plants that express 35S:*AG*. (A) Wild type. (B) *ap2*-1 mutant, a rather mild allele with reduced perianth development. Petals are shorter than wild type. (C,D) Transgenic Arabidopsis expressing 35S:*AG*. These flowers are from plants with mild effects and are similar to the *ap2* mutant. Note that first whorl organs are leaflike with stigmatic papillae, s, at their tips. Bars = 500 µm. (From Mizukami, Y., and Ma, H. 1992. Ectopic expression of the floral homeotic gene *AGAMOUS* in transgenic Arabidopsis plants alters floral organ identity. Cell 71:122. Reprinted by permission of the author and Cell Press.)

35S:*AG* plants showed reduced perianth development, and stigmatic papillae were produced on the tips of first whorl organs (Fig. 8.9C and D). The ectopic expression of *AG,* therefore, suppresses class A function in whorls 1 and 2. The consequences of ectopic *AG* expression in vegetative tissue were quite subtle.

Other floral pattern genes have only cadasteral and no floral organ identification activity. One such gene called *LEUNIG* (*LUG*) influences the territory of *AP2* function (Fig. 8.8). *AP2* is a class A gene that functions in whorls 1 and 2. *AP2* RNA, however, is found in all four whorls (Jofuku et al. 1994). Because of that, Liu and Meyerowitz (1995) proposed that the domain-specific function of *AP2* might be determined by other factors. A search was therefore undertaken to find mutations in other genes that might enhance or suppress a weak *ap2* mutation. They found two enhancers that turned out to be allelic to *lug*. *Lug* mutations are recessive, and the mutants frequently show organ identity transformations in the whorls 1 and 2. Whorl 1 organs are often petal-, stamen- or carpelloid, whereas whorl 2 organs are stamenoid. In wild type flowers, *AG* is normally expressed in whorls 3 and 4. In *lug* mutants, *AG* RNA was detected in all four whorls. Ectopic *AG* expression in whorls 1 and 2 is more pronounced in *lug*-1 *ap1*-1 double mutants than it is in the *lug*-1 single mutant. It appears, therefore, that *LUG* acts in concert with *AP1* to exclude *AG* function from whorls 1 and 2, and that the limits of *AG* expression are probably determined by the territory of *LUG* expression (Fig. 8.8). If that is the case, then one would predict that the organs in whorls 1 and 2 would be properly identified in double *lug*-1 *ag*-1 mutants. That is indeed the case, which suggests that *LUG* in the absence of *AG* has no function in whorl 1 and 2 with respect to organ identification.

POSITIVE REGULATION OF HOMEOTIC GENE FUNCTION

The type 1 intraclass interactions between floral homeotic genes within classes A and B are cooperative or positive interactions (Fig. 8.8). Where it has been investigated, the expression of one gene within a class reinforces and, in fact, is required for the expression of the other. The positive regulatory interaction between *AP3* and *PI,* which are both class B genes, was demonstrated by expressing *AP3* in transgenic Arabidopsis plants under the control of the CaMV 35S promoter (35S:*AP3*) (Jack et al. 1994). They found that carpels were converted to stamens in the fourth whorl, which is a result of the combined expression of class B and C genes in that whorl (Fig. 8.10B). Because only one of the class B genes

FIGURE 8.10

Phenotype of flowers from transgenic Arabidopsis plants expressing 35S:*AP3*. (A) Wild type, (B) 35S: *AP3*. Note that stamens
instead of carpels are produced in the center or 4th whorl. (From
Jack, T., Fox, G. L., and Meyerowitz, E. M. 1994. Arabidopsis
homeotic gene *APETALA3* ectopic expression: Transcriptional and
posttranscriptional regulation determine floral organ identity. Cell
76:704. Reprinted by permission of the author and Cell Press.)

was ectopically expressed in the transgenic plants, the question was
raised whether expression of the endogenous gene in the same class (*PI*)
was required for organ identification in the fourth whorl. To address this
question, the *AP3* transgene was introduced into *pi-1* mutant plants, and
carpels in these plants were not converted into stamens in the fourth
whorl because of the absence of *PI* function. Thus, in 35S:*AP3* plants,
ectopic expression of *AP3* in the fourth whorl appears to activate
endogenous *PI* expression in that same whorl, which indicates that both
genes must be expressed for class B function in the fourth whorl.

Similar findings have been made in studies of the class B genes
DEFICIENS A (*DEF A*) and *GLOBOSA* (*GLO*) in *Antirrhinum*. Tröbner et
al. (1992) noted that the normal high-level expression of *DEF A* and
GLO in whorls 2 and 3 of older buds was not seen in transposon insertion mutations in the *DEF A* and *GLO* structural genes. The high-level
expression of *DEF A* was reduced in *glo* mutants; similarly, the high-
level expression of *GLO* was reduced in *def A* mutants. As discussed earlier, these genes encode transcription factors. To explain the coordinate
regulation, Tröbner et al. (1992) proposed that the gene products are
autoregulatory, interacting to regulate their own transcription after initial activation (Fig. 8.11). More about the molecular basis for this control will be discussed shortly.

Not all positive control of homeotic gene expression is transcrip-

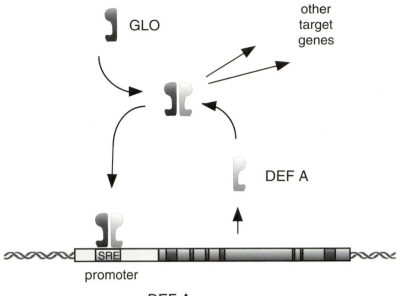

FIGURE 8.11

Model for the interaction of DEFICIENS and GLOBOSA proteins and autoregulation of *DEFICIENS A* (*DEF A*), a Class B gene in *Antirrhinum*. *DEF A* gene is shown with a promoter containing a MADS box-protein binding site (serum response element, SRE). *DEF A* encodes a protein that forms a heterodimer with the protein encoded by *GLOBOSA* (*GLO*) to form an active transcription factor. The transcription factor is thought to activate the expression of *DEF A* gene and other downstream target genes. (Redrawn from Schwarz-Sommer, Z., et al. 1992. Characterization of the Antirrhinum floral homeotic MADS-box gene deficiens—Evidence for DNA binding and autoregulation of its persistent expression throughout flower development. EMBO J 11:261. By permission of the publisher.)

tional. For example, Jack et al. (1994) found that ectopic expression of *AP3* driven by the CaMV 35S promoter (35S:*AP3*) led to misidentification of organs in whorl 4, where *AP3* is not ordinarily expressed. In this whorl, carpels were converted to stamens. Ectopic *AP3* expression, however, did not result in organ misidentification in whorl 1, which is where it might be expected that sepals would be converted to petals. Because *AP3* was driven by a strong constitutive promoter (the 35S promoter), the investigators assumed that *AP3* would be expressed throughout the developing flower in all whorls, including whorl 1. Their assumption was not borne out, however. When they attempted to demonstrate the presence of the AP3 protein in whorl 1 using immunolocalization techniques, they failed to do so. This indicated that posttranscriptional processes were at play that prevented either the synthesis or accumulation of AP3 protein in whorl 1. The mechanism is not yet understood, but it illustrates the complexities that control the regional expression and function of floral homeotic genes.

MOLECULAR FUNCTIONS OF THE MADS BOX GENES

Because the interactions among floral homeotic genes (Fig. 8.8) have such profound effects on pattern formation, there is real interest in understanding the molecular basis for the interactions. Significant progress has been made in investigations into type 1 interactions (cooperative intraclass interactions), but the molecular basis for type 2 interactions (competitive interclass interactions) and type 3 interactions (combinatorial interclass interactions) have been more difficult to demonstrate. There is evidence for the direct interactions between homeotic gene products in type 1 intraclass interactions. As discussed earlier, Schwarz-Sommer et al. (1992) investigated the intraclass interactions between two class B gene products, DEF and GLO, in snapdragons (*Antirrhinum*). Both DEF and GLO are MADS box transcription factors. Using an *in vitro* DNA binding assay, they showed that DEF and GLO together bind to a target oligonucleotide with a MADS box binding site (a SRE element with a CArG sequence motif) (Fig. 8.11). (The gene products were synthesized in an *in vitro* protein synthesis system.) Neither gene product was able to bind to the target oligonucleotide on its own. Thus, it was argued that the class B gene products in snapdragons form heterodimers with each other in order to bind to the target site. Tröbner et al. (1992) envisioned that the interaction between DEF and GLO could explain the reinforcement of one gene on the other's expression. The promoters in both DEF and GLO genes have MADS box binding sites, and it has been proposed that DEF and GLO may interact with each other to stimulate their own transcription.

MADS box transcription factors bind to promoter elements with a core CArG domain (Schwarz-Sommer et al. 1992). Each of these factors, however, plays a unique role in flower development and probably activates the expression of a different (but possibly overlapping) set of downstream genes. What is it, then, that confers the specificity of these factors? In an effort to find out, Mizukami et al. (1996) dissected the MADS box factor AGAMOUS by creating truncated forms of the protein using site-directed mutagenesis. They focused on five domains of the protein: the N region, which is a segment at the amino-terminus of the protein; the M or MADS box region, which is a moderately conserved; the K region that is similar to keratin proteins in animals; an intervening I region that is

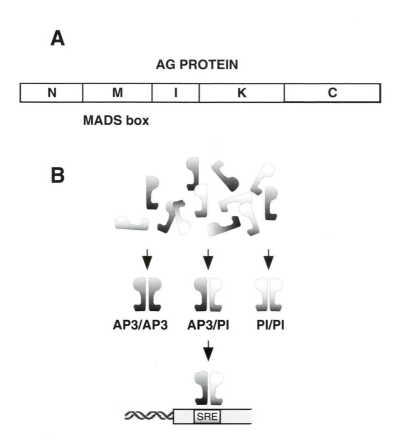

FIGURE 8.12

Structure of the MADS box transcription factor protein. (A) Transcription factor is composed of an amino-terminal N region; the M or MADS box region; keratinlike or K region; an intervening I region; a carboxy-terminal C region. (B) Dimerization among class B MADS box factors, AP3 and PI. Class B factor subunits form homo- and heterodimers; however, only the heterodimers are able to bind to SRE with CArG motifs. (Part A based on Mizukami, Y., et al. 1996. Functional domains of the floral regulator *AGAMOUS*: Characterization of the DNA binding domain and analysis of dominant negative mutations. Plant Cell 8:831–845. Part B based on Riechmann, J. L., Krizek, B. A., and Meyerowitz, E. M. 1996. Dimerization specificity of Arabidopsis MADS domain homeotic proteins *APETALA1, APETALA3, PISTILLATA,* and *AGAMOUS.* Proc. Natl. Acad. Sci. USA 93:4793–4798.)

not conserved; and a nonconserved C or carboxy-terminal region (Fig. 8.12). They observed that the MADS box and the I region were necessary for binding DNA in *in vitro* binding assays. They also found that both regions were involved in the formation of protein dimers, except that the N-terminal quarter of the MADS region was not required for dimerization. Hence, most of MADS box and I region were required for both dimerization and DNA binding.

The interaction between the MADS box factors from Arabidopsis has also been examined *in vitro* by Riechmann et al. (1996), who used an immunoprecipitation approach. In these experiments, an epitope-labeled MADS box factor was cotranslated with each of the other unlabeled MADS factors. The ability to coimmunoprecipitate the unlabeled MADS box factors was examined. It was found that all of the Arabidopsis MADS box factors – AP1, AP3, PI, and AG – interacted with each other in solution, where they formed dimers and coimmunoprecipitated with each other. There were, however, preferred partners when the factors were challenged to bind to the SRE-containing DNA. In general, the members of a class (intraclass combinations) interacted with each other. Similar to the findings of Schwarz-Sommer et al. (1992), Riechmann et al. (1996) observed that AP3/PI heterodimers (class B) bound to the SRE-containing DNA, but AP3 and PI homodimers did not bind (Fig. 8.12). AP1 (class A) and AG (class C) homodimers also bound DNA. (AP1 is the only class A MADS box factor. AP2 is not a MADS box factor.) Although all the MADS box factors dimerized with each other, only the combinations within a class form functional dimers. Again, the formation of functional heterodimers between the two B gene products suggests a basis for the cooperative or positive interactions between class B genes in whorls 2 and 3. The nonfunctional interaction between class A/B and class B/C gene products did not support the idea that heterodimer formation between MADS box factors of different classes can explain the type 3 combinatorial interactions.

The preferred partner interactions in the *in vitro* binding assays were further examined by swapping domains. Riechmann et al. (1996) found that the specificity of interactions lies in the nonconserved I region (Fig. 8.12A). The domain-swapping experiments did not reveal functions for the N, K, and C domains of the MADS box factors even though these regions of the protein are also thought to contact DNA and/or be involved in protein–protein interactions. Other roles for these domains have arisen from the analysis of various mutant alleles in MADS box factor genes, such as *AG* in Arabidopsis (Sieburth et al. 1995). On its own, *AG* directs carpel formation and determinate growth and, in combination with class B genes, specifies stamen formation. Sieburth et al. (1995) described two *ag* alleles with partial activity that affected carpel formation and determinacy separately. One was *ag*-4, a splicing site defect that results in a partial loss of the C-terminus of the K domain. The mutant *ag*-4 had stamens in the 3rd whorl (normal *AG* function) but sepals in the fourth whorl and indeterminate flower development (defective *AG* function). Another allele, *AG*-Met205, had a single amino acid change also in the same region of the K domain. This allele also produced stamens in the 3rd whorl (normal *AG* function), but had carpels in the fourth whorl (normal *AG* function), and showed indeterminate growth (defective *AG* function). (Indeterminate growth involved the production of four, instead of two carpels, and the formation within the carpels of additional whorls of stamens and carpels.) The K domain in AG, therefore, appears to have separable organ identification and growth determinacy functions.

Mizukami et al. (1996) pursued the functions of the N, K, and C domains further by analyzing *AG* constructs introduced into transgenic plants. Recall that when the intact 35S:*AG* construct is introduced into transgenic plants, an *ap2*-like phenotype is produced because *AP2* ordinarily limits the expression of *AG*. When *AG* is driven by the CaMV 35S promoter, however, *AG* expression overrides the territorial controls exerted by *AP2*. The result is a plant with flowers that look like *ap2* mutants in which sepals and petals have been converted to carpelloid and stamenoid organs respectively (similar to Fig. 8.9B). When the C terminal region was removed from 35S:*AG*, the transgene lost its ability to produce the *ap2*-like phenotype in transgenic plants. In fact, a number of the 35S:*AG* transgenic plant lines displayed *ag* mutant phenotypes in which reproductive organs in the third and fourth whorls were converted to perianth organs, and flowers were indeterminate in growth. These findings suggested that the C-terminal region was required for proper function, and that the mutated transgene acted as a dominant negative mutation that inactivated the expression of the endogenous *AG* gene. In 35S:*AG* constructs in which both the K and C regions were removed, the flowers of the resulting transgenic plants had normal reproductive organs, except that they were more numerous than in the nontransgenic control. The effects of these constructs, therefore, were different from the constructs bearing the C-terminal deletions; however, the plants and their flowers were not normal. This suggests that the K region also plays a role in normal *AG* gene expression. Because of its similarity to keratins, the K region is thought to play some role in protein–protein interactions.

GENES THAT MEDIATE THE INTERACTIONS BETWEEN FLORAL MERISTEM AND FLORAL ORGAN IDENTITY GENES

Although floral genes with cadasteral functions define the territory of expression of other homeotic genes, the question can be raised, What turns on the expression of these genes in the first place? As discussed in the Chapter 7, some of the homeotic genes are activated by the prior

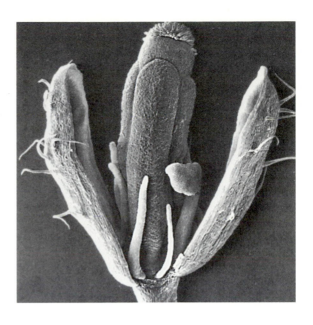

FIGURE 8.13

Flower in *unusual floral organs* (*ufo*) mutant in Arabidopsis. Whorls 2 and 3 form chimeric organs that result from partial or complete homeotic transformations of petals and stamens. One sepal has been removed for viewing inner organs of the flower. Sepals in outer whorl are generally unaffected by the mutation. (From Bowman, J. 1994. Arabidopsis: An Atlas of Morphology and Development. Springer-Verlag, New York, pp. 250 and 258. Reprinted by permission of Springer-Verlag.)

expression of floral meristem-determining genes, such as *LFY* and *AP1* in Arabidopsis (Weigel and Meyerowitz 1993). The homeotic genes are not affected equally (Fig. 7.2). *LFY* is a major activator of class B genes, *AP3* and *PI*. Even though *LFY* and *AP1* activate the expression of the floral homeotic genes, they probably have little to do with determining their spatial pattern of expression. That is reasoned from the experiments described in Chapter 7 in which *LFY* and *AP1* transgenes were used to accelerate the time to flowering. In these experiments, the transgenes were driven by the CaMV 35S promoter, which is expressed generally in the flower, nonetheless, normally patterned flowers were produced in the transgenic plants. On the other hand, *lfy* mutants, such as *lfy-6*, can be rescued by ectopic 35S:*AP3* and 35S:*PI* expression (Krizek and Meyerowitz 1996). While ectopic *AP3* and *PI* expression restores class B gene function, it does not restore the normal pattern of organ formation. Flowers in *lfy-6* 35S:*AP3* 35S:*PI* plants consist mostly of petal and stamenoid organs.

Other floral genes appear to play roles in mediating the interactions between floral meristem and floral organ identity genes. One such gene involved in mediation is *fimbriata* (*fim*) in *Antirrhinum* and its ortholog in Arabidopsis *UNUSUAL FLORAL ORGANS* (*UFO*) (Ingram et al. 1995). This gene fulfills a variety of roles in flower development, much like *AP1*. It appears to be involved in floral organ formation and several other processes in inflorescence and flower development. Mutations in this gene interfere with flower and floral organ formation to various extents (Fig. 8.13). In *Antirrhinum, fim* is expressed at a developmental stage after the meristem identity genes, *floricaula (flo)* and *squamosa (squa)*, and before the floral organ genes, *deficiens (def)* and *plena (ple)*, are expressed in wild type flowers (Simon et al. 1994). Meristem identity genes are normally expressed at node 0 and organ identity genes expressed at nodes 10–11. (Nodes are numbered sequentially starting at the top with node 0, which is the youngest visible node.) *Fim* is first expressed at node 9 in wild type flowers, but *fim* is not expressed in *flo* mutants, which suggests that *flo* is required for *fim* expression. Despite the dependence of *fim* on *flo*, the genes appear to exclude each other's expression in the early floral meristem. *Fim* is expressed in the central domain of the early floral meristem, and *flo* is expressed in the area around *fim*. At the time of floral organ formation, domains of *def* and *ple* expression take over the central two whorls of the flower, whereas *flo* expression is seen in whorl 1, and *fim* expression is limited to whorl 2. *Def* and *ple* expression is reduced in *fim* mutants, which suggests that prior expression of *fim* is required for the activation of organ identity genes in the central whorls. The boundary of *ple* expression appears to coincide and may be determined by *fim* expression. Thus, at the stage of floral organ formation, the four whorls correspond to the domains of *flo, fim, def*, and *ple* expression and to the temporal order in which these genes are activated.

The Arabidopsis ortholog, *UFO*, appears to influence several aspects of floral meristem and floral organ formation in Arabidopsis, as *fim* does in *Antirrhinum* (Levin and Meyerowitz 1995). *UFO* influences the expression of flower homeotic genes, particularly the class B genes, *AP3* and *PI* (Fig. 7.2). *In situ* hybridization experiments showed that *AP3* and *PI* expression was reduced in *ufo* mutants compared with wild type. In addition, intermediate *ufo* alleles have defects in Class B functions in that petals generally show some sepaloid transformation and stamens usually appear as filaments that frequently fuse with carpels. Strong alleles, such as *ufo*-2, show a reduction in the number of petals and stamens. As in *lfy* mutants, the class B defects in *ufo*-2 can be rescued by ectopic 35S:*AP3* and 35S:*PI* expression (Krizek and Meyerowitz 1996). These transformants have petaloid organs in the outer whorls and stamens in the inner whorls.

UFO also has a somewhat similar role to *LFY* in activating the expression of floral homeotic genes (Levin and Meyerowitz 1995). Double *ufo lfy* mutants were constructed to determine whether these genes interact or have overlapping functions. The alleles appeared to interact synergistically in that the phenotype of the double mutant is similar to a strong *lfy* mutant. It was also determined whether *UFO* and *AP1* interact. As discussed earlier, *LFY* and *AP1* interact in determining the identity of the floral meristem; therefore, it was asked whether *UFO* and *AP1* interact similarly. Double *ufo ap1* mutants show a partial conversion of floral meristems to inflorescences, which suggests that the two genes have overlapping functions. The flowers of the double mutants appeared to be made up of leaflike organs and were arranged in a spiral, not whorled phyllotaxy.

Flowers in either *fim* or *ufo* mutants were quite variable. *Ufo* flowers are made up of various filaments and fused organs (Fig. 8.13), and it was unusual to find two flowers that have the same set of floral organs. Levin and Meyerowitz (1995) speculate that *UFO* may play some role in deciding the boundaries of primordia for various floral organs. If the boundaries are too narrow and too few cells are involved in the formation of a primordium, then an organ frequently might only become a filament, or it might not be made at all. On the other hand, if the boundaries are too wide, then the organ might be fused with the neighboring organ.

CONTROL OF WHORL FORMATION AND NUMBER OF ORGANS PER WHORL

The pattern formation genes described earlier affect the identity of organs in various whorls. What determines whether a whorl will be formed and how are the number of organs in a whorl determined? One of two models is usually put forward to explain the control of whorl formation. One is a spatial model; the other a sequential model (Day et al. 1995). In the spatial model, it is argued that whorl formation is regulated by positional information and that whorls and their organs develop independently of surrounding tissue. In the sequential model, the formation of one whorl is thought to induce the formation of the next. Because the outermost whorl forms first, it is thought to induce the next inner whorl, and so forth.

Various microsurgical experiments tend to support a sequential model. For example, McHughen (1980) resected peripheral areas in the floral meristem in tobacco (*Nicotiana tabacum*), and examined the fate of inner floral organs. He found that no petals, stamens, and carpels were formed when sepal-forming areas were resected. No stamens and carpels were formed when petal primordia were resected, and so forth. Despite the elegance of the microsurgical experiments, the results from the genetic studies do not support the conclusion that formation of one type of floral organ depends on another. Genetic studies described earlier in the chapter demonstrate that one can change the identity of the organs in any whorl and thus the order in which the organs are formed. In flower development, therefore, the formation of carpels does not depend on the prior formation of stamens and so forth. The microsurgical experiments are only consistent with the idea that formation of organs in the inner whorls of the flower depends on the formation of organs in outer whorls, regardless of the type of organ produced in any whorl.

More recent studies using genetic ablation techniques, however, favor a spatial rather than a sequential model for the control of whorl formation. Day et al. (1995) ablated organ primordia in the second and third whorls in Arabidopsis and tobacco (*Nicotiana tabacum*) flowers using a whorl-specific "suicide gene" that consisted of the *AP3* promoter linked to diphtheria toxin A chain gene (*AP3:DT-A*). Recall that *AP3* is a class B gene expressed in whorls 2 and 3 of Arabidopsis flowers at a time when sepal primordia initiate (Jack et al. 1992). The suicide gene (*AP3:DT-A*) was therefore used to drive the production of the toxin and ablate cells involved in petal and stamen primordia formation in whorls 2 and 3. The flowers on Arabidopsis plants bearing *AP3:DT-A* failed to form petals and stamens and produced only two kinds of floral organs – sepals and carpels. Thus, the results were inconsistent with a sequential model which assumes that organs are formed in whorls one after another. The results in tobacco were not as clear cut because carpel formation was delayed in the severe phenotypes in which development in whorls 2 and 3 was fully suppressed. Normal pistils, however, were eventually produced in these flowers, suggesting that interference in normal development of the second and third whorls might have had some general effect on flower development, such as impairment in growth regulator production or transport, and not that second and third whorl formation has a specific inductive effect on fourth whorl formation. In any case, even the tobacco experiments generally disputed a sequential model.

Although the results of genetic ablation studies appear to be at odds with those of the microsurgical experiments, it is possible to reconcile the two studies. Both findings are consistent with the idea that the inductive events in the sequential model transpire early in floral meristem development before the time when AP3 expressed. After that time, whorls act independently and produce primordia without inductive signals from other whorls.

There are several Arabidopsis mutants that affect the number and/or positions of organs in a whorl. As discussed in Chapter 5 (Fig. 5.8), *Clavata* (*clv*) mutants affect the overall size and function of vegetative, inflorescence, and floral meristems. These mutants have enlarged meristems with more organs in each whorl as well as also more whorls

(Clark et al. 1993; Clark et al. 1995). Other mutations affect the number of organs in specific whorls. For example, the class B homeotic genes appear to regulate the number of organs, particularly in the third whorl. Loss-of-function *ap3* or *pi* mutations reduce the number of stamens in the third whorl, and ectopic 35S:*AP3* and 35S:*PI* expression leads to additional whorls of stamens in the center of the flower (Krizek and Meyerowitz 1996).

Another mutation called *superman* (*sup*) produces numerous stamens in the third and fourth whorls (Bowman et al. 1992). Thus, the normal *SUP* gene appears to antagonize *AP3* and *PI* and was originally proposed to be a negative regulator that prevented class B from functioning in the fourth whorl. Through cloning of *SUP* and *in situ* hybridization analysis, however, it has been found that the early floral domain of *SUP* expression is at the boundary of whorls 3 and 4 (Sakai et al. 1995). The phenotype of the *sup* mutant is consistent with the idea that the normal *SUP* gene restrains cell divisions (particularly at the boundaries) in the third whorl, and this restraint might have non-cell autonomous effects in promoting cell proliferation in the fourth whorl. In the *sup* mutant, excess divisions in the third whorl are thought to produce supernumerary stamens and suppress divisions in the adjacent fourth whorl (Sakai et al. 1995). *FLORAL ORGAN NUMBER1* (*FON1*) is another locus in Arabidopsis that affects the number of organs per whorl in the third and fourth whorl and is thought to interact with *SUP* (Huang and Ma 1997). *Fon1* mutants appear to maintain meristematic activity in the center of the floral meristem longer than wild type, which generates additional stamen primordia inner to the third whorl as well as an increased number of carpels. The characteristics of *fon1* mutants indicate that the function of the wild type gene must be to limit stamen proliferation to the third whorl. Double mutants generated with *fon1* and the *clv* mutants have novel or additive phenotypes, which suggests that the genes operate on different pathways. *FON1* and *SUP* appear to interact because double heterozygotes between *fon1* and *sup* have a mutant *fon1*-like phenotype. (Both are recessive mutations, and the double heterozygote would be expected to be like wild type, if there was no interaction between loci.)

Perianthia (*pan*) mutants in Arabidopsis have an unusual effect on the number of organs in a whorl. Most of the flowers in *pan* mutants have a pentamer or pentamerous pattern of organs (i.e., 5 sepals, 5 petals, 5 stamens, and 2 carpels) (Running and Meyerowitz 1996) (Fig. 8.14). The pentamer pattern is interesting because it is the most common flower organ pattern in dicots outside the mustard family. Arabidopsis normally has a tetramerous pattern with four sepals, four

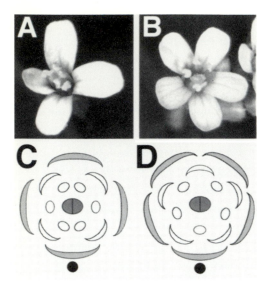

FIGURE 8.14

Flowers on *perianthia*-1 (*pan*-1) mutant in Arabidopsis. (A) Wild type, (B) *pan*-1 mutant. Note that *pan*-1 flower is pentamerous with five petals and stamens. Diagram of the most frequently encountered arrangement of organs in (C) wild type and (D) *pan*-1 mutant. (From Running, M. P., and Meyerowitz, E. M. 1996. Mutations in the *PERIANTHIA* gene of Arabidopsis specifically alter floral organ number and initiation pattern. Development 122:1262. Reprinted by permission of the Company of Biologists Ltd.)

petals, four plus two stamens, and four carpels. The mutant pattern can be seen in the emergence of the floral primordia. In wild type flowers, sepals emerge at medial and lateral positions. Petals arise next in an inner whorl at positions that alternate with the positions of sepals. In *pan* mutants, the adaxial sepal arises in a normal position, but the two lateral sepals emerge somewhat offset toward the adaxial position. Instead of one sepal, two arise in the abaxial position. Petals arise in the next inner whorl, in positions that alternate from the sepals, as in wild type. The authors argue that *PAN* acts differently from genes such as *CLV* in which floral organ number increase through an enlargement in meristem size (Running and Meyerowitz 1996). In addition, the mutations in double *clv pan* mutants are independent and additive. It is thought that *PAN* acts on a process by which primordia of the floral meristem assess their radial position in a whorl. In any case, the action of this gene may be responsible for the unusual arrangement of floral organs in cruciferous plants.

SEX DETERMINATION – MONECIOUS AND DIOECIOUS PLANTS

Most higher plants are hermaphroditic, bearing perfect flowers with both male and female organs. In other plants, unisexual flowers of both sexes are found on the same plant (monecious plants) or on different plants (dioecious plants). Unisexuality promotes outbreeding, which has adaptive advantages because it provides for genetic variability and exchange. The highest outbreeding rates occur among dioecious plants because outbreeding is obligatory, but rates can be variable in monecious plants in which outbreeding is not mandatory (Dellaporta and Calderon-Urrea 1993). Plants also have a variety of other mechanisms that encourage outbreeding such as self-incompatibility discussed in Chapter 10. Although only about 10 percent of plants are known to be strictly monecious or dioecious, unisexuality arose many times during plant evolution and is found in 75 percent of plant families. Because unisexuality arose independently many times, there is no consistent genetic basis for sex determination in plants. In fact, most of the mechanisms found in animals, such as X to autosome ratios, active Y, and autosome determined systems, are also found in plants.

It might seem from the preceding discussion of flower development that unisexual flowers would result from inactivation of floral homeotic genes in either (but not both) the third or fourth whorl. Few situations have been described so far, however, where that seems to be the case. Unisexuality usually arises instead by arresting or aborting the development of sex organ primordia once they have formed (Dellaporta and Calderon-Urrea 1993). Some exceptions are hemp (*Cannabis*) and the herb mercury (*Mercurialis*), which do not appear to form organ primordia of the opposite sex. In most cases, therefore, the genetic basis of unisexuality probably does not involve selective inactivation of floral homeotic genes.

The development of unisexuality has been studied extensively in maize (*Zea mays*). Maize is a monecious plant. The male flower is the tassel (staminate inflorescence) and the female flower is the ear (pistillate inflorescence). Unisexuality in maize results from the selective arrest in the development of the organs of one sex or the other (Dellaporta and Calderon-Urrea 1993). Immature inflorescence meristems are bisexual and virtually identical. In the primary florets of the ear, however, the stamen initials arrest and the gynoecium continues to develop. (Maize forms two florets, primary and secondary florets, in each spikelet, both in the tassel and in the ear.) In secondary ear florets, both stamen and gynoecia initials abort, leaving a single pistillate floret in each ear spikelet. Just the opposite happens in the tassel: Gynoecium initials degenerate and the stamen initials continue to develop. Thus, sex determination in maize appears to involve programmed arrest of cells in organs of the opposite sex.

Sex determination in maize tassel and ear development has been analyzed genetically, and mutations have been identified that block the feminizing genes in the ear or the masculinizing genes in the tassel. One of the functions of the feminizing genes is to arrest development of the stamens. In *dwarf* (*d*) mutants, stamen development is not arrested, and

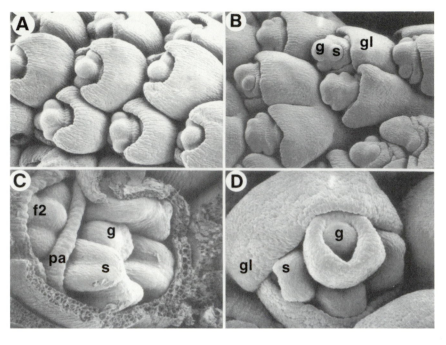

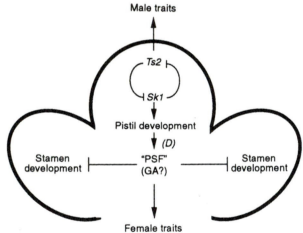

FIGURE 8.15

Development of florets on tassel inflorescences in *tasselseed2* (*ts2*) mutant of maize (*Zea mays*). (A,C) Nonmutant, *ts2-m1/Ts2*; (B,D) *ts2* mutant, *ts2-m1/ts2-R*. (A,B) Both nonmutant and *ts2* mutant initiate production of bisexual florets from which primordia for the female gynoecium, g, and male stamens, s, emerge. (C) Stamens grow in nonmutant spiklet, while the development of the gynoecium is arrested. Glume, gl, has been partially removed to reveal reproductive organs. (D) Both stamens and gynoecium develop in *ts2* mutant, and gynoecia will develop into silks in the mature tassel. Glumes are much shorter in mutant. Visualized by scanning electron microscopy. (Below) Model of sex determination in maize. *Silkless1* (*Sk1*) promotes development of pistils that produce a pistil-specific factor, PSF, which blocks stamen development and promotes the feminization of floral tissue.

(continued)

both pistils and stamens develop in the primary ear florets. *Dwarf* mutants affect both the feminization of the ear and the stature of the plant. *Dwarf* mutants *d1, d2, d3,* and *d5* have defects in gibberellic acid (GA) biosynthesis, and the feminization defect can be partially rescued by treating ears with GA at the appropriate stages in development. These mutants suggest that GAs play a role in programming cell death in male organs in the ear.

The development of primordia for the female gynoecium are blocked at an early stage in normal tassel development (Fig. 8.15C). Mutations blocking the effects of masculinizing genes fail to prevent the development of female organs in the tassel. In the case of the recessive *tasselseed* mutations (such as *ts1* and *ts2*), gynoecium development proceeds (Fig. 8.15D) and functional pistillate flowers formed in the tassel are capable of setting seed. *Ts1* and *ts2* therefore appear to have defects in the mechanisms that block the development of female organs in the tassel. (In severe *ts1* and *ts2* alleles, there is a complete sex reversal in the tassel. Female organ development in the tassel appears to cause ectopic stamen abortion.) *Ts2* has been cloned and encodes a protein that is similar to steroid dehydrogenases, which suggests that steroids might be cell death factors. Because *Ts2* acts antagonistically on the feminizing effects of gibberellic acid in the ear, it might seem that *Ts2* has a direct effect on GA production. The behavior of the *ts2* and *d1* double mutants, however, mitigate against the possibility that *Ts2* may have a direct effect on GA biosynthesis. Another mutation that affects feminizing functions is *silkless* (*sk1*). This mutation differs from *dwarf* in that stamen development is still arrested in ears, but *sk1* interferes with pistil formation, which is a feminizing function.

Dellaporta and Calderon-Urrea (1993) put forward a model of sex determination that involves the interaction of the genes described earlier (Fig. 8.15, diagram). In their model, *Sk1* promotes pistil development in the primary ear florets by blocking *Ts2* action. In a reciprocal manner, *Ts2* suppresses *Sk1* action in the tassel florets blocking pistil development. The model is supported by the action of the *ts2* and *sk1*

FIGURE 8.15 *(continued)*

Sk1 may also prevent the action of *Ts2* in aborting gynoecial development in primary ear florets. *Ts2* is thought to block *Sk1* action in promoting pistil development and PSF production. *Ts2* also promotes stamen development and masculinization of floral tissue. (A–D from DeLong, A., Calderon-Urrea, A., and Dellaporta, S. L. 1993. Sex determination gene *TASSELSEED2* of maize encodes a short-chain alcohol dehydrogenase required for stage-specific floral organ abortion. Cell 74:760. Reprinted by permission of the author and Cell Press. Panel below from Dellaporta, S. L., and Calderon Urrea, A. 1994. The sex determination process in maize. Science 266:1504. Reprinted by permission of the author and Science.)

double mutants, whereby *ts2* is epistatic to *sk1* in the ear, which suggests that *Sk1* may suppress the function of *Ts2* in the normal ear. Likewise, in the tassels, *sk1* is partially epistatic to the feminization effect of *ts2*.

REFERENCES

Bossinger, G. and Smyth, D. R. 1996. Initiation patterns of flower and floral organ development in *Arabidopsis thaliana*. Development 122: 1093–1102.

Bowman, J. 1994. Arabidopsis: An Atlas of Morphology and Development. New York: Springer-Verlag.

Bowman, J. L., Sakai, H., Jack, T., Weigel, D., Mayer, U., and Meyerowitz, E. M. 1992. *SUPERMAN,* a regulator of floral homeotic genes in Arabidopsis. Development 114: 599.

Bowman, J. L., Smyth, D. R., and Meyerowitz, E. M. 1989. Genes directing flower development in *Arabidopsis*. Plant Cell 1: 37–52.

Bowman, J. L., Smyth, D. R., and Meyerowitz, E. M. 1991. Genetic interactions among floral homeotic genes of Arabidopsis. Development 112: 1–20.

Bradley, D., Ratcliffe, O., Vincent, C., Carpenter, R., and Coen, E. 1997. Inflorescence commitment and architecture in Arabidopsis. Science 275: 80–83.

Clark, S. E., Running, M. P., and Meyerowitz, E. M. 1993. *CLAVATA1,* a regulator of meristem and flower development in Arabidopsis. Development 119: 397–418.

Clark, S. E., Running, M. P., and Meyerowitz, E. M. 1995. *CLAVATA3* is a specific regulator of shoot and floral meristem development affecting the same processes as *CLAVATA1*. Development 121: 2057–2067.

Day, C. D., Galgoci, B. F., and Irish, V. F. 1995. Genetic ablation of petal and stamen primordia to elucidate cell interactions during floral development. Development 121: 2887–2895.

Dellaporta, S. L. and Calderon Urrea, A. 1994. The sex determination process in maize. Science 266: 1501–1505.

Dellaporta, S. L. and Calderon-Urrea, A. 1993. Sex determination in flowering plants. Plant Cell 5: 1241–1251.

DeLong, A., Calderon-Urrea, A., and Dellaporta, S. L. 1993. Sex determination gene *TASSELSEED2* of maize encodes a short-chain alcohol dehydrogenase required for stage-specific floral organ abortion. Cell 74: 757–768.

Gustafson Brown, C., Savidge, B., and Yanofsky, M. F. 1994. Regulation of the Arabidopsis floral homeotic gene *APETALA1*. Cell 76: 131–143.

Huang, H. and Ma, H. 1997. *FON1,* an Arabidopsis gene that terminates floral meristem activity and controls flower organ number. Plant Cell 9: 115–134.

Ingram, G. C., Goodrich, J., Wilkinson, M. D., Simon, R., Haughn, G. W., and Coen, E. S. 1995. Parallels between *UNUSUAL FLORAL ORGANS* and *FIMBRIATA,* genes controlling flower development in Arabidopsis and Antirrhinum. Plant Cell 7: 1501–1510.

Jack, T., Brockman, L. L., and Meyerowitz, E. M. 1992. The homeotic gene APETALA3 of *Arabidopsis thaliana* encodes a MADS box and is expressed in petals and stamens. Cell 68: 683–697.

Jack, T., Fox, G. L., and Meyerowitz, E. M. 1994. Arabidopsis homeotic gene *APETALA3* ecotopic expression: Transcriptional and postranscriptional regulation determine floral organ identity. Cell 76: 703–716.

Jofuku, K. D., Den Boer, B. G. W., Van Montagu, M., and Okamura, J. K. 1994. Control of Arabidopsis flower and seed development by the homeotic gene *APETALA2*. Plant Cell 6: 1211–1225.

Krizek, B. A. and Meyerowitz, E. M. 1996. The Arabidopsis homeotic gene *APETALA3* and *PISTILLATA* are sufficient to provide the B class organ identity function. Development 122: 11–22.

Levin, J. Z. and Meyerowitz, E. M. 1995. *UFO:* An Arabidopsis gene involved in both floral meristem and floral organ development. Plant Cell 7: 529–548.

Liu, Z. and Meyerowitz, E. M. 1995. *LEUNIG* regulates *AGAMOUS* expression in Arabidopsis flowers. Development 121: 975–991.

McHughen, A. 1980. The regulation of floral organ initiation. Bot. Gaz. 141: 389–395.

Mizukami, Y., Huang, H., Tudor, M., Hu, Y., and Ma, H. 1996. Functional domains of the floral regulator *AGAMOUS:* Characterization of the DNA binding domain and analysis of dominant negative mutations. Plant Cell 8: 831–845.

Mizukami, Y. and Ma, H. 1992. Ectopic expression of the floral homeotic gene *AGAMOUS* in transgenic Arabidopsis plants alters floral organ identity. Cell 71: 119–131.

Pellegrini, L., Tan, S., and Richmond, T. J. 1995. Structure of serum response factor core bound to DNA. Nature 376: 490–497.

Riechmann, J. L., Krizek, B. A., and Meyerowitz, E. M. 1996. Dimerization specificity of Arabidopsis MADS domain homeotic proteins *APETALA1, APETALA3, PISTILLATA,* and *AGAMOUS*. Proc. Natl. Acad. Sci. USA 93: 4793–4798.

Running, M. P. and Meyerowitz, E. M. 1996. Mutations in the *PERIANTHIA* gene of Arabidopsis specifically alter floral organ number and initiation pattern. Development 122: 1261–1269.

Sakai, H., Medrano, L. J., and Meyerowitz, E. M. 1995. Role of *SUPERMAN* in maintaining floral whorl boundaries. Nature 378: 199–203.

Schwarz-Sommer, Z., Hue, I., Huijser, P., Flor, P. J., Hansen, R., Tetens, F., Lonnig, W. E., Saedler, H., and Sommer, H. 1992. Characterization of the antirrhinum floral homeotic MADS-box gene deficiens – Evidence for DNA binding and autoregulation of its persistent expression throughout flower development. EMBO J 11: 251–263.

Sieburth, L. E., Running, M. P., and Meyerowitz, E. M. 1995. Genetic separation of third and fourth whorl functions of *AGAMOUS*. Plant Cell 7: 1249–1258.

Simon, R., Carpenter, R., Doyle, S., and Coen, E. 1994. *Fimbriata* controls flower development by mediating between meristem and organ identity genes. Cell 78: 99–107.

Sommer, H., Beltran, J. P., Huijser, P., Pape, H., Lonnig, W.-E., Saedler, H., and Schwarz-Sommer, Z. 1990. *Deficiens,* a homeotic gene involved in the control of flower morphogenesis in *Antirrhinum majus:* The protein shows homology to transcription factors. EMBO J. 9: 605–613.

Tröbner, W., Ramirez, L., Motte, P., Hue, I., Huijser, P., Lönnig, W. E., Saedler, H., Sommer, H., and Schwarz-Sommer, Z. 1992. *GLOBOSA:* A homeotic gene which interacts with *DEFICIENS* in the control of Antirrhinum floral organogenesis. EMBO J. 11: 4693–4704.

Tucker, S. C. 1988. Heteromorphic flower development in *Neptunia pubescens,* a mimsoid legume. Am. J. Bot. 75: 201–224.

Weigel, D. and Meyerowitz, E. M. 1993. Activation of floral homeotic genes in Arabidopsis. Science 261: 1723–1726.

Weigel, D. and Meyerowitz, E. M. 1994. The ABCs of floral homeotic genes. Cell 78: 203–209.

Yanofsky, M. F., Ma, H., M., Bowman, J. L., Drews, G. N., Felmann, K. A., and Meyerowitz, E. M. 1990. The protein encoded by the *Arabidopsis* homeotic gene *agamous* resembles transcription factors. Nature 346: 35–39.

9

Development of Floral Reproductive Organs and Gametophytes

Sexual reproduction in higher plants involves the production of gametes, either eggs or sperm, that are borne in gametophytes. The female gametophyte in angiosperms is an embryo sac that develops within the ovule of the ovary, and the male gametophyte is a pollen grain. In angiosperms, gametophytes are small but multicellular. Pollen grains (Fig. 9.1A) are made up of two or three cells, a small generative cell that is often enclosed within a large vegetative cell (see Fig. 9.4 also). The generative cell divides to give rise to two sperm cells, which are the male gametes. The female gametophyte or embryo sac is also multicellular and is made up of cells that include the egg cell and a central cell that contains two polar nuclei (Fig. 9.1B). Fertilization in higher plants is a "double fertilization" process in which one of the sperm cells from a pollen grain fuses with the egg cell, giving rise to the zygote, whereas the other fuses with the central cell that produces the endosperm (Fig. 9.2).

The simple gametophytes in higher plants are relics of the more elaborate gametophytes in lower plants. Gametophytes are haploid structures that are often well-developed, multicellular plant forms in lower plants. (Gametophytes are haploid or 1N in plants in which the sporophyte is diploid or 2N.) For example, a moss gametophyte is a leafy plant that supports a less prominent sporophyte. In higher plants, the situation is reversed; the haploid phase is less prominent, and the sporophyte dominates the life cycle.

In higher plants, male gametophytes are derived from premeiotic cells called **microsporocytes** or **pollen mother cells** (Fig. 9.2). The microsporocyte undergoes meiotic divisions to produce four meiotic products or microspores. The four microspores are encased together in a microspore mother cell, and the four-celled structure is called a tetrad. Tetrads in most plants fall apart during pollen formation and form individual pollen grains. The nucleus usually divides once (pollen mitosis I),

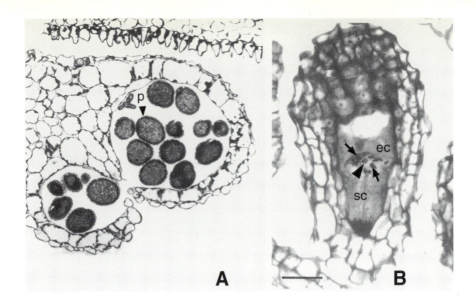

FIGURE 9.1

Male and female gametophytes of Arabidopsis. (A) Mature pollen grains (male gametophytes) in locules of an anther. Individual microspores released in the locule of the anther where they undergo the first mitotic division to form the generative and vegetative cells. Generative cell undergoes second mitotic division to form sperm cells. (See also Fig. 9.4.) (B) Egg cell, ec, seen in a section through the egg or embryo sac (female gametophyte) of Arabidopsis. The egg cell lies between the synergid cells, sc, and the central cell. Egg cell nucleus is indicated by the arrowhead, starch grains are pointed out by small arrows. Light microscopic section stained with periodic acid and Schiff's stain. (From Bowman, J. 1994. Arabidopsis: An Atlas of Morphology and Development. Springer-Verlag, New York, pp. 281 and 323. Reprinted by permission of Springer-Verlag.)

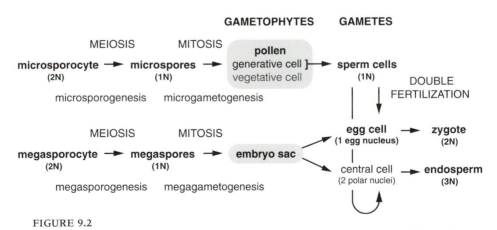

FIGURE 9.2

Diagram of events in gametophyte development and fertilization. Pollen development in male floral organs involves microsporogenesis and microgametogenesis, whereas embryo sac development in female floral organs results from megasporogenesis and megagaemtogenesis. Double fertilization in angiosperms is brought about by fusion of one sperm cell with the egg cell to form the zygote and another sperm cell with the two polar nuclei in the central cell to form a triploid endosperm. Ploidy numbers are given for a diploid plant.

during pollen formation, forming vegetative and generative cells. The generative cell divides again (pollen mitosis II), usually following pollen germination, to form two sperm cells.

The female gametophyte is derived from a premeiotic cell in the ovule called the **megasporocyte** or megaspore mother cell (Fig. 9.2). Megasporocytes undergo meiotic divisions in a process called megas-porogenesis to form four meiotic products or megaspores. Only one of the four meiotic products survives (in monosporic or Polygonum-type megasporogenesis), and the surviving megaspore gives rise to the game-tophyte during a subsequent process called megagametogenesis. The megaspore undergoes a series of mitotic divisions to produce usually eight nuclei within the embryo sac. The embryo sac cellularizes, com-partmentalizing the nuclei in separate cells, such as the egg cell and the synergids (Fig. 9.1B). The central cell contains two polar nuclei that often fuse prior to fertilization. During fertilization, the egg cell fuses with one of the sperm cells from the pollen to form the diploid zygote (in a diploid plant). The other sperm cell fertilizes the central cell (with fused polar nuclei), which gives rise to triploid tissue that will form the endosperm. (A more detailed description will be found in Chap. 11.)

ANTHER DEVELOPMENT AND POLLEN FORMATION

Male gametophytes or pollen grains develop in anthers, and anthers are structures borne on stamens, which are the male reproductive organs of the flower. Anthers are formed from different cell layers, and the cell layer origin of various tissues in the anther was determined in Jimson weed (*Datura*) by Satina and Blakeslee (1941). Epidermal cells were derived from L1; microsporocytes, middle and outer tapetum from L2; and vascular tissue and inner tapetum from L3. Microsporocytes origi-nate from diploid archesporial cells that give rise to it and a tapetal ini-tial in single division (McCormick 1993). In maize, microsporocytes also appear to be derived from L2. The origin of microsporocytes from a sin-gle cell layer creates an interesting situation in chimeric plants where a somatic cell layer in the tassel, such as L1, can be a different genotype than the sporogenous tissue. In their studies of cell lineages in maize anthers, Dawe and Freeling (1990) examined pigmented sectors created by the excision of the *Ds* element (of the *Ac-Ds* transposable element system) from the *bz2-m* or *a1-m* loci, which are two pigmentation genes in maize. (Excision of a *Ds* element can lead to activation of the gene in which the element was inserted.) They found that anther walls were bilayered and could be colored by activation of the pigment genes in

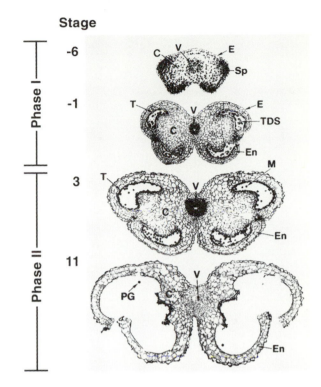

Stage

FIGURE 9.3

Anther development in tobacco (*N. tabacum*). Development is divided into two phases. The nonsporogenic tissue grows and develops in phase I and degenerates in phase II, providing materials for pollen development and allowing for pollen release. Transverse sections through anther at various stages of development. Connective, C, epidermis, E; endothecium, En; microspores, M; pollen grain, PG; sporogenous cells, SP; tapetum, T; tetrads, TDS; vascular bundle, V. (From Goldberg, R. B., Beals, T. P., and Sanders, P. M. 1993. Anther development: Basic principles and practical applications. Plant Cell 5:1221. Reprinted by permission of the American Society of Plant Physiologists.)

either L1 or L2; however, the activated gene was only transmitted through pollen when the excision event occurred in L2.

Anther development is thought to occur in two phases correlated with events in pollen development (Goldberg et al. 1993). During phase I, sporogenic cells in the anther engage in microsporogenesis, whereas nonsporogenic cells form the epidermis, tapetum, and so on (Fig. 9.3). The tapetum is a tissue that surrounds sporogenic cells and provides materials for developing pollen. During phase II, the anther enlarges and the filament (another part of the stamen) elongates. At this time pollen grains form, dehiscence occurs and pollen grains are released. During phase I, the nonsporogenic tissue provides structural support for sporogenic cells. Some of the same tissues degenerate in phase II, allowing for pollen release.

Microsporocytes undergo meiosis and produce tetrads of microspores that are individually compartmentalized, but which are nonetheless encased together in a callose cell wall. During pollen development, an enzyme, callase, is secreted by the tapetum. Callase digests the tetrad wall, releasing individual microspores. Upon release, the microspore undergoes an unequal or asymmetric mitotic division, called pollen mitosis I, that gives rise to a small generative cell and a large vegetative cell that often surrounds and encloses the generative cell in mature pollen

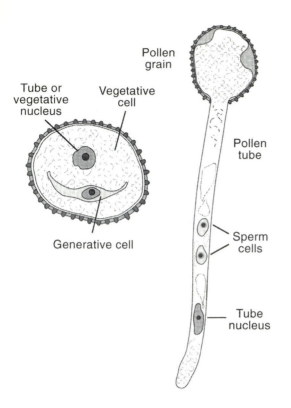

Pollen grain

Tube or vegetative nucleus

Vegetative cell

Generative cell

Pollen tube

Sperm cells

Tube nucleus

FIGURE 9.4

Mature pollen grain from lily (*Lilium*). (Left) Pollen grain is a two-celled gametophyte that contains a generative cell within a vegetative cell. The generative cell undergoes mitosis to form two sperm cells. (Right) Pollen grain germinates and extends a pollen tube. Vegetative cell cytoplasm, tube nucleus, and sperm cells are carried along by the growing tube.

(Fig. 9.4). The vegetative cell contains the tube or vegetative nucleus and the stored metabolites that power pollen germination and the elongation of the pollen tube. The small generative cell usually divides again to produce two sperm cells during elongation of the pollen tube. (In the mustard family, Cruciferae, and grasses, Gramineae, the generative nucleus divides and produces two sperm cells before pollen release.) The tube nucleus is transported through the elongating pollen tube, and the sperm are delivered after the pollen tube enters the micropyle.

Eady et al. (1995) examined the process by which the generative and vegetative cells differentiate from each other by studying the expression of the tomato *lat52* gene, which is transcriptionally activated only in vegetative cells during late pollen development. They determined whether the preferential expression of *lat52* in the vegetative cells was a consequence of unequal mitotic division. To do so, they produced transgenic tobacco plants that bear a *lat52* promoter-GUS fusion (*lat52*:GUS) and cultured microspores from the transgenic plant in high levels of colchicine (a mitotic spindle poison) to block pollen mitosis I. They found that blocking division did not prevent *lat52*:GUS expression. *Lat52*:GUS was instead expressed in the undivided cells at about the same time that it would have been expressed in untreated cells. Nonetheless, they found that unequal division was required for the

FIGURE 9.5

Arabidopsis pollen grains with highly sculpted exine coat. Visualized by scanning electron microscopy. (From Bowman, J. 1994. Arabidopsis: An Atlas of Morphology and Development. Springer-Verlag, New York, p. 341. Reprinted by permission of Springer-Verlag.)

asymmetric expression of *lat52*:GUS in daughter cells. Treatment of microspores with lower levels of colchicine permitted division, but prevented it from occurring asymmetrically in some pollen. In these cases, two equal daughter cells were produced that each contained a large vegetative cell nucleus; *lat52*:GUS was expressed in both cells. They hypothesized that the asymmetric pollen mitosis I unequally partitions factors required for generative cell differentiation and that the factors either activate *lat52*:GUS expression in the vegetative cell nucleus or repress the gene in the generative nucleus. This is an elegant example of the impact of unequal cell division on the differentiation of daughter cells.

The formation of pollen cell walls is initiated before the individual microspores are released from tetrads. The pollen wall is made up of two layers: an inner intine layer and an outer exine layer. The intine layer is initiated before the disintegration of the tetrad callose walls. The exine layer is largely derived from the materials deposited by the tapetum and is mostly composed of a polymer called sporopollenin. The exine layer is intricately sculpted (Fig. 9.5). The sculpting is species-specific, and the pattern is determined by the sporophytic tissue. As a result, the sculpting pattern is the same for individual meiotic products even if the pollen segregate for pollen pattern determinants. From the disintegrating tapetum, other compounds, such as flavonols and lipids, are also deposited to form an outer tryphine coat. These compounds are thought to be involved in pollen germination and in the interactions between pollen and stigma (Preuss et al. 1993).

The onset of phase II in anther development is marked by a change from growth to degeneration of the supporting tissue, followed by dehiscence. Degeneration of supporting tissue (tapetum) is accompanied by pollen development; however, the events that lead to dehiscence do not appear to be set in motion by activities in the pollen grains. Tobacco, made male-sterile by the action of a pollen-suicide transgene, dehisces normally (Mariani et al. 1990). Dehiscence involves an orchestrated set of events in the anther that includes breakdown of the tapetum, degeneration of the wall interconnecting the two pollen sacs, and, finally, rupture of the anther wall.

It is not known what signals the transition between phase I and II; however, molecular markers may be helpful in sorting out that process. Anther-specific cDNAs have been isolated and their temporal and spatial expression followed (Koltunow et al. 1990). An example is a cDNA called TA56 that is expressed in nonsporogenic tissue and encodes a thiol endopeptidase. The expression of this gene precedes cell-specific degeneration events. The gene is first expressed in the tissues that separate the two pollen sac compartments. The activation of TA56 gene expression appears to be transcriptionally controlled because a TA56 promoter-reporter gene construct shows a similar pattern of expression. Other cDNAs expressed in the tapetum demonstrate specific temporal patterns of expression in phase II; many are expressed at high levels prior to dehiscence.

Formation of pollen grains also involves gene expression in gametophytic tissue (McCormick 1993). The example of *LAT52* gene expression in microspores, described earlier, is a case in point. Muschietti et al. (1994) demonstrated that *LAT52*, which encodes a protein similar to Kunitz trypsin inhibitor, is expressed late in pollen development. They also demonstrated through the use of an antisense construct that *LAT52* gene expression was required for functional pollen development.

GENETICS OF ANTHER DEVELOPMENT AND POLLEN FORMATION

Male-sterile mutants have been studied intensively because they provide a basis for hybrid seed production. Male and female flowers or floral organs develop independently, and it is possible to obtain plants that are male sterile, but female fertile. These plants are not self-fertile, but they can be fertilized by pollen from related but not identical plants to produce hybrids. Male fertility is sensitive to many different types of mutations. Male sterility results from mitochondrial mutations (cytoplasmic male sterility) or nuclear mutations expressed either sporophytically (by maternal tissue) or gametophytically (by gametes).

Nuclear Male Sterility

Among the nuclear genes that confer male fertility, some act on anther development, whereas others affect microsporogenesis, pollen development, release, or function. The formation of microspore tetrads is easily scored and can be used to identify and characterize mutants. Mutants with abnormal tetrads generally have defects in microsporogenesis or in premeiotic development of the archesporial cells, whereas those with normal tetrads often have defects in later stages of pollen development, release, or function (Chaudhury et al. 1994).

Mutants have been described that disrupt the development of anthers. For example, a nuclear recessive mutant in maize, called *antherless* (*at*), has normal filaments, but lacks anthers (Kaul 1988). Other maize mutants, such as the dwarf gibberellic acid–deficient mutants, *d2, d3,* and *d5,* form small anthers that fail to produce pollen. In tomato, a mutant called *stamenless-2* has been described that produces short stamens and nonfunctional microspores. The tomato mutant also appears to be defective in gibberellic acid (GA) production and can be rescued by application of GA (Sawhney and Bhadula 1988). Both maize and tomato mutants demonstrate the importance of GA in stamen development.

Other nuclear mutants have been described that specifically affect pollen development. In Arabidopsis, four groups of sporophytically expressed nuclear genes have been identified among *male sterile* (*ms*) mutants that have premeiotic developmental lesions (Chaudhury et al. 1994). For example, *ms3* and *ms4* mutants have normal sporogenous cells, but have aberrant tetrads indicating that premeiotic or meiotic microspore development is perturbed or arrested (Fig. 9.6B and C). In maize, premeiotic or meiotic mutants appear infrequently among nuclear male sterile mutants. In a study by Albertsen et al. (1981), only two out of thirteen male sterile mutants affected early stages of microsporocyte development. One maize mutant, called *ms8,* showed abnormalities during early prophase stages in microsporocyte meiosis. Most of the *ms* mutants are postmeiotic mutants that demonstrate the important role of sporophytic genes in controlling pollen development

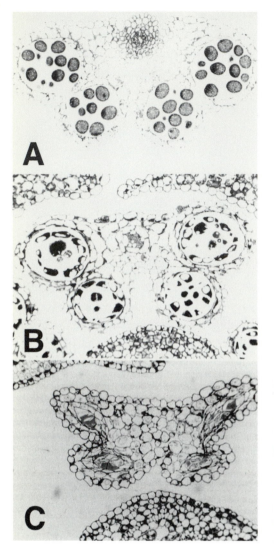

FIGURE 9.6

Defects in male-sterile mutants in Arabidopsis.
Cross-section through anthers of (A) normal,
male-fertile mutant, (B) *ms4* mutant in which
premeiotic or meiotic microspore development
is blocked, and (C) *ms3* mutant in which
locules collapse at the time of microspore
release. (From Chaudhury, A. M. 1993.
Nuclear genes controlling male fertility. Plant
Cell 5:1279. Reprinted by permission of the
American Society of Plant Physiologists.)

in maize. For example, normal tetrads are formed in mutants *ms2* and
ms7, but after the tetrad stage, chromosomes condense prematurely and
pollen development aborts (Albertsen and Phillips 1981).

Sporophytic male sterile mutants have been described in Arabidopsis, but they affect later pollen functions, such as pollen-stigma interactions (Preuss et al. 1993). These mutants will be discussed in Chapter
10. Arabidopsis mutants have been found that affect pollen development, but were not selected on the basis of male sterility. These mutants
were fertile and were found by screening for pollen morphology. Two
mutants, called *quartet* (*qrt*) mutants, were identified that produce
pollen which fail to separate into individual pollen grains (Preuss et al.

1994). In *qrt* mutants, the outer exine coat of the four meiotic products became fused, and pollen grains were released as tetrads. Callose cell walls normally keep individual microspores apart. In any case, an unusual advantage of these mutants is that they are viable and fertile, and they allow for tetrad analysis of meiotic products in Arabidopsis.

Cytoplasmic Male Sterility

Cytoplasmic male sterility (CMS) has been reported in many different plant species, but it has been most intensively studied in maize (*Zea mays*) and petunia (*Petunia hybrida*) (Levings 1993). The molecular lesions in mitochondrial DNA that result in CMS have been described, but the reasons why the lesions lead to male sterility are still not clear. T-type CMS in maize results from the expression of a nonessential chimeric gene in the mitochondrial genome called T-*urf13*. T-*urf13* is expressed in most tissues of CMS maize plants without apparent consequence, but its expression in anthers leads to degeneration of tapetal cells and aborted pollen formation. T-*urf13* encodes a 13 kD polypeptide (URF13) that is associated with the inner mitochondrial membrane. T-*urf13* expression renders maize plants susceptible to the toxin (T-toxin) produced by the fungal pathogen, *Bipolaris maydis* race T, which is the causative agent of Southern corn leaf blight (Williams and Levings 1992). The toxin appears to interact with URF13 and forms membrane pores (Levings 1993). It has been speculated that male sterility in maize results from the production of a toxinlike substance in anthers which kills cells that accumulate *URF13* (Flavell 1974). Such a substance, however, has not been found. Male fertility in CMS-T plants can be restored by the joint action of two dominant nuclear restorer genes, *Rf1* and *Rf2*. *Rf1* reduces T-*urf13* gene expression. *Rf2* has been cloned by transposon tagging and found to encode a protein that appears to be a mitochondrial aldehyde dehydrogenase (Cui et al. 1996). This was quite unexpected and led to several theories to explain the finding. One is that the RF2 protein might be required to rescue a metabolic disorder caused by T-*urf13* expression, such as scavenging or detoxifying aldehydes. On the other hand, the predicted enzymatic properties of RF2 might be unrelated to its effect. For example, RF2 might physically interact with URF13, ameliorating its effects.

In petunia, a similar type of CMS has been described that involves the expression of a nonessential chimeric gene in the mitochondrial genome called *pcf* (Hanson 1991). *Pcf* encodes a 25 kDa polypeptide found in mitochondria, and the abundance of the *pcf* gene product is reduced in nuclear restored lines. CMS in *pcf*-expressing lines is associated with a change in the partitioning of electrons through the normal oxidative and alternative oxidase electron transport pathways. In CMS lines, the activity of the alternative oxidase pathway is reduced, but it is restored in fertility-restored lines. Whether changes in the activity of the alternative oxidase pathway cause CMS in these lines is not known.

PISTIL, OVULE, AND EMBRYO SAC DEVELOPMENT

In angiosperms, the female organs in the flower are referred to as gynoecia, or pistils, which can be composed of one or more carpels that are usually fused together in higher plants. At anthesis (i.e., the time when the flower opens and fertilization occurs), the pistil has three functional parts; the ovary, which contains ovules; the style, through which pollen tubes grow; and the stigma, which is at the top of style where pollen grains land and germinate (Fig. 9.7).

The ovary is an enclosed space that is often divided into separate locules and is formed by tissue fusions at the margins of the carpels. In Arabidopsis, the carpels grow as a primordial cylinder and two locules are formed by the extension and fusion of tissues from either side of the cylinder (Gasser and Robinson-Beers 1993). The space defines the ovary and is enclosed by fusions at the margins of the extending cylinder, and by the development of an inner wall layer called the placenta. In many other plants, carpels fuse after they are formed (in a so-called postgenital fusion) (Verbecke 1992). Fusion involves the redifferentiation of cells at the surfaces where fusion occurs, as described in Chapter 2 (Fig. 2.8). After the carpels have fused and the ovary is enclosed, tissues at the top of the ovary extend to form one or more styles. Styles are short in Arabidopsis, longer in tomato (*Lycopersicon*), and very long in maize (*Zea mays*), where the styles are silks that emerge from the end of an ear of corn (Fig. 9.7). Tissues at the top of the style differentiate into the stigma. Stigma formation involves the proliferation and extension of papillary cells that secrete compounds to which pollen adheres and which promote or prevent the growth of pollen tubes.

Development of the ovule and embryo sac has been described by Reisner and Fischer (1993) and Schneitz et al. (1995). The ovule consists of a nucellus in which the megasporangium (structure that bears the megasporocyte) develops, one or more integument layers, and a supporting stalk called the funiculus (Fig. 9.8). The ovule is derived from an ovule primordium and is formed from the placenta of the inner

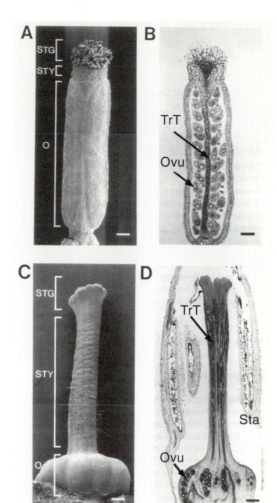

FIGURE 9.7

Flower pistils in Arabidopsis and tomato (*Lycopersicon*) at anthesis, which is the time when the flower opens. (A,B) Arabidopsis pistil has a short style and (C,D) tomato (*Lycopersicon*) pistil has a longer style. Stigma (STG), style (STY), ovary (O), ovule (Ovu), transmitting tract (TrT), stamen (Sta). Pistils are visualized by scanning electron microscopy (A,C) and by light microscopy of longitudinal sections (B,D). Bars = 100 μm in A and B, 400 μm in C, and 370 μm in C. (From Gasser, C. S., and Robinson-Beers, K. 1993. Pistil development. Plant Cell 5:1232. Reprinted by permission of the American Society of Plant Physiologists.)

ovary wall (Fig. 9.9A). Shortly after ovule initiation, a single subdermal cell, immediately below the apex of the nucellus, enlarges. The enlarged cell, called the archesporium, functions directly as a megasporocyte in some species, whereas in other species, such as soybean, the subdermal cell divides to form a multicellular archesporium out of which one cell will become the megasporocyte. During enlargement of the ovule primordium, three different regions along the proximal–distal axis of the primordium can be discerned: the nucellus containing the archesporium, the chalaza flanked by integuments, and the funiculus that anchors the ovule to the ovary wall (Schneitz et al. 1995) (Fig. 9.8). The development and specification of identity of these three regions is the important issue in pattern formation in ovule development.

Like the anther, the ovule is polyclonal in origin, and is derived from

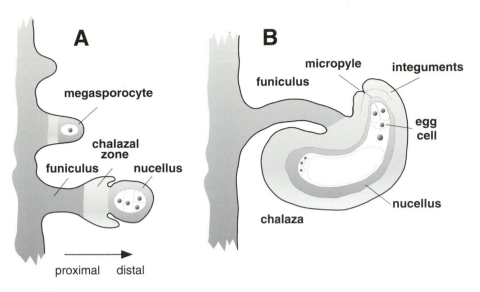

FIGURE 9.8

Ovule development viewed as a pattern formation process along the proximal–
distal axis of the developing ovule. (A) First step (upper) is the emergence of the
ovule primordium, followed by formation and development of three pattern elements: the distal element that contains the megasporocyte becomes the nucellus,
the middle element that forms the chalaza, and the basal element that forms the
funiculus. (B) Representation of a mature, unfertilized ovule. Embryo sac (white
area) is surrounded by sporophytic tissue, including the integuments. Egg cell is
located near the micropylar end of the embryo sac. Embryo sac is largely occupied by the central cell and large vacuole. Three antipodal cells, found at the chalazal end of the embryo sac, usually degenerate before or during fertilization.
(Based on Schneitz, K., Huelskamp, M., and Pruitt, R. E. 1995. Wild-type ovule
development in Arabidopsis thaliana: A light microscope study of cleared whole-
mount tissue. Plant Journal 7:731–749.)

more than one cell layer. Analysis of periclinal chimeras in Jimson weed
(*Datura*) has shown that the megasporocyte is derived from the L2 layer
(Satina 1945). The integuments arise from tissue surrounding the nucellus. Many of the dicots and most of the monocots have two integument
layers. The inner integument is derived from L1 and the outer integuments from L1 and L2 (Reisner and Fischer 1993). The integuments
grow during embryo sac formation, surrounding and overgrowing the
nucellus (Fig. 9.9C). Before the embryo sac matures, the nucellus degenerates in many species, leaving the embryo sac in direct contact with the
inner integuments. In these species, the integument layer in contact
with the embryo sac may differentiate into a specialized tissue called the
endothelium. The endothelium has cytological characteristics similar to
the anther tapetum and may function like the tapetum in the production and secretion of substances for the developing reproductive cells.

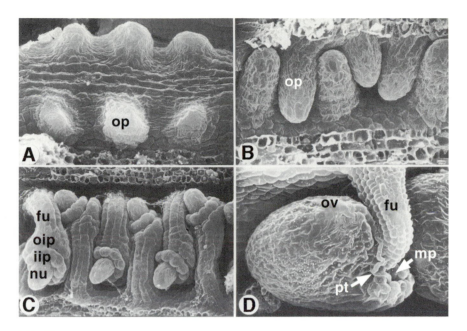

FIGURE 9.9

Development of ovules in wild type Arabidopsis. (A) Ovule primordia, op, emerge from placenta. (B) Ovule primordia enlarge. Note that ovule primordia from different rows within a locule arise in staggered positions. (C) Inner integument primordia, iip, and outer integument primordia, oip, grow and begin to cover nucellus, nu. (D) Ovule, ov, at maturity and in the process of being fertilized. Outer integument overgrows and covers the inner integument and nucellus and in doing so forms a micropyle, mp. Ovule is folded on itself so that the micropyle lies next to the funiculus, fu. Pollen tube, pt, can be seen entering the micropyle opening. Visualized by scanning electron microscopy. (From Bowman, J. 1994. Arabidopsis: An Atlas of Morphology and Development. Springer-Verlag, New York, p. 303. Reprinted by permission of Springer-Verlag.)

The polarity of the ovary is described in terms of its micropylar and chalazal ends. (The micropyle is the site where the integuments terminate and where the pollen tube penetrates the ovule. The chalazal end is where the funiculus attaches.) During its enlargement the ovule bends so that the micropyle lies close to the placenta along which the pollen tubes grow (Fig. 9.9D). During fertilization the pollen tube emerges from the placenta and enters the micropyle.

The megasporocyte gives rise to the megaspore via meiotic divisions during megasporogenesis (Fig. 9.2). Following that, during megagametogenesis, the megaspore undergoes mitotic divisions to produce the embryo sac (Fig. 9.10). The embryo sac contains an egg nucleus that fuses with the pollen sperm nucleus to give rise to a zygote. In standard cases, the set of nuclei in the embryo sac is derived from two meiotic and three or more mitotic divisions, although the number and timing of mitotic divisions, in particular, varies from species to species. Arabidopsis and maize have a monosporic pattern (or Polygonum-type) of megasporogenesis in which the megasporocyte undergoes two successive meiotic divisions and usually forms a linear array of megaspores that are aligned along the micropyle–chalazal axis. The megaspore nearest the chalaza enlarges, and the three other megaspores degenerate and are crushed by the enlarging megaspore. It is not known how a single megaspore is selected, but it is thought to relate to the formation of callose walls around the megaspores. The nonfunctional megaspores form thicker callose cell walls, which may possibly exclude a continuous supply of nutrients.

The functional megaspore in many monosporic plants undergoes three rounds of mitoses, giving rise to eight nuclei (Fig. 9.10). The eight nuclei are distributed among seven cells in the developing embryo sac, with two polar nuclei occupying a single central cell. (In certain species, such as maize, the antipodal nuclei may undergo additional rounds of division.) The nuclei are characteristically distributed in the embryo sac because, at the four nuclei stage, two nuclei migrate to the chalazal end of the embryo sac, and two migrate to the micropylar end. Of the pair that migrate to the micropylar end, the more distal nucleus divides to form the synergids, and the more proximal divides to form the egg nucleus and one of the two polar nuclei. The egg cell lies adjacent to the two synergid cells at the micropylar end of the embryo sac, forming a unit called the egg apparatus.

The mitotic divisions that give rise to the embryo sac nuclei are not followed immediately by cell wall formation. The embryo sac is instead multinucleate (coenocytic) until the final division; cellularization occurs thereafter. Cellularization appears to take place between sister nuclei by normal formation of cell plates following division. Such is the case in the formation of the common cell wall between synergids. It is not known, however, how the walls between nonsister nuclei that were not recently separated by division are formed. The walls of the synergid cells that face the micropyle develop distinctive ingrowths that form the filiform apparatus, which is a structure that is specialized to allow for penetration by the pollen tube (Russell 1993).

Fertilization in flowering plants is a double fertilization process in which the egg cell is fertilized to form the diploid zygote (in diploid plants) and the central cell is fertilized to form triploid endosperm. At fertilization, the pollen tube penetrates the embryo sac at the micropylar end and discharges its contents into one of the synergid cells (Russell 1993). Both sperm nuclei migrate to the chalazal end of the synergid

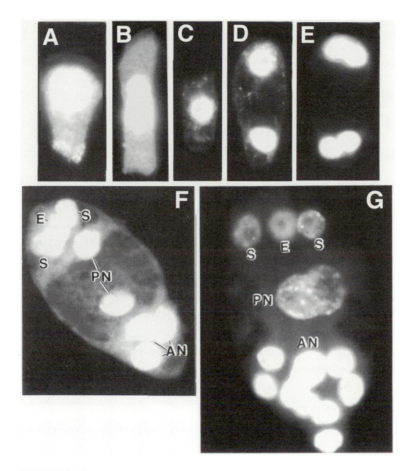

FIGURE 9.10

Megasporocyte and embryo sac development in maize (*Zea mays*). (A) Megas-porocyte at premeiotic stage. (B) Megasporocyte at prophase I of meiosis. (C) Megaspore. (D) Embryo sac, two-nucleate stage. (E) Embryo sac, four-nucleate stage. (F) Embryo sac, eight-nucleate stage. (G) Mature embryo sac with ten or more antipodal cells. Egg, E; synergid, S; polar nucleus, PN; antipo-dal cell nuclei, AN. Embryo sacs were isolated by enzyme digestion of dissected ovaries and stained with Hoechst 33258 stain and viewed in a fluorescent microscope. (From Huang, B. Q., and Sheridan, W. F. 1994. Female gameto-phyte development in maize: Microtubular organization and embryo sac polar-ity. Plant Cell 6:846. Reprinted by permission of the American Society of Plant Physiologists.)

cell. One sperm nucleus fuses with the nucleus in the egg cell, which gives rise to the zygote. The other sperm nucleus fuses with the polar nuclei in the central cell, which gives rise to tissue that will form the endosperm. It is not known how the sperm nuclei move; however, for-mation of cell walls is incomplete in the nonsister cells between which sperm nuclei move – between the synergids and egg cell and between the synergids and central cell.

An interesting question is whether the synergid cells release a signal that attracts the pollen tube once it has penetrated the ovule. (Guidance of the pollen tube to the micropyle will be discussed in Chap. 11.) The receptive synergid degenerates either before or after the pollen tube arrives and is easily recognized because the plasma and vacuole (tonoplast) membranes break down when it degenerates. The synergid cell vacuole sequesters large quantities of calcium, and it has been suspected that some released material might act as a pollen tube attractant (Russell 1993).

MUTANTS THAT AFFECT PISTIL, OVULE, AND EMBRYO SAC DEVELOPMENT

Mutations in ovule and female gametophyte development have been identified by examining plants with reduced female fertility. Most of these mutants are sporophytic mutants (i.e., they are determined by the diploid genotype of the maternal tissue and affect all gametes regardless of their haploid genotype). Gametophytic mutants are more difficult to recognize because they are not transmitted through the female and can only be scored as heterozygotes that are often not completely sterile (Angenent and Colombo 1996). In Arabidopsis, sporophytic mutants such as *short integument1* (*sin1*), *bell1* (*bell*), and *ovule* (*ovm2* and *ovm3*), have specific effects on one or more processes, such as development of the nucellus, formation of the integuments, megasporogenesis or megagametogenesis.

Bell mutants of Arabidopsis show an interesting homeotic transformation of ovules. In *bell,* the integuments are transformed into carpelloid structures (Fig. 9.11) (Modrusan et al. 1994; Ray et al. 1994). The inner integument fails to initiate, the outer integument becomes carpelloid, and megagametogenesis is blocked at a late stage of development. *BEL1* therefore appears to be required for the formation of the inner integument and for the identification of the outer integument. *BEL1* has been cloned and found to encode a homeodomain transcription factor (Reiser et al. 1995). In the flower, *BEL1* is expressed throughout the ovule primordium; however, as the primordium begins to grow, *BEL1* expression is confined to a region between the nucellus and funiculus where the inner and outer integuments will form. Thus, the expression of *BEL1* predicts the region of integument development.

Ray et al. (1994) found that the ovules resembled those of *bell* mutants in transgenic plants that constitutively express *AGAMOUS* (*AG*) from a 35S:*AG* transgene. Thus, the expression of *AG* appeared to suppress the function of *BEL1*. It was speculated that *BEL1* and *AG* might

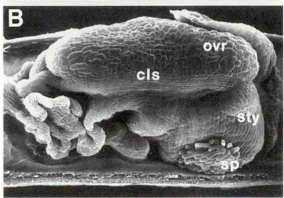

FIGURE 9.11

Abnormal ovule development in *bell1*
(*bell*) mutants of Arabidopsis. (A) Ovule
develops into a carpelloid structure (cls)
while two adjacent ovules degenerate.
(B) Further development of carpelloid
structure with recognizable tissue representing ovule, ovr; style, sty; and stigmatic papillar cells. Visualized by scanning electron microscopy. (From
Modrusan, Z., et al. 1994. Homeotic
transformation of ovules into carpel-like
structures in Arabidopsis. Plant Cell
6:340. Reprinted by permission of the
American Society of Plant Physiologists.)

counteract each other's expression, and that *BEL1* might direct normal
integument development, in part, by suppressing *AG* expression in the
integuments. In the absence of normal *BEL1* function, *AG* expression is
thought to convert the outer integuments to carpelloid structures. The
pattern of AG expression until right before bud opening in *bel-3*
mutants, however, does not appear to be different from wild type. It is
possible that *BEL1* and *AG* might counteract each other's function and
not expression (Reiser et al. 1995).

Other Arabidopsis mutants, such as *aintegumenta* (*ant*) and *short
integument* (*sin1*), also block both integument and megagametophyte
development. *ANT* appears to function differently than *BEL1* in that the
latter is required for initiation of the inner integument and identification
of the outer integument (Elliott et al. 1996; Klucher et al. 1996). Neither
integument is initiated in stronger alleles of *ant,* and the growth of the
integument is variable in weaker alleles, although there is no obvious
homeotic transformation of the integuments. *ANT* has been cloned and
is related to the *APETALA2* (*AP2*) domain family of transcription factors
(Elliott et al. 1996; Klucher et al. 1996). *ANT* is expressed in the develop-

239

ing ovule primordium, particularly in the regions of the primordium where integuments begin to form. *ANT,* however, is also expressed in vegetative primordia, except roots, and in different organs of the flower. *Ant* mutants have a number of pleiotropic effects that may reflect the role of *ANT* in various vegetative primordia as well as ovule primordia.

Angenent et al. (1995) described an unusual way to find genes that affect early ovule formation and ovule identity. They generated a cDNA library from pistils of petunia (*Petunia hybrida*) and screened the library with a hybridization probe to detect MADS box sequences, based on the assumption that controlling genes in ovule development might be MADS box-containing transcription factors. Two cDNAs called *floral binding protein 7 and 11* (fbp7 and fbp11), were identified. It was demonstrated by *in situ* hybridization that the genes represented by the cDNAs were expressed in the center of the gynoecium and later in the developing ovules. To determine whether these genes were involved in ovule development, the cDNAs were linked to the CaMV 35S promoter, and the constructs were introduced into transgenic plants. Bizarre plants with ovules that had a spaghettilike structure were observed among the transgenic plants (Fig. 9.12B). The *spaghetti* plants arose through cosuppression (down-regulation of expression) of the corresponding genes because RNA from neither gene (represented by fbp7 and fbp11) accumulated in the ovaries of the transformants. Detailed analysis of the aberrant ovules indicated that they had a carpelloid structure with papillae on top.

Plants in which the transgene was overexpressed (rather than ones in which the gene was cosuppressed) were also observed among the other transgenic plants in the group produced by Angenent et al. (1995). Some plants transformed with the CaMV 35 promoter-fbp11 constructs (35S:fbp11) developed ovulelike structures on the adaxial side of sepals and the abaxial side of petals (Fig. 9.13B) (Colombo et al. 1995). The aberrant formation of ovules indicates, nonetheless, that carpel development is not a prerequisite for ovule formation. This is somewhat similar to ovule development in gymnosperms that form free ovules on ovuliferous scales, and not within an ovary.

A number of mutants have also been described that affect gametophyte development with obvious defects in the sporophytic development of the ovule. *Meiotic* (*mei*) mutants in maize block stages of meiosis and ultimately interfere with megasporogenesis or megagametogenesis (Golubovskaya et al. 1992). Meiosis, however, is not an absolute prerequisite for female gametophyte formation. In plant species that are apomictic, meiosis is either abnormal or bypassed altogether, and the embryo sac is formed from nonarchesporial tissue. These examples sug-

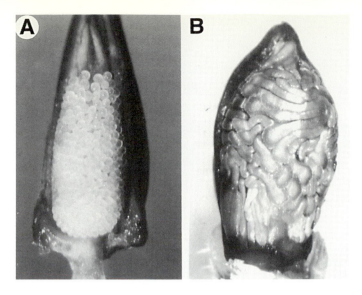

FIGURE 9.12

Ovaries of (A) normal petunia and (B) transgenic petunia that bears MADS box–containing cDNA constructs. Introduction of cDNA constructs results in cosuppression of the corresponding genes and in the development in ovaries of spaghettilike structures that contain carpelloid tissues. (From Angenent, G. C., et al. 1995. A novel class of MADS box genes is involved in ovule development in petunia. Plant Cell 7:1574. Reprinted by permission of the American Society of Plant Physiologists.)

FIGURE 9.13

Ectopic formation of ovulelike structures on the sepals of transgenic petunia (*Petunia hybrida*). Transgenic plant overexpresses the floral binding protein MADS box gene (*FBP11*) through the action of a 35S:*FBP11* construct. (A) Placenta, pl, in ovary of non-transgenic petunia. Some of the ovules, ov, have been removed so that the structure of individual ovules and funiculi, fu, can be seen. (C) Close-up of ovulelike structures, ov*, on a sepal of transgenic plant. Visualized by scanning electron microscopy. (From Colombo, L., et al. 1995. The petunia MADS box gene FBP11 determines ovule identity. Plant Cell 7:1862. Reprinted by permission of the American Society of Plant Physiologists.)

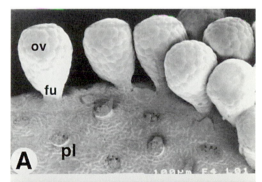

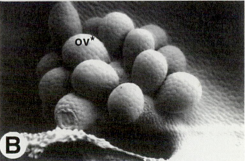

gest that embryo sac development can be uncoupled from meiosis, and that the megasporocyte is not the only cell that can acquire competence to form an embryo sac. Fewer embryo sac mutants have been described that are gametophytic (i.e., determined by the haploid genotype of the gametes). The lack of mutants may be due to problems in gamete transmission rather than the fact that few genes are needed for ovule or embryo sac development.

REFERENCES

Albertsen, M. C. and Phillips, R. L. 1981. Developmental cytology of 13 genetic male-sterile loci in maize. Can. J. Genet. Cytol. 23: 195–208.

Angenent, G. C. and Colombo, L. 1996. Molecular control of ovule development. Trends Plant Science 1: 228–232.

Angenent, G. C., Franken, J., Busscher, M., Van Dijken, A., Van Went, J. L., Dons, H.J.M., and Van Tunen, A. J. 1995. A novel class of MADS box genes is involved in ovule development in petunia. Plant Cell 7: 1569–1582.

Bowman, J. 1994. Arabidopsis: An Atlas of Morphology and Development. New York: Springer-Verlag.

Chaudhury, A. M. 1993. Nuclear genes controlling male fertility. Plant Cell 5: 1277–1283.

Chaudhury, A. M., Lavithis, M., Taylor, P. E., Craig, S., Singh, M. B., Signer, E. R., Knox, R. B., and Dennis, E. S. 1994. Genetic control of male fertility in *Arabidopsis thaliana*: Structural analysis of premeiotic developmental mutants. Sexual Plant Reproduction 7: 17–28.

Colombo, L., Franken, J., Koetje, E., Van Went, J., Dons, H.J.M., Angenent, G. C., and Van Tunen, A. J. 1995. The petunia MADS box gene FBP11 determines ovule identity. Plant Cell 7: 1859–1868.

Cui, X., Wise, R. P., and Schnable, P. S. 1996. The *rf2* nuclear restorer gene of male-sterile T-cytoplasm maize. Science 272: 1334–1336.

Dawe, R. K. and Freeling, M. 1990. Clonal analysis of the cell lineages in the male flower of maize. Devel. Biol. 142: 233–245.

Eady, C., Lindsey, K., and Twell, D. 1995. The significance of microspore division and division symmetry for vegetative cell-specific transcription and generative cell differentiation. Plant Cell 7: 65–74.

Elliott, R. C., Betzner, A. S., Huttner, E., Oakes, M. P., Tucker, W.Q.J., Gerentes, D., Perez, P., and Smyth, D. R. 1996. *AINTEGUMENTA,* an *APETALA2*-like gene of Arabidopsis with pleiotropic roles in ovule development and floral organ growth. Plant Cell 8: 155–168.

Flavell, R. 1974. A model for the mechanism of cytoplasmic male sterility in plants, with special reference to maize. Plant Sci. Lett. 3: 259–263.

Gasser, C. S. and Robinson-Beers, K. 1993. Pistil development. Plant Cell 5: 1231–1239.

Goldberg, R. B., Beals, T. P., and Sanders, P. M. 1993. Anther development: Basic principles and practical applications. Plant Cell 5: 1217–1229.

Golubovskaya, I., Avalkina, N. A., and Sheridan, W. F. 1992. Effects of several meiotic mutations on female meiosis in maize. Dev. Genet. 13: 411–424.

Hanson, M. R. 1991. Plant mitochondrial mutations and male sterility. Annu. Rev. Genet. 25: 461–486.

Huang, B. Q. and Sheridan, W. F. 1994. Female gametophyte development in maize: Microtubular organization and embryo sac polarity. Plant Cell 6: 845–861.

Kaul, M. L. H. 1988. Male Sterility in Higher Plants. Berlin: Springer-Verlag.

Klucher, K. M., Chow, H., Reiser, L., and Fischer, R. L. 1996. The *AINTEGU-MENTA* gene of Arabidopsis required for ovule and female gametophyte development is related to the floral homeotic gene *APETALA2*. Plant Cell 8: 137–153.

Koltunow, A. M., Truettner, J., Cox, K. H., Wallroth, M., and Goldberg, R. B. 1990. Different temporal and spatial gene expression patterns occur during anther development. Plant Cell 2: 1201–1224.

Levings, C. S. 1993. Thoughts on cytoplasmic male sterility in cms-T maize. Plant Cell 5: 1285–1290.

Mariani, C., De Beuckeleer, M., Truttner, J., Leemans, J., and Goldberg, R. B. 1990. Induction of male sterility in plants by a chimeric ribonuclease gene. Nature 347: 737–741.

McCormick, S. 1993. Male gametophyte development. Plant Cell 5: 1265–1275.

Modrusan, Z., Reiser, L., Feldmann, K. A., Fischer, R. L., and Haughn, G. W. 1994. Homeotic transformation of ovules into carpel-like structures in Arabidopsis. Plant Cell 6: 333–349.

Muschietti, J., Dircks, L., Vancanneyt, G., and McCormick, S. 1994. LAT52 protein is essential for tomato pollen development: Pollen expressing antisense LAT52 RNA hydrates and germinates abnormally and cannot achieve fertilization. Plant J. 6: 321–338.

Preuss, D., Lemieux, B., Yen, G., and Davis, R. W. 1993. A conditional sterile mutation eliminates surface components from Arabidopsis pollen and disrupts cell signaling during fertilization. Genes Dev. 7: 974–985.

Preuss, D., Rhee, S. Y., and Davis, R. W. 1994. Tetrad analysis possible in Arabidopsis with mutation of the *QUARTET* (*QRT*) genes. Science 264: 1458–1460.

Ray, A., Robinson Beers, K., Ray, S., Baker, S. C., Lang, J. D., Preuss, D., Milligan, S. B., and Gasser, C. S. 1994. Arabidopsis floral homeotic gene *BELL* (*BEL1*) controls ovule development through negative regulation of *AGAMOUS* gene (*AG*). Proc. Natl. Acad. Sci. USA 91: 5761–5765.

Reiser, L., Modrusan, Z., Margossian, L., Samach, A., Ohad, N., Haughn, G. W., and Fischer, R. L. 1995. The *BELL1* gene encodes a homeodomain protein involved in pattern formation in the Arabidopsis ovule primordium. Cell 83: 735–742.

Reisner, L. and Fischer, R. L. 1993. The ovule and the embryo sac. Plant Cell 5: 1291–1301.

Russell, S. D. 1993. The egg cell: Development and role in fertilization and early embryogenesis. Plant Cell 5: 1349–1359.

Satina, S. 1945. Periclinal cimeras in *Datura* in relation to the development and structure of the ovule. Am. J. Bot. 32: 72–81.

Satina, S. and Blakeslee, A. F. 1941. Periclinal chimaeras in *Datura stramonium* in relation to development of leaf and flower. Am. J. Bot. 28: 862–871.

Sawhney, V. K. and Bhadula, S. K. 1988. Microsporogenesis in the normal and male-sterile stamenless mutant of tomato (*Lycopersicon esculentum*). Can. J. Bot. 66: 2013–2021.

Schneitz, K., Huelskamp, M., and Pruitt, R. E. 1995. Wild-type ovule development in Arabidopsis thaliana: A light microscope study of cleared whole-mount tissue. Plant J. 7: 731–749.

Verbecke, J. A. 1992. Fusion events during floral morphogenesis. Annu. Rev. Plant Physiol. Plant Mol. Biol. 43: 583–598.

Williams, M. E. and Levings, C. S. 1992. Molecular biology of cytoplasmic male sterility. In Plant Breeding Reviews, ed. Janick, J., pp. 23–51. New York: John Wiley.

10

Pollination and Apomixis

Pollination in plants is a fascinating process to study because pollination embraces so many concepts in modern biology: cell–cell communication, self-recognition, growth guidance, plant–insect interaction, and so on. Pollination is a key step in plant sexual reproduction because it involves the transfer of male gametes from the male to the female organs of the flower and the union of male gametes with female gametes. Male gametes are carried by pollen grains (male gameto-phytes) to the female floral organ (pistil), and growth of a pollen tube down the style delivers sperm cells to an embryo sac (female gameto-phyte) that lies within an ovule.

Pollination begins when a pollen grain lands on the surface of the stigma, which can be a meadow of fingerlike papillar cells in some species (Fig. 10.1). In species with dry pollen, pollen grains are hydrated by the papillae, the pollen germinate, and pollen tubes emerge and grow. The path taken by a pollen tube down the style can be readily visualized by staining callose in the pollen tube with the fluorescent dye, aniline blue (Fig. 10.2). Four stages of pollen tube growth follow: (1) The pollen tube penetrates through the cuticle and grows through the cell wall of the papillar cell; (2) the tube enters the transmitting tract at the base of the papillar cell and grows basipetally down through the transmitting tract of the style; (3) the pollen tube emerges from the septum or placenta near the funiculus of an ovule; (4) the tube grows along the surface of the septum or funiculus and enters the micropyle of the ovule.

Successful pollination requires guidance of the pollen tube from the stigma to the embryo sac. Guidance involves the exchange of signals between the pistil and the pollen at many steps along the way (Wilhelmi and Preuss 1997). The exchange of signals serves to direct pollen tube growth and assures delivery of gametes from the right plant

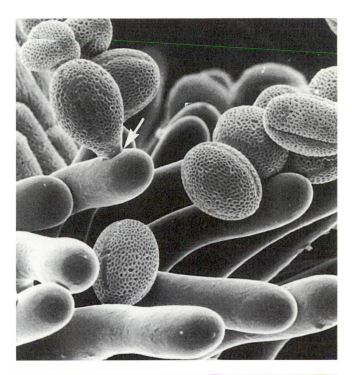

FIGURE 10.1

Pollen binding and germination on the papillae of the stigma in Arabidopsis. Pollen forms a foot (arrow) as it penetrates the cuticle of the stigma and grows in the cell wall of the papillar cell. (From Bowman, J. 1994. Arabidopsis: An Atlas of Morphology and Development. Springer-Verlag, New York, p. 343. Reprinted by permission of Springer-Verlag.)

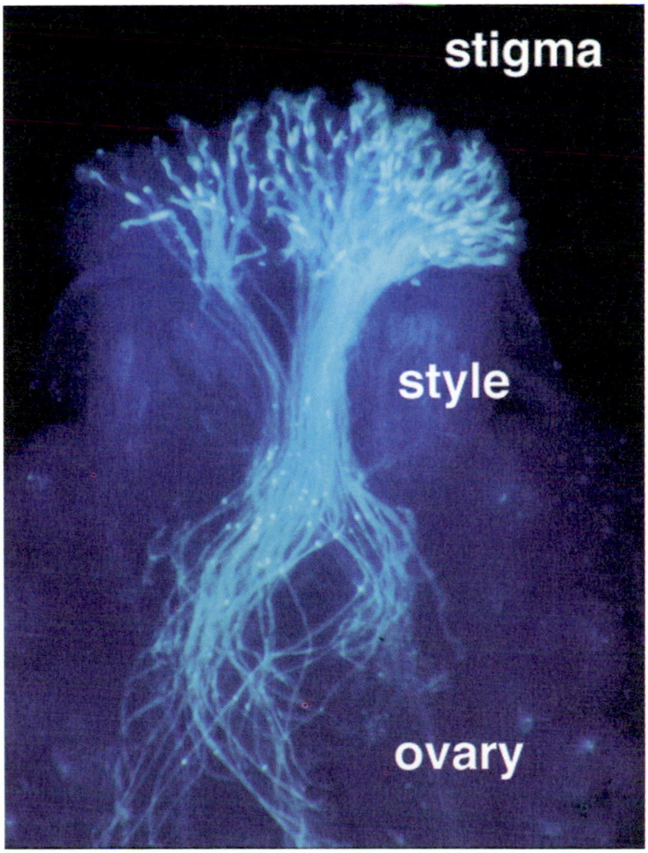

FIGURE 10.2

Pollen tube growth in the pistil of Arabidopsis. Pollen grains land and germinate on the stigma, and pollen tubes grow down through the transmitting tract of the style and ovary. Callose formed during pollen tube elongation is visualized by staining with Aniline blue. Bright spots are callose plugs formed at intervals along the pollen tube. Whole mount viewed with ultraviolet fluorescent microscopy. (Kindly provided by June and Mikhail Nasrallah, Cornell University.)

species to the embryo sac. Because plant environments are often filled with different pollen, safeguards are built into every step of the pollination process to insure species specificity.

POLLEN GERMINATION AND POLLEN TUBE GROWTH

Hydration of pollen is the usual trigger for pollen germination in species with dry pollen. Stigma surfaces can either be wet or dry depending on the species and environmental conditions. On wet and sticky stigma surfaces, such as in tobacco (*Nicotiana tabacum*), there are no barriers to pollen hydration. On dry stigmas, pollen germination can be controlled by pollen–stigma interactions and may involve the delivery of water and other substances from the stigma (Wilhelmi and Preuss 1997). A conditional, male sterile mutant in Arabidopsis was found that prevents hydration by interfering with pollen–pistil interactions (Preuss et al. 1993). Pollen from the mutant, *pollen–pistil interactions 1* (*pop1*), incites callose production in stigmas of Arabidopsis flowers (usually a defense response to foreign pollen). *Pop1* pollen germinates *in vitro*, but it fails to germinate on the stigma surface because callose deposits block access to the stigma fluids that are needed for pollen germination. The mutation is conditional because in humid conditions, *pop1* pollen will take up sufficient moisture to germinate despite the formation of the callose barrier. The defense response in the stigma to *pop1* pollen occurs because *pop1* pollen lack extracellular lipids in their tryphine coat (i.e., the coat deposited by the breakdown of the tapetum). *Pop1* also has defects in vegetative characters, such as in the production of epicuticular waxes that are composed of long-chain lipids. In fact, *pop1* is allelic to *cer6*, which is a mutant described by others as wax defective. These findings demonstrate that the tryphine coat (or materials in the tryphine coat) is critical for pollen–stigma interactions and prevents the stigma from mounting a defense to compatible pollen.

Pollen germination in marked by the emergence of the pollen tube (Fig. 9.4). Pollen tube growth has been studied extensively because pollen from many species will germinate and grow pollen tubes *in vitro* (see Mascarenhas 1993). Maize pollen germinates rapidly *in vitro* (within 5 minutes of hydration), and the pollen tubes grow through stylar tissue at a fast pace (1 cm/hour). (Pollen tube growth in other species, such as orchids, can be very slow, taking months for a pollen tube to reach an ovule.) Gene expression in the male gametophyte appears to contribute significantly to pollen tube development. Ten to twenty percent of the RNAs in pollen are pollen-specific, and most transcription occurs around meiosis and the mitotic divisions that give rise to the generative and vegetative cells (Mascarenhas 1993). Pollen germination in many plant species does not require new RNA or protein synthesis; however, pollen tube growth does require new synthesis, particularly late pollen growth. These conclusions have been reached largely by applying RNA and protein synthesis inhibitors *in vitro* to pollen from plants such as spiderwort (*Tradescantia*).

The pollen tube is a tip-growing cell, and growth of the pollen tube has many parallels to hyphal growth in fungi. The tip is thought to grow by the fusion of secretory vesicles to the plasma membrane at the tip and the deposition of cell wall substances to the outside. Like other tip growing-cells, the tip contains a large number of secretory vesicles that are delivered to the tip by cytoplasmic streaming. Streaming appears to involve an actomyosin-based system. An actin network has been observed at the pollen tube tip, and cytochalasin treatment (a drug that promotes actin depolymerization) inhibits tip growth (see the review by Pierson and Cresti, 1992). In other organisms, such as yeast, the *rho* family of GTPases regulates a number of actin-based cellular processes. These processes include cell morphogenesis and the establishment of polarity. A member of the rho GTPase family, called *Rop1*, is specifically expressed in pollen and pollen tubes in pea (*Pisum sativa*) (Lin et al. 1996). Rop1 protein was shown by immunolocalization techniques to be concentrated near the cortical region of the tip and associated with the generative cell. Injection of anti-Rop1 antibodies into the tips of pea pollen tubes induced rapid growth arrest but did not block cytoplasmic streaming (Lin and Yang 1977). Thus, it has been suggested that Rop1 may play a pivotal role in tip growth, perhaps, in regulating vesicle docking/fusion in the growing tip.

The pollen tip contains an apical growth zone and a subapical, organelle-rich zone. Behind is a nuclear zone that contains the tube nucleus and the generative or sperm cells (Fig. 9.4). As the pollen tube grows, the tube nucleus and generative or sperms cells move down the pollen tube as a unit (Palevitz and Tiezzi 1992). Thus, growth of the pollen tube is unique in that the cell body moves along with the growing front of the pollen tube (Heslop-Harrison 1987; Steer and Steer 1989). In fact, once the pollen tip has grown into the style, the pollen grain and the spent pollen tube can be eliminated without interfering with fertilization (Jauh and Lord 1995). During pollen tube growth, callose is deposited as plugs formed periodically in the pollen tube behind the cytoplasmically rich zone near the tip. These plugs cordon off the cytoplasm from the noncytoplasmic regions behind (Fig. 10.2). Pollen grains in different species contain either two or three cells, depending on whether the generative nucleus has divided to form two sperm cells prior to pollen release. If not, the generative nucleus usually undergoes division during the migration of the nuclei down the pollen tube (Fig. 9.4).

POLLEN TUBE GUIDANCE

For successful pollination, pollen tubes must grow down through the style, into the ovary, and emerge from the placenta near the site of an ovule with an unfertilized egg (Fig. 10.3). Pollen tubes grow down the transmitting tract of the style and take rather direct paths to their targets. The pollen tube tip emerges from the placenta, makes its way to a nearby ovule, and penetrates the micropyle. The problem of homing in on the target has been likened to axon guidance in nervous system development in animals (Wilhelmi and Preuss 1996). The challenge in these systems is to identify the guiding signals and to understand the mechanisms by which the pollen tube or axon responds.

Pollen tubes grow down the style in a transmitting tract that contains an extracellular matrix (ECM) rich in polysaccharides, glycoprotein, and glycolipids. An abundant class of the glycoproteins in the transmitting tract are arabinogalactan-rich glycoproteins (AGPs), and these glycoproteins are thought to play an important role in the support and guidance of the pollen tube. Cheung et al. (1995) extracted the AGPs from styles of tobacco (*Nicotiana tabacum*) flowers and showed that the glycoproteins

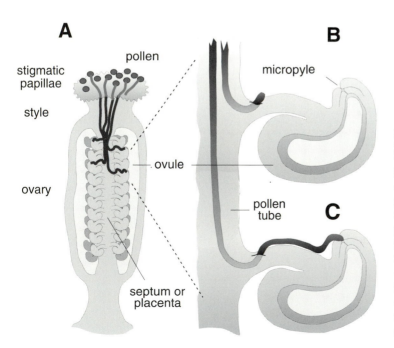

FIGURE 10.3

Pathway followed by growing pollen tubes in Arabidopsis. (A) Pollen germinates on the stigmatic papillae, and pollen tubes grow down through the transmitting tract of the short style. (B) Pollen tube emerges from the septum or placenta, and (C) grows to the nearest ovule and enters the micropyle.

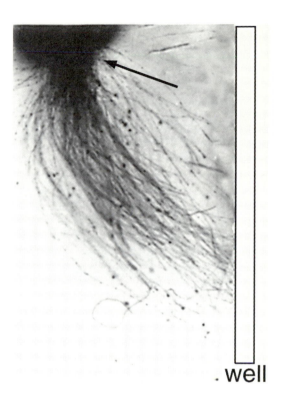

FIGURE 10.4

Assay to detect pollen tube attractant extracted from styles of tobacco (*Nicotiana tabacum*) flowers. Flowers were pollinated, and styles were cut and placed on an agar plate containing pollen tube growth medium. In the absence of an attractant, the pollen tubes fan out when they emerge from the base of the style (arrow). In this case, stylar transmitting tract proteins (0.5 µg/µl) were introduced in a well to the right of the style and emerging pollen tubes grew toward the well. (From Cheung, A. Y., Wang, H., and Wu, H. M. 1995. A floral transmitting tissue-specific glycoprotein attracts pollen tubes and stimulates their growth. Cell 82:387. By permission of the author and Cell Press.)

stimulated the growth of tobacco pollen *in vitro*. The proteins themselves are about 38 kD, while the glycosylated forms are between 50 and 100 kD. These glycoproteins were progressively deglycosylated (reduced in size down to the 26 kD form) upon incubation with growing pollen (Wu et al. 1995). Cheung et al. (1995) used a novel assay to show that these glycoproteins directed pollen tube growth. In the assay, a pollinated pistil was cut close to the bottom of the style and placed on an agarose plate with pollen tube growth medium (Fig. 10.4). Pollen tubes that emerged from the cut end were all oriented, but they tended to fan out in subsequent growth across the agarose surface. If the glycoprotein extract from the tobacco transmitting tract was implanted in the medium to one side or another of the growing pollen tube front, then the pollen tubes grew preferentially toward the implanted glycoprotein extracts. The pollen tubes were not attracted to extracts that had been deglycosylated. Several controls were carried out to demonstrate that the attraction was not just metabolic support of pollen tube growth; however, that possibility is difficult to dismiss. Wu et al. (1995) argued that the growth of pollen down the transmitting tract might be due to a gradient of AGP glycosylation. They found that the transmitting tract glycoproteins were more highly glycosylated (and were larger in size) at the base of the style rather than at the top. The gradient of more highly glycosylated glycoproteins, therefore, might be a general directional cue for the growth of pollen tubes.

An effort was undertaken to block the production of these glycoproteins in transgenic plants to examine their role in pollen guidance *in planta*. Copy DNAs were identified (called TTS cDNAs) that encode forms of the glycoproteins specifically expressed in transmitting tracts. Antisense forms of the cDNAs (35S:antisense TTS) were introduced into transgenic plants to interfere with the production of transmitting tract arabinogalactan-rich glycoproteins (Cheung et al. 1995). In tobacco lines with reduced glycoprotein production, the rate of pollen tube growth in the style was slowed, and these lines had reduced fertility.

Once the pollen has grown down the style, how is the site for pollen tip emergence selected? A pollen tube usually emerges near the ovule that it will penetrate. Hülskamp et al. (1995) examined whether pollen tubes emerged at random sites in Arabidopsis. In flowers with few pollen tubes, they found that there was a higher probability that the pollen tube would emerge at the upper (apical) ovules. Pollen tubes only emerged at lower ovules when greater numbers of pollen tubes were found in a given flower. Thus, site selection does not appear to be random, and competition for sites of emergence (when there are greater numbers of pollen tubes) may direct pollen tubes to less probable sites for emergence and prevent the incidence of polyspermy (fertilization of an egg by more than one sperm).

Pollen tube guidance to the ovule may be influenced by the release of Ca^{2+}. There is a steep intracellular Ca^{2+} gradient at the growing tip of the pollen tube that is derived from the inward movement of Ca^{2+} at the very apex of the tube (Pierson et al. 1996). Localized Ca^{2+} influx is thought to result from the activation or deposition of new Ca^{2+} channels and their inactivation as they are continually swept back from growth at the tip. The growth of pollen tubes can be reoriented by locally manipulating free cytosolic Ca^{2+} levels to one side or the other of the growing tip using microinjection or flash photolysis of caged Ca^{2+}(Malho and Trewavas 1996). By elevating the cytosolic Ca^{2+} levels, the growing tip bends toward the side of Ca^{2+} release. Despite the involvement of Ca^{2+} in the mechanism of pollen reorientation, there is still controversy whether external gradients of Ca^{2+} are signals for the reorientation of pollen tube growth. It is just as likely that other substances actually serve as signals, but that the mechanism of reorientation involves some relocalization of Ca^{2+} channels at the tip.

Hülskamp et al. (1995) also examined the targeting of the pollen tube after it emerged from the septum or placenta (Fig. 10.5). They found that most of the pollen tubes grow to the nearest ovule; however,

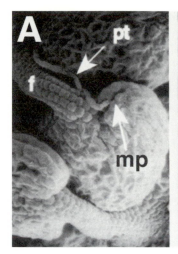

FIGURE 10.5

Directed growth of the pollen tube in Arabidopsis after emerging from the placenta. (A) Pollen tube, pt, grows along the funiculus, f, and penetrates the micropyle opening, mp, in wild type ovary. (B) Pollen tube growth is undirected in a mutant with defective embryo sac development, 47H4. Visualized by scanning electron microscopy. (From Hülskamp, M., Schneitz, K., and Pruitt, R. E. 1995. Genetic evidence for a long-range activity that directs pollen tube guidance in Arabidopsis. Plant Cell 7:59. Reprinted by permission of the American Society of Plant Physiologists.)

a number target an ovule one or two ovules away. The behavior of pollen tubes after they emerge is not dependent on the number of pollen tubes. Roving of the pollen tube suggested that the ovule might produce a factor to attract the pollen tube. To test this hypothesis, the authors studied the behavior of pollen tubes in mutants defective in ovule development. They found that in mutants defective in embryo sac formation (*short integument, sin1,* and 47H4), pollen tubes did not emerge preferentially at the upper ovules; rather, they emerged with nearly equal probability anywhere along the septum. When the pollen tube emerged in these mutants, it did not home in on the nearest ovule; rather, it grew randomly over all available surfaces (Fig. 10.5B). The interpretation of these experiments was that the ovule or ovule development affects the emergence of the pollen tube. It is also possible, however, that the mutations affect some properties of the transmitting tract.

To determine what tissue inside the ovule was required for pollen tube targeting, Hülskamp et al. (1995) examined 54D12, an Arabidopsis mutant in which ovules develop to various extents. Those ovules in which the embryo sacs appear to undergo normal development generally attracted pollen tubes with high frequency, whereas those with less well-developed embryo sacs attracted fewer pollen tubes. Those ovules that failed to develop embryo sacs at all did not attract pollen tubes. The authors concluded, therefore, that the embryo sac itself, or that some aspect of embryo sac development, is required to attract pollen.

Another mutant has been identified in Arabidopsis that is defective in pollen tube guidance yet forms normal embryo sacs (Wilhelmi and Preuss 1996). Seed yield was much reduced in the mutant which bears two mutations: *pop2*, a recessive mutation, and *pop3*, a dominant mutation. It is interesting that the mutant was self-sterile in that selfing crosses were sterile, whereas reciprocal crosses of the mutant to wild type were fully fertile. The defect depended on the parental genotype rather than on that of the gametes. For example, *pop2/pop2, pop3/POP3+* mutant plants were self-sterile even though only half of the male and female gametes carried the *POP3+* allele. This pattern of inheritance was sporophytic, meaning that the mutant phenotype was not determined by the gametes themselves, but by the parental tissue from which the gametes were derived. (More discussion about sporophytic inheritance is found in the next section.) The sporophytic inheritance pattern was difficult to reconcile with the observation that the defect in the mutant involved a late step in pollination, after the pollen tube emerged from the septum. These pollen tubes did not home in on the nearest ovule; rather, they grew randomly in the ovary locule. From reciprocal crosses among homozygous single mutants, double mutants, and wild type, it was found that *POP2* and *POP3* function in both the male and female parents, which suggests that these genes encode molecules that are present on both pollen tubes and pistil cells (Wilhelmi and Preuss 1996). The simple self-sterility trait seen in this mutant may represent the sort of characteristic from which more elaborate self-incompatibility systems in plants evolved.

SELF-INCOMPATIBILITY

Plants with perfect flowers (both male and female organs in the same flower) or plants with unisexual flowers of both sexes on the same individual (monecious plants) are usually capable of self-pollination. To prevent inbreeding, however, some plants have evolved self-incompatibility mechanisms to interfere with pollination. In self-incompatible plants, "self" pollen is recognized and rejected. Self-incompatibility mechanisms play upon the intimate interactions between pollen and pistil during pollination and generate barriers to prevent fertilization. Pollen rejection in the incompatible interaction is one of the best examples of specificity in cell recognition in plants.

Gametophytic system

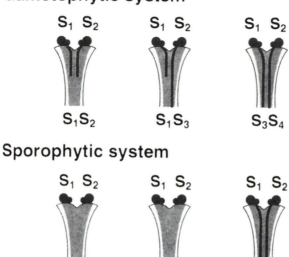

Sporophytic system

FIGURE 10.6

Schematic representation of the differences between gametophytic and sporophytic incompatibility systems in plants. Different incompatibility alleles are indicated as S_1, S_2 . . . and so on. In gametophytic incompatibility, the genotypes of the individual male gametophytes (pollen) determine whether the interaction is compatible or not. If the genotype of a pollen grain (shown above the pollen grains) matches either allele of the diploid style tissue (indicated below the pistil), then elongation of the pollen tube is arrested in the style. In sporophytic incompatibility, the genotype of the pollen parent determines the compatibility of the interaction. In this case, the genotype of the pollen parent is S_1S_2. If either one of the pollen parent alleles matches that of the diploid style tissue, then pollen germination arrests at the stigma surface. The case shown here is applicable when the S alleles are dominant or codominant. (From Newbigin, E., Anderson, M. A., and Clarke, A. E. 1993. Gametophytic self-incompatibility systems. Plant Cell 5:1317. Reprinted by permission of the American Society of Plant Physiologists.)

In all plants where self-incompatibility has been genetically characterized, the trait is controlled by a single locus called the S-locus with many alleles. Pollen rejection occurs when pollen from a plant with one S allele lands on a pistil of the same plant or on another plant with the same S allele. There are two major types of self-incompatibility systems in plants: gametophytic and sporophytic (Fig. 10.6). In gametophytic self-incompatibility, the genotype of the male gametophyte (pollen) determines whether pollination will be successful. In gametophytic self-incompatibility, pollen growth is blocked as the pollen tube grows down through the transmitting tract in the style. In sporophytic self-incompatibility, the genotype of the plant (sporophyte) from which the pollen is derived determines whether the pollen will be rejected or not. In this type of self-incompatibility, development of the pollen is usually blocked on the stigma surface. Both gametophytic and sporophytic self-incompatibility mechanisms have independently evolved a number of times in angiosperms, and they appear in many plant families. All the self-incompatible species within a given family, however, are either gametophytic or sporophytic.

Sporophytic Self-Incompatibility

Sporophytic self-incompatibility has been most extensively studied in the Brassicas (rape, broccoli, etc.), where the S locus has been found to be a cluster of tightly linked genes (see Nasrallah et al. 1994). The gene cluster, which encompasses more than 200 kb, specifies a particular self-incompatibility phenotype called the **S haplotype.** The S locus is highly polymorphic, and different allelic forms of genes in the S locus cluster constitute different S haplotypes. In sporophytic self-incompatibility, pollen rejection occurs when pollen from a plant of one S haplotype lands on a pistil of the same plant or on a plant of the same S haplotype (providing that the matching S haplotype is dominant or codominant in the anther that produces the pollen and in the style of the pollinated flower) (Fig. 10.6). Haplotypes in the Brassicas have been classified with respect to their dominance. In general, class I haplotypes are stronger, and they are dominant over class II haplotypes.

Two S-locus genes appear to be the major determinants of the S-haplotype: the *S-locus glycoprotein* (*SLG*) and the *S-locus receptor kinase* (*SRK*) (Fig. 10. 7A). SLG is a major glycoprotein that is secreted into the cell walls of the papillar cells, and SRK is a receptor protein kinase with an extracellular domain at the N terminus, a single-pass transmembrane domain, and a protein kinase domain at its C-terminus. The extracellular domain of SRK is similar in sequence to SLG. Through sequence comparison studies, Tantikanjana et al. (1993) proposed that *SLG* was derived from *SRK* by gene duplication and that the two genes coevolved as a gene pair in various haplotypes. The argument was made on the grounds that the sequences of *SLG* and *SRK* within a haplotype are more similar to each other than they are from one haplotype to another.

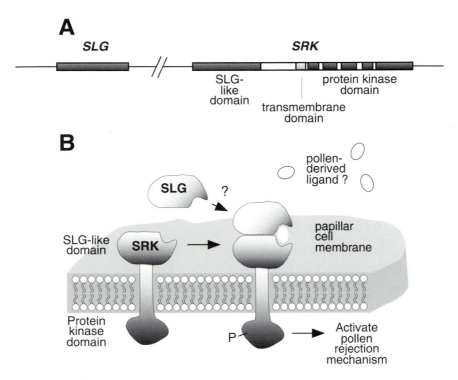

FIGURE 10.7

S locus determines sporophytic self-incompatibility type in Brassicas. (A) S-locus contains genes that encode the S-locus glycoprotein (SLG) and S-locus receptor kinase (SRK). The S-locus is very large (~ 200 kb) and the particular *SLG* and *SRK* alleles (along with other genes) constitute a "haplotype." (B) SLG encodes an abundant soluble glycoprotein expressed in the stigma papillar cells. SRK is thought to be located in the plasma membrane of stigma papillar cells. Factor contributed by the pollen is unknown; however, it is hypothesized to be a ligand that binds to the SRK receptor (and perhaps to SLG as a co-receptor), which activates the protein kinase activity and, presumably, a cascade of events that lead to self-pollen rejection. (Based on Nasrallah, J. B., and Nasrallah, M. E. 1993. Pollen-stigma signaling in the sporophytic self-incompatibility response. Plant Cell 5: 1325–1335. Nasrallah, J. B., et al. 1994. Signaling the arrest of pollen tube development in self-incompatible plants. Science 266:1505–1508.)

SLG and *SRK* are both needed for the pollen rejection. Self-compatible mutant strains of Brassicas have been analyzed, and in certain strains mutations at the S-locus disrupt the *SRK* gene. Other self-compatible mutations that are unlinked to the S-locus down-regulate the expression of *SLG*. Pollen rejection is thought to be a response that occurs in the stigma because both *SRK* and *SLG* are coordinately expressed in stigmatal papillar cells (Stein et al. 1996). The interaction of self-pollen with the papillar cells is thought to activate *SRK*, which leads to a response that rejects self-pollen (Fig. 10.7B). It is not known what SRK recognizes because *SLG* and *SRK* are not expressed in pollen. It has been proposed that there is an unidentified ligand present on pollen, such as the product of the *SLA* gene, and another S locus gene, which is expressed specifically in the anther.

It is also not known how self-pollen are rejected after they are recognized. The response is fast, however, and it is localized to the site of contact between the pollen grain and the stigma because pollination by compatible pollen is not blocked by mixed pollination with both compatible and incompatible pollen grains. Ikeda et al. (1997) described an unlinked mutation in *Brassica campestris* at the *mod* locus that compromises the self-incompatibility response. It is interesting that the *mod* locus was found to encode a protein similar to an aquaporin or water channel protein. The function of this protein in the rejection response is not known; however, it is speculated that the channel might be activated in a self-recognition response and may remove water from the site of initial pollen elongation.

Gametophytic Self-Incompatibility

Gametophytic self-incompatibility has been described in ornamental tobacco (*Nicotiana alata*), petunia (*Petunia inflata*), and tomato cultivars (*Lycopersicon peruvianum*). In an incompatible interaction, pollen tube growth is usually blocked in the transmitting tube tract where pollen tube growth becomes irregular, the walls thicken, and the tip frequently bursts.

As in sporophytic self-incompatibility, pollen rejection is controlled by the action of a multiallelic S-locus. The S locus encodes an abundant glycoprotein located in the transmitting tract of the style. S-locus glycoproteins are found in different isoforms (different electrophoretic mobility forms) in different S-types. S-locus glycoproteins have five conserved domains, two of which are similar to fungal ribonucleases. The sequence similarity led to the discovery that S-locus glycoproteins have nonspecific ribonuclease activity, which is essential for the rejection of incompatible pollen (McClure et al. 1989). Convincing arguments for the role of the ribonuclease activity (S-RNase) in pollen rejection came from studies in which various S-alleles (S_1, S_2, S_3, etc.), in sense or antisense form, were introduced as transgenes into transgenic plants. In antisense constructs, some or all of the transcribed portion of a gene is inserted in reverse orientation with respect to the promoter. In such constructs, the promoter drives RNA synthesis from the opposite strand of the gene, which generates a complementary or **antisense mRNA.**

The strategy succeeds if the antisense RNA is able to extinguish the expression of the endogenous gene. In petunia, Lee et al. (1994) demonstrated that antisense inhibition of S_2 and S_3 expression in plants of S_2S_3 genotype resulted in transgenic plants that failed to reject S_2 and S_3 pollen. They further demonstrated that expression of the S_3 transgene in plants of S_1S_2 genotype conferred the ability to reject S_3 pollen.

To show that pollen rejection was dependent on ribonuclease activity, Huang et al. (1994) substituted two histidine residues at the catalytic site in the protein derived from the S_3 gene. The histidine-substituted form of the S_3 gene was introduced into plants of the S_1S_2 genotype, and the resulting transgenic plants that expressed the mutated S_3 gene failed to reject self-pollen. McCubbin et al. (1997) used the same construct to determine whether a mutated S_3 allele would act as a **dominant negative mutation** in S_3-containing genotypes (i.e., whether the mutant allele could suppress the effect of the normal endogenous S-allele). The mutated S_3 allele was introduced into plants with an S_2S_3 genotype, and transformants that expressed both S-type ribonucleases and the mutant forms were selected. It was found that the transgenic plants were self-incompatible but unable to reject pollen from S_3S_3 tester plants. The self-incompatibility derived from the operation of the S_2 recognition system that was still intact in the transgenic plants. (The transgenic plant rejected pollen from the S_2S_2 tester.) The transgenic plants bearing the mutated S_3 gene still produced the S_3 RNase (which was active when tested in an in-gel assay); however, the S_3 RNase was ineffective in self-incompatibility functions. The mechanism by which the mutated S_3 gene acts in a dominant negative mutation in an allele-specific manner is still in dispute. It seems unlikely that the mutated S-gene product forms a heterodimer that poisons the function of the normal S_3 RNase because there is no evidence for dimer formation. A favored model is that self S-RNases are recognized by a receptor(s) on the pollen tube and, if taken up, the degradative activity of the RNase prevents further growth. A mutated S_3 RNase might compete with the normal S_3 RNase at the receptor site, reducing or blocking the uptake of the active RNase (McCubbin et al. 1997). That would mean that the mutated S_3 RNase would still retain its recognition function even though it had lost its RNase activity.

The self-incompatibility properties of the pollen were not affected in either the antisense or dominant negative constructs described earlier. In general, S-RNase genes are highly expressed in the pistil and only modestly so in pollen, which is consistent with the finding that S-RNase activity is involved in the behavior of the pistil and not the pollen. Genetic evidence demonstrates that the self-incompatibility type of the pollen is determined by the S-locus; however, the genetic determinant in the S-locus that is responsible for the pollen character is not known.

APOMIXIS

In sexual reproduction, seeds are produced by a developmental program that is initiated following fertilization. In some plants, such as the common dandelion (*Taraxacum vulgare*), viable seeds can be produced without fertilization of the egg in a process called apomixis (Asker and Jerling 1992). There are many different apomictic mechanisms that have been observed in different plant species. Most, however, involve processes that bypass some of the normal steps in gametophyte formation. Three general classes of apomictic mechanisms are recognized – diplospory, apospory, and adventitious embryony (reviewed by Koltunow 1993) (Fig. 10.8).

In diplospory, the embryo arises from a megasporocyte (megaspore mother cell) that has failed to initiate or to execute the reductional divisions of meiosis (meiosis I). In diplosporic apomicts, the subsequent steps of gametophytic development are quite normal, and an embryo sac that contains eight unreduced nuclei is formed in which the egg cell undergoes embryo development without fertilization. In apospory, the embryo derives from nucellar cells and not the megaspore, but, nonetheless, it does produce an embryo sac with unreduced nuclei. Thus, both diplosporic and aposporic apomicts, which are referred to as gametophytic apomicts, produce gametophytes or embryo sacs. Another form of apomixis is called sporophytic apomixis. This usually occurs when nucellar cells undergo embryogenesis and form adventitious embryos outside the embryo sac.

In normal sexual reproduction, fertilization is required to initiate both embryo and endosperm development. In diplosporic apomicts, formation of the endosperm usually occurs autonomously (i.e., without fertilization). It is not known how this happens, and the ploidy of the resulting endosperm can be variable. Diplosporic apomixis bypasses the normal requirement that male and female genomes contribute to the formation of endosperm (to be discussed in Chapter 11). In aposporic apomicts, autonomous endosperm formation rarely occurs, and the initiation of endosperm development usually depends on fertilization of the polar nuclei. It is not known why the egg cell is not also fertilized in these apomicts. Several embryo sacs can be formed per ovule in aposporic apomicts, although only one is usually fertilized. Thus, in these apomicts the limits on the formation of one embryo sac per ovule

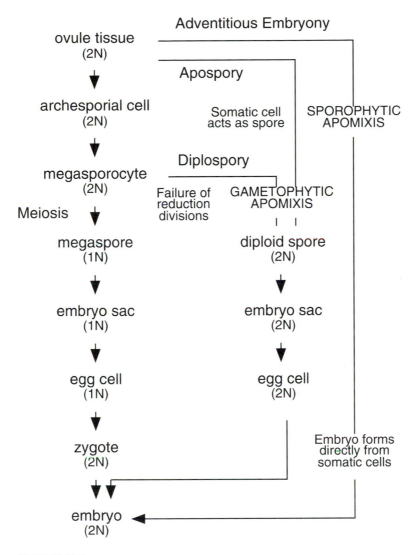

FIGURE 10.8

Sexual and apomictic pathways that lead to embryo development in angiosperm ovules. Normal sexual reproduction (left) is compared with two gametophytic apomixis pathways and a sporophytic apomixis pathway of adventitious embryony (right). (Based on Koltunow, A. M., Bicknell, R. A., and Chaudhury, A. M. 1995. Apomixis: Molecular strategies for the generation of genetically identical seeds without fertilization. Plant Physiol. 108:1346.)

have been compromised. Sporophytic apomicts do not form embryos from an embryo sac; therefore, they depend on the fertilization of a gametophyte in the same ovule for endosperm formation. Apomixis and sexual reproduction are not mutually exclusive. In facultative apomicts, both apomixis and sexual reproduction occur in the same plant.

The control of apomixis in plant breeding programs would be quite desirable because embryos that result from apomictic processes are entirely maternal in origin and, therefore, are genetically identical. In apomicts, the genotype is fixed because the male gametophyte makes no contribution to the embryo. Because of this, controlled apomixis would be useful in hybrid seed production, particularly in propagating F_1 hybrids (Koltunow et al. 1995). Normal hybrids lose vigor when sexually propagated because maternal and paternal characters segregate through sexual propagation. The genotypes in apomictic progeny from F_1 hybrids would be fixed and the vigor associated with hybrid formation might be maintained. Of the forms of apomixis, facultative apomixis might be most useful for hybrid seed propagation purposes. Facultative apomicts in which a small portion of the seeds arise from sexual reproduction would allow for the production of hybrids, but not for their propagation in bulk by apomixis.

The genetics of apomixis have been studied using the apomicts as the pollen parents. In a number of cases, it has been found that the predisposition for apomictic development through the various pathways is conditioned by single dominant genes (Koltunow 1993). Of interest is whether the dominant characters are gain-of-function traits in the execution of apomixis or represent loss-of-function in sexual reproduction. For example, diplosporic apomixis might result from a defect in meiosis I that fails to arrest further gametophyte development. Koltunow et al. (1995) have proposed a scheme to select for Arabidopsis mutants that produce seed without fertilization. Because Arabidopsis is a self-fertilizing plant, selection for fertile mutants could be made in male-sterile lines. For example, selections for fertile apomicts could be carried out in lines bearing *apetela3* (*ap3*) or *pistillata* (*pi*) mutations that are male sterile, but female fertile.

REFERENCES

Asker, S. E. and Jerling, L. 1992. Apomixis in plants. Boca Raton, FL: CRC Press.

Bowman, J. 1994. Arabidopsis: An atlas of morphology and development. New York: Springer-Verlag.

Cheung, A. Y., Wang, H., and Wu, H. M. 1995. A floral transmitting tissue-specific glycoprotein attracts pollen tubes and stimulates their growth. Cell 82: 383–393.

Heslop-Harrison 1987. Pollen germination and pollen tube growth. Int. Rev. Cytol. 107: 1–78.

Huang, S., Lee, H. S., Karunanandaa, B., and Kao, T. H. 1994. Ribonuclease activity of *Petunia inflata* S proteins is essential for rejection of self-pollen. Plant Cell 6: 1021–1028.

Hülskamp, M., Schneitz, K., and Pruitt, R. E. 1995. Genetic evidence for a long-range activity that directs pollen tube guidance in Arabidopsis. Plant Cell 7: 57–64.

Ikeda, S., Nasrallah, J. B., Dixit, R., Preiss, S., and Nasrallah, M. E. 1997. An aquaporin-like gene required for the Brassica self-incompatibility response. Science 276: 1564–1586.

Jauh, G. Y. and Lord, E. M. 1995. Movement of the tube cell in the lily style in the absence of the pollen grain and the spent pollen tube. Sex. Plant Repro. 8: 168–172.

Koltunow, A. M. 1993. Apomixis: Embryo sacs and embryos formed without meiosis or fertilization in ovules. Plant Cell 5: 1425–1437.

Koltunow, A. M., Bicknell, R. A., and Chaudhury, A. M. 1995. Apomixis: Molecular strategies for the generation of genetically identical seeds without fertilization. Plant Physiol. 108: 1345–1352.

Lee, H. S., Huang, S., and Kao, T. H. 1994. S proteins control rejection of incompatible pollen in *Petunia inflata*. Nature 367: 560–563.

Lin, Y., Wang, Y., Zhu, J. K., and Yang, Z. 1996. Localization of a Rho GTPase implies a role in tip growth and movement of the generative cell in pollen tubes. Plant Cell 8: 293–303.

Lin, Y. and Yang, Z. 1997. Inhibition of pollen tube elongation by microinjected anti-Rop1Ps antibodies suggests a crucial role for Rho-type GTPases in the control of tip growth. Plant Cell 9: 1647–1659.

Malho, R. and Trewavas, A. J. 1996. Localized apical increases of cytosolic calcium control pollen tube orientation. Plant Cell 8: 1935–1949.

Mascarenhas, J. P. 1993. Molecular mechanisms of pollen tube growth and differentiation. Plant Cell 5: 1303–1314.

McClure, B. A., Haring, V., Ebert, P. R., Anderson, M. A., Simpson, R. J., Sakiyama, F., and Clarke, A. E. 1989. Style self-incompatibility gene products of *Nicotiana alata* are RNase. Nature 342: 955–957.

McCubbin, A. G., Chung, Y.-Y., and Kao, T. 1997. A mutant S_3 RNase of *Petunia inflata* lacking RNase activity has an allele-specific dominant negative effect on self-incompatibility interactions. Plant Cell 9: 85–95.

Nasrallah, J. B. and Nasrallah, M. E. 1993. Pollen-stigma signaling in the sporophytic self-incompatibility response. Plant Cell 5: 1325–1335.

Nasrallah, J. B., Stein, J. C., Kandasamy, M. K., and Nasrallah, M. E. 1994. Signaling the arrest of pollen tube development in self-incompatible plants. Science 266: 1505–1508.

Newbigin, E., Anderson, M. A., and Clarke, A. E. 1993. Gametophytic self-incompatibility systems. Plant Cell 5: 1315–1324.

Palevitz, B. A. and Tiezzi, A. 1992. Organization, composition and function of the generative cell and sperm cytoskeleton. Int. Rev. Cytol. 140: 149–185.

Pierson, E. S. and Cresti, M. 1992. Cytoskeleton and cytoplasmic organization of pollen and pollen tubes. Int. Rev. Cytol. 140: 73–125.

Pierson, E. S., Miller, D. D., Callaham, D. A., Van Aken, J., Hackett, G., and Hepler, P. K. 1996. Tip-localized calcium entry fluctuates during pollen tube growth. Dev. Biol. 174: 160–173.

Preuss, D., Lemieux, B., Yen, G., and Davis, R. W. 1993. A conditional sterile mutation eliminates surface components from Arabidopsis pollen and disrupts cell signaling during fertilization. Genes Dev 7: 974–985.

Steer, M. W. and Steer, J. M. 1989. Pollen tube tip growth. New Phytol. 111: 323–358.

Stein, J. C., Dixit, R., Nasrallah, M. E., and Nasrallah, J. B. 1996. SRK, the stigma-specific S locus receptor kinase of Brassica, is targeted to the plasma membrane in transgenic tobacco. Plant Cell 8: 429–445.

Tantikanjana, T., Nasrallah, M. E., Stein, J. C., Chen, C. H., and Nasrallah, J. B. 1993. An alternative transcript of the S locus glycoprotein gene in a class II pollen-recessive self-incompatibility haplotype of *Brassica oleracea* encodes a membrane-anchored protein. Plant Cell 5: 657–666.

Wilhelmi, L. K. and Preuss, D. 1996. Self-sterility in Arabidopsis due to defective pollen tube guidance. Science 274: 1535–1537.

Wilhelmi, L. K. and Preuss, D. 1997. Blazing new trails: Pollen tube guidance in flowering plants. Plant Physiol. 113: 307–312.

Wu, H. M., Wang, H., and Cheung, A. Y. 1995. A pollen tube growth stimulatory glycoprotein is deglycosylated by pollen tubes and displays a glycosylation gradient in the flower. Cell 82: 395–403.

11

Seed and Fruit Development

SEED DEVELOPMENT

Seeds are a highly successful adaptation that insure the spread and survival of higher plants (angiosperms). A true seed is protected by a coat and contains food reserves to nurture the embryo during germination and early development. Angiosperms develop seeds within an ovary, whereas gymnosperms produce "naked" seeds, which are not developed within an ovary, but are borne instead on ovuliferous scales. Both angiosperm and gymnosperm seeds are protected by a coat that is derived from ovule integuments; however, in angiosperms the embryo and specialized extraembryonic tissues serve food storage functions. Food reserves in gymnosperms are stored in the highly developed female gametophyte. In this chapter, the development of the embryo and extraembryonic tissue in angiosperms will be discussed in the context of a developing seed.

Fertilization sets in motion the events of seed development (except in forms of apomixis in which fertilization is not obligatory). As discussed in Chapter 9, fertilization in angiosperms is a double fertilization event in which one of two sperms in a pollen tube fertilizes the egg nucleus, which gives rise to the zygote, whereas the other fertilizes the polar nuclei, which leads to the formation of the endosperm. Most angiosperm seeds undergo desiccation and are dormant until they germinate. Maintenance of viable embryos in a dried state provides plants with extraordinary means of survival and dispersal. In preparation for dormancy, seeds accumulate osmoprotectants and other solutes (sugars, proteins, etc.) that help to maintain the viability of the embryo during desiccation.

Seed development in soybean (*Glycine max*), for example, has been divided into four stages, the first three of which are shown in

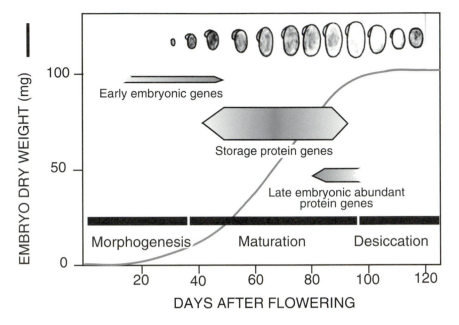

FIGURE 11.1

Time course of events during seed development in soybean (*Glycine max*). Time is expressed in days after flowering and embryo dry weight is plotted. Stages of seed development are indicated with black bars, and the time when major groups of genes are expressed is indicated. Seed development is depicted above. (Adapted from Goldberg, R. B., Barker, S. J., and Perez-Grau, L. 1989. Regulation of gene expression during plant embryogenesis. Cell 56:151. By permission of the author and Cell Press.)

Figure 11.1 (Goldberg et al. 1989). First, **morphogenesis** is an early stage during which the embryo and extraembryonic tissues are formed. (In Chap. 3, early embryogenesis was discussed separately.) Second, the **maturation** stage follows, during which storage materials accumulate and the seed grows. Third, **desiccation** or the postabscission stage occurs after the ovule abscises and the seed is cut off from maternal tissue. The seed dries and the embryo becomes dormant. Fourth, **germination** breaks seed dormancy. The embryo grows, utilizing storage materials from within the seed. The timing of these stages must be carefully orchestrated for successful seed development. Mutants that interfere with seed development events have provided some insight into how these events are coordinated. For example, *viviparous* (*vp*) mutants germinate precociously (usually during late maturation stage). These mutants are generally recessive loss-of-function mutants, which demonstrates that the action of the wild type genes is to suppress germination during seed development.

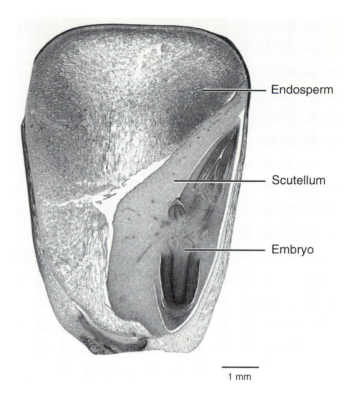

Endosperm

Scutellum

Embryo

1 mm

FIGURE 11.2

Mature kernel of corn (*Zea mays*). Longitudinal section shows endosperm, scutellum and embryo. Scutellum is equivalent to a cotyledon. (From Raven, P. H., Evert, R. F., and Eichhorn, S. E. 1986. Biology of Plants. Worth Publishers, New York, 4th ed., p. 383. Reprinted by permission of the author and Worth Publishers, Inc.)

MORPHOGENESIS STAGE

Development of the Endosperm

Double fertilization in angiosperms gives rise to the embryo and endosperm. In gymnosperms, the enlarged female gametophyte serves as an endosperm. Therefore, the endosperm, which is derived from a sexual process, is a new development in the evolution of seed plants. Morphogenesis of the embryo was described in Chapter 3, and development of the endosperm will be discussed here. The endosperm in monocot seeds, such as maize (*Zea mays*), plays a very prominent role in seed development and can be the principal nutrient storage tissue in seeds (Fig. 11.2) (Lopes and Larkins 1993). In a number of dicots, however, the endosperm is a transient structure that may be completely absorbed before the seed matures. In soybean, food reserves for the germinating embryo are stored in the cotyledons, which are organs of the embryo, and not in extraembryonic tissues, such as the endosperm.

Endosperm and embryos develop quite differently. In general, there are three types of endosperm development: cellular, nuclear, and helobial (Brink and Cooper 1947). In cellular types, which are found in

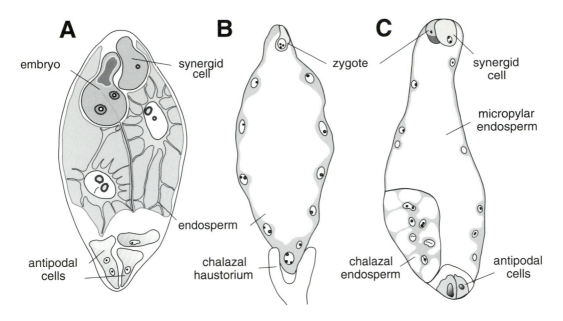

FIGURE 11.3

Various endosperm types result from differences in cellularization following division of endosperm nuclei. Drawings illustrate (A) cellular type of endosperm in *Centranthus macrosiphon*. Cellularization follows most divisions and endosperm is composed of some uninucleate cells. (B) Nuclear type in *Phacelia tanacetifolia*. Endosperm is formed from divisions without cellularization, which results in a multinucleate endosperm. (C) Helobial type in *Ixolirion montanum*. First endosperm division occurs with cellularization creating micropylar and chalazal endosperm cells. Further nuclear divisions without cellularization result in two multinucleate endosperm cells. (Redrawn from Schnarf, K. 1929. Embryologie der Angiospermen. In: Handbuch der Pflanzenanatomie II, part 2, Gebrüder Borntraeger, Berlin, pp. 338, 345, and 329.)

some tomato (*Lycopersicon*) species and in the example shown here in *Centranthus* (Fig. 11.3A), cellularization occurs after each round of nuclear division. Cereals, such as maize and as illustrated in *Phacelia* (Fig. 11.3B), have nuclear forms in which nuclei divide in the endosperm without cellularization (i.e., without cell wall formation). The nuclei migrate to the periphery of the central cell, and cellularization follows thereafter. In helobial types (Fig. 11.3C), the primary endosperm nucleus divides to form cells of unequal size. The larger micropylar cell undergoes cellular development, whereas the smaller chalazal cell remains undivided or forms a multinucleate cell.

The patterns of cell divisions in endosperm have been traced by sector analysis in maize (McClintock 1978). The *waxy* (*Wx*) locus controls the accumulation of amylose, and the excision of the unstable *Ac* transposon from *Wx* generates sectors that can be visualized by iodine stain-

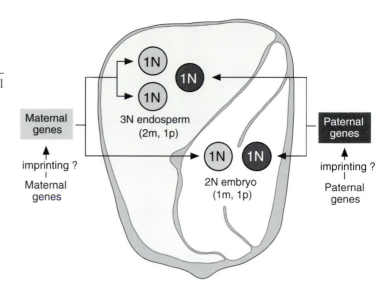

FIGURE 11.4

Balance between maternal and paternal genomes in the production of the triploid endosperm and diploid embryo in maize (*Zea mays*). Successful endosperm development requires a 2:1 ratio between maternal and paternal genomes (2m:1p). The differences between maternal and paternal genomes are attributed to imprinting.

ing. Sector analysis demonstrated that early divisions establish the left and right halves of the endosperm, and later divisions generate conical sectors. This pattern is consistent with growth of the endosperm by simple, radial enlargement.

Development of the seed involves an interplay between maternal tissue, embryo, and endosperm – tissues of different genetic constitution. Endosperm is usually triploid (or 3N) because a sperm nucleus from the pollen fuses with the two polar nuclei in the central cell (Fig. 11.4). In monosporic plants, the genetic makeup of the endosperm and the embryo are qualitatively similar in that both are derived from the same meiotic product in the male and the same meiotic product in the female. (As discussed in Chap. 9, all the nuclei in the female gametophyte are derived from the same meiotic product in monosporic plants.) The genetic makeup of endosperm and embryo, however, is quantitatively different in that the ratio of maternal to paternal genomes differs in the two tissues. Some have argued that the quantitative differences contribute to their different developmental fates of the tissues (Kermicle and Alleman 1993).

A number of studies have demonstrated that disruption in the balance of maternal and paternal genomes between embryo, endosperm, and maternal tissue can lead to "seed breakdown," which is an arrest in seed development. The success of the embryo hinges on the development of the endosperm because failure in endosperm development leads to embryo abortion. Genetic differences between parents frequently interfere with endosperm development, and such incompatibilities are thought to be an important factor in the reproductive isolation of angiosperm populations. Ploidy differences within a species represent a common reproductive barrier attributable to problems in endosperm development. Johnston and Hanneman (1982) pointed out in potato (*Solanum tuberosum*) that crosses between diploid and tetraploid potato species typically fail because of problems in endosperm development, whereas crosses among diploids or tetraploids usually succeed. A few tetraploid potato species, however, will cross with diploid species. These tetraploids cannot be crossed to other tetraploids unless their chromosome number is doubled ($4X \rightarrow 8X$, where X is the ploidy level in a haploid genome). These observations gave rise to the concept of **endosperm balance number** (EBN), which states that any species can be assigned an effective ploidy number, which may not be the actual ploidy number. Successful crosses between species usually occur when the endosperm has a maternal–paternal ploidy ratio of $2m{:}1p$ (2 maternal genomes: 1 paternal genome) (Fig. 11.4). Depending on their genetic origin, various tissues in the seed have different maternal–paternal ploidy ratios. In a typical seed from a cross of two diploids, maternal–paternal ploidy ratios are $2m{:}0p$ in maternal tissues, $2m{:}1p$ in endosperm, and $1m{:}1p$ in the embryo.

Maternal–paternal ploidy ratios have been more extensively studied in maize, where ploidy ratios in the embryo and in the endosperm can be manipulated. Lin (1984) studied the mutation *indeterminate gametophyte* (*ig*), which produces abnormal numbers of polar nuclei (see Chap. 9) and leads to variation in the maternal ploidy levels in endosperm. Lin used pollen from either diploids or tetraploids to vary the paternal genome contribution and, again, it was found that endosperm development succeeded only when maternal–paternal ploidy ratio in the endosperm was $2m{:}1p$. It was clear from his study that the overall ploidy level was not critical to the success of the endosperm, but that the relative contributions from the male and female parent were important. For example, among $6X$ endosperms, $4m{:}2p$ were normal, whereas others with $5m{:}1p$, for example, were aborted. From these observations, Lin proposed that endosperm failure was due to parental imprinting in which the genomes derived from the male and female parent operate differently in the context of the endosperm. Imprinting is a phenomenon in animals which ensures that propagation occurs by sexual reproduction. Derivation of gametic alleles from both male and female parents is needed for successful embryo development. In flowering plants, it appears that a different imprinting process required for endosperm development is a checkpoint for sexual reproduction (Walbot 1996). The proper balance of maternal and paternal genomes is needed for successful endosperm development.

The fragile relationship between embryo, endosperm, and maternal tissue when the endosperm ratio is unbalanced may be a consequence of the intimate contact between these tissues in seed development. Charlton et al. (1995) examined the structural interaction between these tissues in aborted maize kernels when the endosperm ratio was 2*m*:2*p*. They found that the formation of the transfer cell layer in the endosperm, which normally occurs about 8 days after pollination, was completely suppressed. The transfer cells are thought to be responsible for the transport of nutrients from the maternal tissue to endosperm; thus, it would be expected that the unbalanced endosperm would be starved of nutrients. The reason transfer cells do not form is not yet known.

A number of genes are specifically expressed in the endosperm. Many of these appear to be duplicates of genes expressed elsewhere, but which have been specialized for endosperm-specific expression (Lopes and Larkins 1993). They encode products involved in the synthesis of stored nutrients, such as starch. Loss-of-function mutations in these genes have a profound effect on endosperm development. For example, *shrunken-2* (*sh2*), a starch deficiency mutation in maize kernels, is defective in an endosperm form of ADP-glucose phosphorylase required for starch biosynthesis (Bhave et al. 1990). The endosperm also accumulates lipids and storage proteins. More will be discussed about the regulation of storage protein synthesis in the next section.

There is important interplay between maternal tissue and embryo/endosperm tissue during seed development, although there is no direct or symplasmic connection between them (Thorne 1985). It has been argued that this arrangement prevents the transmission of viruses from maternal to embryonic tissue. Because of that, special accommodations have been made to move nutrients from the plant to the seed during seed maturation or grain filling. Transfer occurs by unloading nutrients into the apoplast from phloem terminals in the maternal tissues at the base of the seed. The nutrients are assimilated at the base of the endosperm, by cells called transfer cells, that have extensive fingerlike projections. Very little is known about the development of the placental structure; however, a cDNA was found that may serve as a marker for these cells during endosperm development (Hueros et al. 1995). The cDNA was isolated from a library derived from maize endosperm RNA and found by *in situ* hybridization to be located in transfer cells. The cDNA, called Bet1, encodes a small cell-wall protein that may be unique to transfer cells.

It will be important ultimately to determine what kind of cell–cell interactions are involved in the development of this contact structure between maternal and endosperm tissue. In addition, the kind of cell differentiation required for apoplastic nutrient transfer at the contact surface needs to be understood, as does the reason why the development of symplastic connections is prevented along this interface.

MATURATION STAGE

During the maturation stage of seed development, seeds accumulate large quantities of storage macromolecules, such as proteins, lipids, and carbohydrates. Some of the stored proteins serve as food sources for the developing embryo, whereas others are involved in biosynthetic functions or in protecting the seed from predation or pathogens. The major storage proteins common to all seed-forming species are called globulins and albumins, whereas others found exclusively in cereals are called prolamines.

Because these storage materials are the major source of food for people and livestock, the synthesis of seed proteins and the control of seed storage protein gene expression have been studied extensively in crop plants. Most storage proteins are encoded by large families of genes, and many seed storage protein genes are expressed during seed maturation. The high-level expression of so many genes results in a massive amount of protein accumulation at that stage. In most cases, storage protein synthesis appears to be transcriptionally regulated (Goldberg et al. 1989). Dure (1985) proposed that genes expressed during seed development could be categorized based on their temporal patterns of expression. Two major groups of genes are those that encode the **seed storage proteins** (synthesized at midmaturation stage) and those that encode the **late embryonic abundant (LEA) proteins** (Fig. 11.1). LEA proteins are synthesized at the late maturation and early postabscission stages, and are thought to serve as protectants from desiccation. The fact that various constellations of genes are switched on at different developmental stages suggests that a hierarchical control system regulates gene expression during seed development and, most likely, that different global regulators or factors control the expression of large groups of genes.

Abscisic acid (ABA) is one of the important regulators of genes that encode seed storage proteins and late embryogenesis proteins. ABA enhances seed storage protein synthesis and suppresses precocious germination in explanted embryos. In wheat, ABA levels in developing seeds rise and fall in a manner expected for a regulator of these developmental processes. ABA levels rise at the time of embryo maturation and seed storage protein accumulation, and the levels fall at the time of desiccation (King 1976). In other systems, however, ABA accumulation patterns do not correlate as well with the pattern of storage protein synthesis.

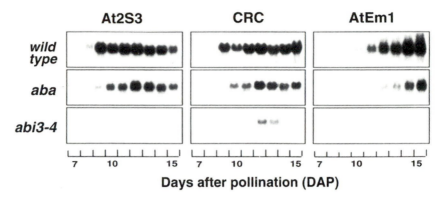

Days after pollination (DAP)

FIGURE 11.5

Expression pattern of genes during Arabidopsis seed development in wild type (Ler, Landsberg erecta ecotype) and abscisic acid (ABA) mutants *aba* and *abi3*-4. RNA samples were subjected to gel electrophoresis and transferred to filters or blots for hybridization. Autoradiographs of blots of RNA extracted from siliques (seed pods) at various days after pollination (DAP) are shown. Gene-specific probes for At2S3 (napin), CRC (cruciferin), and AtEm1 were hybridized to RNA on blots. (From Parcy, F., et al. 1994. Regulation of gene expression programs during Arabidopsis seed development: roles of the *ABI3* locus and of endogenous abscisic acid. Plant Cell 6:1569. Reprinted by permission of the American Society of Plant Physiologists.)

The role of ABA in seed development has been examined in Arabidopsis mutants that are defective in either ABA synthesis or ABA response. ABA mutants were identified by Koornneef et al. (1984), and ABA-insensitive mutants were selected for their ability to germinate on media containing ABA levels that normally inhibit germination. Parcy et al. (1994) monitored the expression pattern of a number of genes during Arabidopsis seed development in *aba-1*, which is a mutant defective in ABA synthesis, and in *abi3*, which is a mutant that is defective in ABA response. They found that the activation of some of the genes were affected in *abi3*, and the genes that were affected did not necessarily belong to the same temporal class. For example, mRNAs encoding the storage proteins, napin (At2S3), and cruciferin (CRC) accumulate during maturation phase in normal Arabidopsis seeds. As expected, accumulation of these mRNAs was strongly curtailed in the *abi3* mutant, which is defective in ABA responses (Fig. 11.5). It was unexpected that the accumulation of these mRNAs was not blocked in *aba-1*, which is a mutant with defects in ABA accumulation. The results suggest that the block in ABA synthesis in *aba-1* is not complete and that ABA levels are above the threshold needed for gene expression. On the other hand, the results suggest that not all seed protein genes are activated by ABA and that *ABI3* has a larger role in seed development that extends beyond its involvement in ABA responses. The role of *ABI3* in other ABA responses will be discussed in a later section in this chapter. In general, the findings indicate that ABA is important, but that it is not the only regulator of seed development. Other regulators must be equally important in controlling gene expression during seed development even within specific temporal classes.

Nonetheless, because ABA is an important regulator of some of the genes expressed during seed development, the molecular basis of ABA action has been studied extensively. In wheat, one of the major storage protein genes induced during maturation is the *Em* gene, and ABA stimulates its transcription (Marcotte et al. 1988). The expression of the *Em* gene promoter has been examined in a transient expression system in protoplasts from cell suspension cultures of nonembryonic tissue. (In a transient expression system, plasmid DNA that contains a gene under study is introduced into protoplasts; that is, into plants cells from which the cell wall has been removed. In this case, the *Em* gene promoter was linked to the *uidA* or GUS reporter gene, creating an *Em* promoter:GUS construct. The expression of *Em*:GUS was assessed 18 hours after the introduction of DNA.) The *Em* promoter was activated in protoplasts in an ABA-dependent manner.

From the analysis of various deletions in the *Em* promoter, a *cis*-acting ABA response element (ABRE) in the promoter was identified (Fig. 11.6). The ABRE is related to G-box elements, which is a general class of elements found in the promoters of many other regulated genes in plants, including light-regulated genes. One factor that binds to this particular G-box element and activates the expression of the *Em* gene has been identified and cloned from wheat. The factor, called EmBP-1, is a basic leucine zipper (bZIP) transcription factor (Guiltinan et al. 1990). It is not known how ABA affects the expression of the *Em* gene through the action of EmBP-1. EmBP-1 is present and binds to DNA in extracts from cell suspension culture cells whether or not they have been treated with hormone. The activation of *Em* by ABA also requires another transcription factor VP1 (a homolog of ABI3) that acts synergistically with EmBP-1. VP1 and the effects of mutations in the *vp1* gene will be discussed more fully in a later section in this chapter.

Different regulatory factors control the expression of other seed storage proteins. A well-studied case is a gene regulatory circuit in maize that controls the expression of a group of genes that encode prolamine storage proteins called **zeins.** The zeins in maize can constitute more than half of the seed storage proteins, and the zeins are encoded by a very large gene family composed of more than 100 members (Lopes

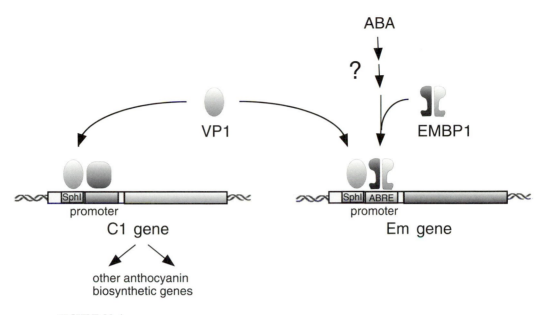

FIGURE 11.6

Scheme for the regulation of *Em* and *C1* genes during seed development in cereals, such as maize. VP1 in maize is a co-activator of both *C1* and *Em* gene expression. *Em* gene expression is activated by abscisic acid (ABA) and requires the action of the b-ZIP transcription factor, EMBP1. The mechanism of ABA activation is not understood. EMBP1 binds to an abscisic acid response element (ABRE) in the *Em* promoter. VP1 is also required for activation and binds to a nearby promoter site (SphI site). VP1 is also thought to bind to the C1 promoter and to interact with other transcription factors. C1 encodes a regulator that activates the expression of other genes on the anthocyanin biosynthetic pathway. (Based on Vasil, V., et al. 1995. Overlap of Viviparous1 (VP1) and abscisic acid response elements in the Em promoter: G-box elements are sufficient but not necessary for VP1 transactivation. Plant Cell 7:1511–1518.)

and Larkins 1993). The zeins fall into five major size classes, of which the most abundant are the 19 and 22 kD proteins. Like other seed storage proteins, the synthesis of zeins is temporally regulated during seed development. Several mutations are known to affect the regulation of groups of *zein* genes, particularly those that affect the 19 and 22 kD zein proteins. One of the better known mutants is *opaque2* (*o2*), in which affected seeds lose their vitreous nature. *Zein* gene expression is reduced by 60 to 80 percent in *o2* null alleles, which is largely due to the reduction in the rate of transcription of 22 kD class of zeins (Kodrzycki et al. 1989). These findings suggested that O2 (the protein encoded by the *o2* gene) might be a transcription factor, and when the gene was cloned by transposon tagging, it was indeed found to encode a bZip transcription factor (Schmidt et al. 1990).

Schmidt et al. (1992) demonstrated that O2 recognizes and binds to a specific target sequence 5′-TCCACGTAGA-3′ in the 22 kD *zein* promoters. To determine whether or not O2 does, indeed, stimulate *zein* expression, the *O2* gene, driven by the strong CaMV 35S promoter (35S: *O2*), was bombarded into maize endosperm slices along with the 22 kD *zein* promoter linked to a reporter gene (from a maize line homozygous for *o2*) (Unger et al. 1993). (Biolistic bombardment was carried out with a "gene gun." DNA is absorbed onto small metal particles and propelled at high speed into plant tissue by the gun.) In these experiments, O2 transactivated the 22 kD *zein* promoter, which resulted in heightened expression of the GUS reporter gene.

In similar experiments, Unger et al. (1993) also demonstrated that mutant forms of O2, particularly, a form in which the basic domain of the protein was deleted, inhibited the activation by normal forms of the O2 protein. Mutations that can interfere with the function of the normal gene are called dominant negative mutations. (Such mutations were described in Chap. 10.) O2 is a bZIP transcription factor and binds to DNA as a dimer. The investigators demonstrated that the mutant protein was able to form dimers with the normal protein, but that the resulting heterodimers were unable to bind to DNA. Thus, the dominant negative mutation interferes with the function of the normal protein by forming inactive heterodimers.

DESICCATION STAGE

Dormancy allows seeds to survive inclement weather and to be dispersed before germination; however, dormancy interrupts the steady progress of development from embryo to seedling. Is dormancy an interlude required for germination and normal seedling development? At the late maturation stage of seed development, embryos are able to germinate precociously and develop into seedlings, but are prevented from doing so in the seed. Late maturation phase embryos (50 days postanthesis) from species, such as rapeseed (*Brassica napus*), germinate precociously when explanted (taken out of the seed) and placed in culture (Finkelstein et al. 1985). Embryos explanted at midmaturation phase (30–40 days postanthesis) do not develop normally, but continue on in embryonic growth, synthesizing storage proteins. At germination, the embryo normally initiates a new program of events and switches from the accumulation of stored reserves to their breakdown.

Does this switch occur during the period of drying and desiccation? Comai and Harada (1990) compared the pattern of transcription in maturing rapeseed embryos, embryos in dry seeds, and germinating seedlings. Because dry seeds are not actively engaged in RNA and protein synthesis, the authors assessed the potential for gene expression in seeds by carrying out "run-on" transcription *in vitro* in nuclei obtained from various embryos. They monitored the pattern of a number of genes that are specifically expressed during seed maturation (largely storage proteins) and others that are expressed following germination. They found that the pattern of gene expression in dry seed was similar to maturing embryos, but that it was different from germinating seedlings. They concluded that no significant changes in gene expression occur during drying, and that the new pattern of transcription is initiated upon germination.

DORMANCY AND THE CONTROL OF GERMINATION

As described earlier, embryos at the late maturation stage of development can germinate if they are explanted, but are prevented from doing so in the seed. ABA appears to play a major role in controlling embryo dormancy in the seed and in suppressing precocious germination (or vivipary). This has been demonstrated in *viviparous* mutants that interfere with ABA synthesis or action. *Viviparous* mutants in maize (*Zea mays*), such as *vp5* and *vp7*, are blocked in ABA biosynthesis. These mutants have low ABA levels in kernels and precociously germinate (Neill et al. 1986). The effect of these mutants can be phenocopied in wild type by application of an ABA synthesis inhibitor (Fong et al. 1983). ABA is synthesized indirectly through the breakdown of carotenoids, and carotene biosynthesis inhibitors, such as fluridone, can block ABA synthesis. Fluridone induces precocious germination in maize (although there is only a narrow developmental window during which the treatment is effective).

VP1 Gene in Maize

Another *viviparous* mutant in maize, *vp1*, germinates precociously (Fig. 11.7), but it is defective in response to ABA rather than in its synthesis (Robichaud et al. 1980). Embryos from these mutants are insensitive to the suppressive effects of ABA on germination in culture. *Vp1* is pleiotropic and affects multiple developmental processes associated with the maturation phase of seed formation (McCarty et al. 1989). *Vp1* also impacts anthocyanin biosynthesis (Fig. 11.7) because the mutant fails to activate the expression of the anthocyanin regulatory gene, *C1*. *C1* encodes a *myb*-like transcription factor that is involved in the regulation

275

FIGURE 11.7

Pleiotropic effects of *vp1* mutation on maize seed development. A maize ear segregating for the *vp1* mutation is shown. Mutant kernels lack anthocyanin pigment and are viviparous (germinate precociously on the cob). Note the sprouts from light-colored kernels. Wild type kernels have anthocyanin pigment and do not germinate precociously. (From McCarty, D. R., et al. 1991. The *viviparous-1* developmental gene of maize encodes a novel transcriptional activator. Cell 66:896. Reprinted by permission of the author and Cell Press.)

of structural genes on the anthocyanin biosynthesis pathway. Thus, VP1 appears to be a master regulator in a regulatory cascade that includes genes that control germination as well as those involved in anthocyanin biosynthesis (Fig. 11.6). *Vp1* was cloned by transposon tagging and encodes what appears to be an acidic transcription factor (McCarty et al. 1991). VP1 was demonstrated to be a transcription factor in transient expression experiments where VP1 transactivated the expression of a *C1* promoter–reporter gene construct in maize protoplasts (Hattori et al. 1992).

As discussed in a previous section, VP1 also activates the expression of other ABA-regulated genes during seed development. VP1 potentiates the effects of ABA on the expression of the maize *Em* gene in the protoplast system described earlier (McCarty et al. 1991). (The *Em* gene was originally identified in wheat. The maize gene is similar to the wheat gene and is expressed during the maturation phase of maize kernel development.) VP1 is a transcriptional co-activator along with EMBP-1 of the *Em* gene and with other undescribed factors of the *C1* gene and other genes (Vasil et al. 1995) (Fig. 11.6). Thus, VP1 has broad ranging effects because it appears to be an essential coactivator with a variety of other transcription factors, some of which require interaction with the hormonal response machinery, and some that do not.

ABI-3, LEC1, and *FUS3* Genes in Arabidopsis

In Arabidopsis, *ABI3* encodes a protein that is similar to VP1 (Giraudat et al. 1992). *Abi3* mutants have reduced seed dormancy, but have no other vegetative phenotype. Consistent with these observations, it was found that *ABI3* is specifically expressed in seeds and not in the vegetative parts of the plant. As discussed earlier, functional *ABI3* is required for the expression of genes that encode various seed proteins during the embryo maturation stage in Arabidopsis seed development. These same genes are not expressed in vegetative tissues after treatment with ABA. Giraudat et al. (1992) determined whether these seed protein genes would be expressed in vegetative tissues in response to ABA if *ABI3* was ectopically expressed in vegetative tissues. They linked the ubiquitously expressed CaMV 35S promoter to *ABI3* (35S:*ABI3*, which is a construct that promotes *ABI3* expression in vegetative tissue) and produced transgenic plants expressing this construct. They found that the seed protein genes were indeed expressed in vegetative tissue in response to ABA. Thus, it appears that ABI3 is a powerful regulator of seed storage protein genes and can drive their expression even in nonembryonic tissue in the presence of ABA.

In addition to the control that *ABI3* exercises over the accumulation of storage proteins and the maturation phase of embryonic development, the gene has an interesting role in late embryogenesis and germination. Like *vp1* in maize, *abi3* seeds in Arabidopsis tend to break dormancy and will occasionally germinate precociously. Wild type Arabidopsis seeds require light to germinate efficiently; however, severe *abi3* alleles germinate in the dark without light stimulus. When *abi*3 is germinated in the dark, cotyledons expand, and small true leaves are produced (Fig. 11.8). As discussed in Chapter 4, that is quite unexpected because dark-grown, wild type seedlings have neither expanded cotyledons nor true leaves. In the dark, the SAM is usually inactive, and leaves are not produced until seedlings are exposed to light. Nambara et al. (1995) argued that cotyledon expansion and precocious activation of the SAM result from the fact that the mutant embryos enter a postgerminative stage of development while still in the seed, and in so doing, the maturation phase is cut short. There are other indicators which support the argument that the maturation phase has been truncated. For example, *abi3* embryos fail to degreen (wild type embryos degreen during maturation phase) and do not develop desiccation tolerance.

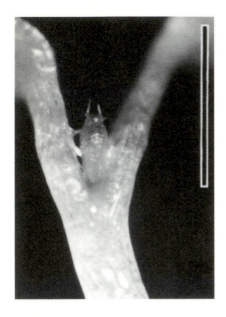

FIGURE 11.8

Precocious shoot apical meristem (SAM) development in dark-grown Arabidopsis *abi3* mutant seedlings. SAM development is usually not activated in dark-grown wild type seedlings. SAM activation is evidenced by the emergence of the first true leaves with trichomes. (From Nambara, E., et al. 1995. A regulatory role for the *ABI3* gene in the establishment of embryo maturation in *Arabidopsis thaliana*. Development 121:633. Reprinted by permission of the Company of Biologists Ltd.)

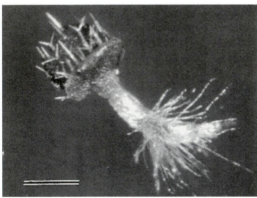

FIGURE 11.9

Leafy cotyledon1 (*lec1*) mutant in Arabidopsis. Cotyledons (upper left) have leaflike characteristics in the mutant, particularly the presence of trichomes. Cotyledons on wild type seedlings do not ordinarily have trichomes. Bar = 0.5 μm. (From Meinke, D. W., et al. 1994. Leafy cotyledon mutants of Arabidopsis. Plant Cell 6:1053. Reprinted by permission of the American Society of Plant Physiologists.)

Homozygous *abi*3 seeds (embryos) can be rescued by harvesting them early before they dry completely.

It is interesting that other mutations, such as *leafy cotyledon1* (*lec1*) and *fusca3* (*fus3*), have similar effects on late embryogenesis and germination, but are, nonetheless, sensitive to ABA. *Lec1* was identified by Meinke et al. (1994) as a mutant seedling that produced trichomes on its cotyledons (Fig. 11.9). Another allele of *lec1* found by West et al. (1994) was embryo lethal, but it could be rescued by explanting embryos. Trichomes are only formed on true leaves in wild type seedlings; therefore, the appearance of trichomes on cotyledons suggested that cotyledons had been transformed into leaves. There were other indicators of a leaflike transformation of cotyledons. Cotyledons in *lec1* embryos did not accumulate storage protein bodies, and the vasculature developed further, more like the leaf, with mature xylem elements rather than just the procambium found in cotyledons. Because cotyledons in *lec1* appeared to be transformed into leaves, the mutant was first interpreted to be a homeotic mutant. Homeotic mutants ordinarily show complete replacement of one organ for another, but the conversion was not complete in the case of *lec1*. For example, the cotyledons did not produce stipules that are characteristic of leaves. Furthermore, storage protein genes such as cruciferin, which are normally not expressed in leaves, are expressed in cotyledons of *lec1* embryos (West et al. 1994).

Other properties of *lec1* suggest that the real problem with the mutant was that control of late embryogenesis and germination were compromised (West et al. 1994). Like *abi3*, *lec1* is desiccation-intolerant, which indicates that *lec1* failed to complete the program of maturation that protects the embryo during desiccation. Genes, such as isocitrate lyase, which is normally expressed after germination, were also turned on in the embryo.

Other mutants similar to *lec1* were found with trichomes on cotyledons (Meinke et al. 1994). One was allelic to *fus3*, and another defined a new locus called *leafy cotyledon2* (*lec2*). *Fus3* had been identified previously as one of the embryo/seedling lethal mutants that accumulates high levels of anthocyanin (similar to *fus6* in Fig. 4.8). Alleles of *fus3* were not as pleiotropic as *lec1* (i.e., did not affect as many postgerminative events as *lec1*). Like *lec1*, *fus3* was desiccation intolerant and did not accumulate protein bodies. Development of the vasculature in *fus3* cotyledons, however, was intermediate between that of cotyledons and leaves. Like *fus3*, *lec2* also had a more limited effect on postgerminative events. Through the analysis of single and double mutants, Meinke et al. (1994) determined whether the genes interacted with each other and whether they represented a pathway of gene action (Fig. 11.10). It was argued that *LEC1* acted on the same pathway as *LEC2* and *FUS3* because the double *lec1 lec2* and *lec1 fus3* mutants did not have new phenotypes. *Lec1*, however, was placed upstream from *lec2* and *fus3* because it was more pleiotropic (i.e., it affects a broader range of postgerminative events).

Because none of the mutants in the *lec1*, *lec2*, or *fus3* group are ABA insensitive, it was reasoned that *abi3* must be on another pathway with overlapping functions. In summary, ABA and *abi3* are major players in late embryogenesis and postgerminative events, but they are not the whole story. The *LEC1/LEC2/FUS3* pathway has even more widespread influence on these events. The pathways are overlapping or redundant, but they undoubtedly provide greater adaptive advantage for seed function or survival.

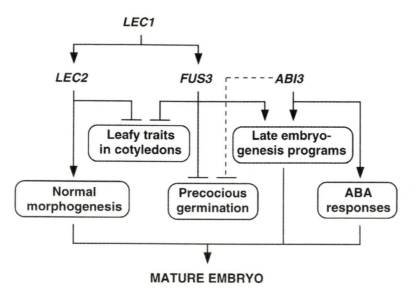

FIGURE 11.10

Genetic regulation of events in late embryogenesis based on characteristics of *leafy cotyledon* (*lec*) and ABA-insensitive (*abi*) mutants. (Activation indicated by an arrow, ↓, whereas inactivation is represented by ⊥.) *ABI3* is only limitedly involved in suppressing vivipary or precocious germination (dotted line). (Adapted from Meinke, D. W., et al. 1994. Leafy cotyledon mutants of Arabidopsis. Plant Cell 6:1049–1064.)

FRUIT DEVELOPMENT

Fruit Growth

Fruits are produced in certain angiosperms and are auxiliary reproductive structures that contain seeds (unless parthenocarpic fruit development has occurred). We think of apples and bananas as fruits, but tomatoes, pea pods, and even Arabidopsis siliques are fruits in botanical terms. Fruits are organs in which seeds develop and which promote seed dispersal. Fruits develop from ovaries and their development is usually triggered by signals from pollination or fertilization. Most fruits are composed of a fleshy, thickened wall called the pericarp, which is derived from the ovary wall. The pericarp is typically composed of a skin, cuticle, or exocarp, a spongy mesocarp, and often a firmer endocarp. Other fruits can also have dry or nonfleshy pericarps that often split when the fruit reaches maturity. In fleshy fruits, such as tomato, the bulk of the fruit is derived from L3. Only the epidermal and subepidermal cell layers are derived from L1 and L2 (see the review by Gillaspy et al. 1993).

The nature of the signals that set the program of fruit development in motion are not known. It is thought that growth factors are produced as a result of pollination and/or fertilization. Some insight into the stimuli for fruit development can be gained from the induction of parthenocarpy or the formation of seedless fruits. Parthenocarpy can be genetically controlled or induced by the exogenous application of hormones, such as gibberellins or auxin (George et al. 1984). It is not known, however, whether these hormones are the initial signal generated by pollination and/or fertilization or whether the hormones only amplify the signal.

The development and ripening of commercial fruit have been staged for growers, processors, and sellers. Fruit development in tomato has been divided into three phases (Gillaspy et al. 1993): Phase I occurs at anthesis and involves ovary development, fertilization and fruit set. Cell division occurs in the developing fruit in phase II, when seeds form and the embryos begin to develop. In phase III, cells of the fruit expand and the embryos mature. Fruit development phases are followed by fruit ripening, in which stages are largely determined by color and softening in tomato.

During the early stages of fruit growth in tomato (Phase II), active cell division occurs in the pericarp and in the placental tissue (the tissue that bears the seeds). The rate and/or extent of cell division appears to be influenced by seed development. The rate of fruit growth during the cell division stage correlates with the number of developing seeds in tomato (Hobson and Davies 1970). There is significant enlargement of the fruit during later stages of fruit growth (Phase III) (Gillaspy et al. 1993). By this time most cell divisions have ceased and the growth of the fruit is fueled by cell expansion. The onset of the cell expansion phase of tomato fruit growth coincides with a peak in auxin accumulation. There is little growth of the embryos in tomato seeds during fruit cell expansion. When fruit cell expansion ceases, however, embryos grow considerably and complete their maturation.

Growth of fruit in plants with large or numerous fruits can place demands on the metabolic capacity of the plant. During the rapid enlargement phase, most fruits accumulate large quantities of sugars and starch, and they can become significant metabolic sinks for the accumulation of storage metabolites. Sink strength in developing fruits is genetically determined, but the regulation may be indirect and affect such properties as the control of cell number in ovaries prior to anthesis. Fruit tissues, such as the pericarp tissue in tomato, are photosynthetically active and have other morphological properties of leaf tissues to which they are ontogenically related. The cells of the green tomato contain chloroplasts, and the bulk of the pericarp is organized much like the palisade layer of the leaf. The photosynthetic capacity of the fruit is modest, however, and it cannot support the development of the tomato.

Size, weight, and percentage of solids are important commercial traits for fruit such as tomatoes. Tomato fruit weight is one of the best-known examples of a complex quantitative trait with continuous variation. Fruit weight is controlled by many genes, and from five to twenty genes are thought to be involved in tomato (see Alpert and Tanksley 1996). Efforts have been undertaken to map **quantitative trait loci (QTLs)** that determine tomato fruit weight and to clone the major loci. One locus called *fw2.2* has been mapped with high precision to a defined region of the tomato genome, which suggests that the major QTL represents a single gene (Alpert and Tanksley 1996).

Fruit Ripening

Fruit ripening involves the events that lead to changes in the color, softness, aroma, and flavor in many different fruits. While the biochemical pathways involved in these changes are different from species to species or cultivar to cultivar, there are some common features about the fruit ripening process. Fruits can be classified according to whether they are climacteric or nonclimacteric. Climacteric fruit, such as tomatoes and bananas, have a respiratory burst at the onset of ripening and they evolve ethylene, which acts in an autocatalytic manner to stimulate the further evolution of ethylene. Nonclimacteric fruits, such as oranges and strawberries, do not have a respiratory burst and do not show a significant increase in ethylene evolution (see Picton et al. 1995).

Ethylene is a key regulator of ripening in climacteric fruit. Ripening begins after cell division and cell expansion have terminated, and it is marked by a burst in respiration and a surge in ethylene production. The importance of ethylene to the ripening process can be demonstrated by the action of inhibitors that block ethylene perception such as norbornadiene or silver ions. Treatment of tomatoes with these inhibitors delays the onset or the progression of the ripening process (Picton et al. 1995). Ethylene has its effect on fruit ripening by activating the expression of a constellation of ethylene inducible genes, many of which have been cloned in cDNA libraries made from RNAs extracted from ripening fruit.

To understand the role of ethylene in fruit ripening, tomato mutants have been sought that are insensitive to ethylene, but which are similar to the Arabidopsis ethylene-insensitive mutants described in Chapter 4. Tomato mutants were identified by screening existing fruit-ripening mutants for ethylene insensitivity at the seedling stage (Lanahan et al. 1994). Recall that ethylene produces a "triple response" in dark-grown seedlings (Chap. 4). A fruit ripening mutant called *Never ripe* (*Nr*) failed to display the triple response after treatment of dark-grown seedlings with ethylene. The *Nr* gene that conditions ethylene insensitivity in tomato was found to cosegregate with a gene that is similar to the *ETR1* gene in Arabidopsis (Wilkinson et al. 1995). It was described in Chapter 4 that *ETR1* encodes a two-componentlike protein kinase that has a receptor-protein kinase and a response regulator domain. The gene that cosegregates with *Nr* is similar to *ETR1* but lacks the response regulator domain, like ERS in Arabidopsis (Fig. 4.11). Nonetheless, it is thought that *Nr* encodes a component of a critical ethylene receptor in tomato. In any case, *Nr* mutants ripen slowly and incompletely, do not redden, and only slightly soften, which demonstrates the diverse roles of ethylene in tomato fruit ripening.

For commercial purposes, various genetic engineering manipulations have been carried out to generate tomatoes that are defective in ethylene production or evolution, but not in ethylene perception. The incentive to do so is to develop tomatoes that can be ripened on demand by controlled application of ethylene, as is done for bananas. The production of ethylene in tomatoes is largely controlled by 1-aminocyclopropane-1-carboxylate (ACC) synthase, which is an enzyme that catalyzes the rate-limiting step in ethylene biosynthesis, the conversion of S-adenosylmethionine to ACC, and by ACC oxidase, which is an enzyme involved in the last step of ethylene biosynthesis in which ACC is converted to ethylene. Copy cDNAs (cDNAs) that encode ACC synthase and oxidase have been identified in tomato and used to develop antisense constructs. (See Chap. 10. Antisense RNAs can be effective in silencing the expression of the endogenous plant gene and of family members in multigene families, as long as there is a sufficient sequence similarity among the genes.)

Antisense ACC synthase constructs (antisense 35S:*ACS2*) were effective in reducing ethylene synthesis in transgenic tomatoes and preventing ripening (Oeller et al. 1991; Theologis et al. 1993) (Fig. 11.11). Treatment of antisense ACC synthase tomatoes with exogenous ethylene produced fruit that were indistinguishable from nontransgenic controls, which indicates that the only defect in the transgenic plants was in production of ethylene. Antisense ACC oxidase constructs were somewhat less effective in reducing ethylene production, but the constructs prevented ripening in fruits detached from transgenic tomatoes (Picton et al. 1993). The application of ethylene to the detached fruits was only partially able to restore normal ripening. Antisense ACC synthase plants were also useful in sorting out which fruit-ripening events were ethylene inducible and which were not (Theologis et al. 1993). For example, it was found that fruit in antisense plants accumulated

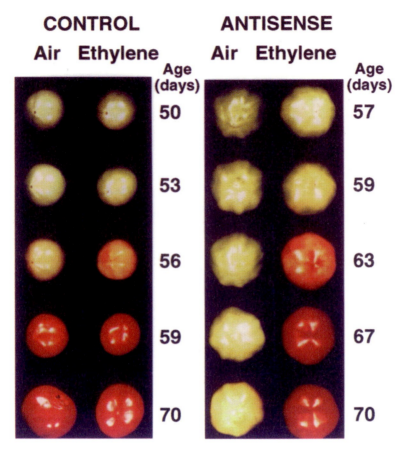

FIGURE 11.11

Inhibition of tomato fruit ripening by antisense ACC synthase constructs. Detached fruits were treated with air or ethylene (1 μl ethylene per milliliter air). (Control) Fruits from nontransgenic plants ripened similarly when treated with air or ethylene. (Antisense) Fruits from transgenic line homozygous for antisense ACC synthase construct (antisense 35S:ACS2) ripen very little when treated with air, however these fruits ripen normally when treated with ethylene. (From Theologis, A., et al. 1993. Use of a tomato mutant constructed with reverse genetics to study fruit ripening a complex developmental process. Devel. Gen. 14:285. Copyright © 1993 John Wiley and Sons. Reprinted by permission of Wiley-Liss, Inc., a subsidiary of John Wiley & Sons, Inc.)

normal levels of mRNA from genes that were thought to be ethylene inducible, such as ACC oxidase itself and polygalacturonase (PG). The fruit did not accumulate PG protein, however, which suggests that ethylene is required for proper translation of PG RNA or accumulation of the protein. In any case, the antisense plants demonstrated that tomato fruit ripening involved molecular events that are ethylene dependent as well as those that are not.

The control of tomato fruit ripening by interfering with ethylene production in other ways has also been successful. For example, another approach that has been used is the introduction of a bacterial gene into tomato which encodes an enzyme that degrades ACC (Klee et al. 1991). The bacterial gene encoding ACC deaminase was expressed under the control of the CaMV 35S promoter (35S:ACC deaminase) such that ACC deaminase was expressed throughout the tomato plant, including the ripening fruit. The effect of the antisense construct was to delay ripening in the tomato fruit, but there was very little other consequence to the plant, which indicates that ethylene may have a negligible contribution to the development of the plant (excluding such functions as senescence, etc.).

Other strategies to control or delay ripening in tomato have been developed, including interference with the expression of the gene that encodes polygalacturonase (PG), which is an enzyme involved in tomato fruit softening. Polygalacturonides are major polysaccharides in the tomato cell wall and PG activity is expressed at high levels during fruit ripening. Copy DNAs for PG were cloned from tomato, and cDNA constructs were introduced in antisense orientation in an effort to reduce the levels of PG expression (Sheehy et al. 1988; Smith et al. 1988). The effectiveness of antisense constructs in controlling ripening varied among different transgenic lines; however, commercially successful lines such as the FLAVR SAVR tomato were developed in which antisense expression improves field-holding and postharvest qualities (Kramer and Redenbaugh 1994). The FLAVR SAVR tomato was the first transgenic crop approved for sale in the United States.

REFERENCES

Alpert, K. B. and Tanksley, S. D. 1996. High-resolution mapping and isolation of a yeast artificial chromosome contig containing *fw2.2:* A major fruit weight quantitative trait locus in tomato. Proc. Natl. Acad. Sci. USA 93: 15503–15507.

Bhave, M. R., Lawrence, S., Barton, C., and Hannah, L. C. 1990. Identification and molecular characterization of *shrunken-2* complementary DNA clones of maize. Plant Cell 2: 581–588.

Brink, R. A. and Cooper, D. C. 1947. The endosperm in seed development. Bot. Rev. 13: 423–541.

Charlton, W. L., Keen, C. L., Merriman, C., Lynch, P., Greenland, A. J., and Dickinson, H. G. 1995. Endosperm development in *Zea mays:* Implication of gametic imprinting and paternal excess in regulation of transfer layer development. Development 121: 3089–3097.

Comai, L. and Harada, J. J. 1990. Transcriptional activities in dry seed nuclei indicate the timing of the transition from embryogeny to germination. Proc. Natl. Acad. Sci. USA 87: 2671–2674.

Dure, L. 1985. Embryogenesis and gene expression during seed formation. Oxford Surv. Plant Mol. Cell Biol. 2: 179–197.

Finkelstein, R. R., Tenmbarge, K. M., Shumway, J. E., and Crouch, M. L. 1985. Role of ABA in maturation of rapeseed embryos. Plant Physiol. 78: 630–636.

Fong, F., Smith, J. D., and Koehler, D. E. 1983. Early events in maize seed development. Plant Physiol. 52: 350–356.

George, W., Scott, J., and Spilttstoesser, W. 1984. Parthenocarpy in tomato. Hort. Rev. 6: 65–84.

Gillaspy, G., Ben-David, H., and Gruissem, W. 1993. Fruits: A developmental perspective. Plant Cell 5: 1439–1451.

Giraudat, J., Hauge, B. M., Valon, C., Smalle, J., Parcy, F., and Goodman, H. M. 1992. Isolation of the Arabidopsis *abi3* gene by positional cloning. Plant Cell 4: 1251–1261.

Goldberg, R. B., Barker, S. J., and Perez-Grau, L. 1989. Regulation of gene expression during plant embryogenesis. Cell 56: 149–160.

Guiltinan, M. J., Marcotte, W. R., and Quatrano, R. S. 1990. A plant leucine zipper protein that recognizes an abscisic acid response element. Science 250: 267–271.

Hattori, T., Vasil, V., Rosenkrans, L., Hannah, L. C., Mccarty, D. R., and Vasil, I. K. 1992. The *viviparous-1* gene and abscisic acid activate the c1 regulatory gene for anthocyanin biosynthesis during seed maturation in maize. Genes Dev 6: 609–618.

Hobson, G. and Davies, J. 1970. The tomato. In The Biochemistry of Fruits and Their Products, ed. Hulme, A. C., pp. 437–482. London: Academic Press.

Hueros, G., Varotto, S., Salamini, F., and Tompson, R. D. 1995. Molecular characterization of *BET1*, a gene expressed in the endosperm transfer cells of maize. Plant Cell 7: 747–757.

Johnston, S. A. and Hanneman, R. E., Jr. 1982. Manipulations of endosperm balance number overcome crossing barriers between diploid *Solanum* species. Science 217: 446–448.

Kermicle, J. L. and Alleman, M. 1993. Genetic imprinting in maize in relation to the angiosperm life cycle. Development (Suppl.): 9–14.

King, R. W. 1976. Abscisic acid in developing wheat grains and its relationship to grain growth and maturation. Planta 132: 43–51.

Klee, H. J., Hayford, M. B., Kretzmer, K. A., Barry, G. F., and Kishore, G. M. 1991. Control of ethylene synthesis by expression of a bacterial enzyme in transgenic tomato plants. Plant Cell 3: 1187–1194.

Kodrzycki, R., Boston, R. S., and Larkins, B. A. 1989. The *opaque-2* mutation of maize differentially reduces *zein* gene transcription. Plant Cell 1: 105–114.

Koornneef, M., Reuling, G., and Karssen, C. M. 1984. The isolation and characterization of abscisic acid-insensitive mutants of *Arabidopsis.* Physiol. Plant. 61: 377–383.

Kramer, M. G. and Redenbaugh, K. 1994. Commercialization of a tomato with an antisense polygalacturonase gene: The FLAVR SAVR-TM tomato story. Euphytica 79: 293–297.

Lanahan, M. B., Yen, H. C., Giovannoni, J. J., and Klee, H. J. 1994. The never ripe mutation blocks ethylene perception in tomato. Plant Cell 6: 521–530.

Lin, B.-Y. 1984. Ploidy barrier to endosperm development in maize. Genet. 107: 103–115.

Lopes, M. A. and Larkins, B. A. 1993. Endosperm origin, development, and function. Plant Cell 5: 1383–1399.

Marcotte, W. R., Bayley, C. C., and Quatrano, R. S. 1988. Regulation of a wheat promoter by abscisic acid in rice protoplasts. Nature 335: 454–457.

McCarty, D. R., Carson, C. B., Stinard, P. S., and Robertson, D. S. 1989. Molecular analysis of *viviparous-1,* an abscisic acid-insensitive mutant of maize. Plant Cell 1: 523–532.

McCarty, D. R., Hattori, T., Carson, C. B., Vasil, V., Lazar, M., and Vasil, I. K. 1991. The *viviparous-1* developmental gene of maize encodes a novel transcriptional activator. Cell 66: 895–906.

McClintock, B. 1978. Development of the maize endosperm as revealed by clones. In The Clonal Basis of Development, ed. Subtelny, S. and Sussex, I. M., pp. 418–471. London: Academic Press.

Meinke, D. W., Franzmann, L. H., Nickle, T. C., and Yeung, E. C. 1994. Leafy cotyledon mutants of Arabidopsis. Plant Cell 6: 1049–1064.

Nambara, E., Keith, K., McCourt, P., and Naito, S. 1995. A regulatory role for the *ABI3* gene in the establishment of embryo maturation in *Arabidopsis thaliana.* Development 121: 629–636.

Neill, S. J., Horgan, R., and Parry, A. D. 1986. The carotenoid and abscisic acid content of viviparous kernels and seedlings of *Zea mays L.* Planta 169: 87–96.

Oeller, P. W., Min Wong, L., Taylor, L. P., Pike, D. A., and Theologis, A. 1991. Reversible inhibition of tomato fruit senescence by antisense RNA. Science 254: 437–439.

Parcy, F., Valon, C., Raynal, M., Gaubier Comella, P., Delseny, M., and Giraudat, J. 1994. Regulation of gene expression programs during Arabidopsis seed development: Roles of the *ABI3* locus and of endogenous abscisic acid. Plant Cell 6: 1567–1582.

Picton, S., Gray, J. E., and Grierson, D. 1995. Ethylene genes and fruit ripening. In Plant hormones: Physiology, biochemistry and molecular biology, ed. Davies, P., pp. 372–394. Dordrecht: Kluwer Academic.

Picton, S. J., Barton, S. L., Bouzayen, M., Hamilton, A. J., and Grierson, D. 1993. Altered fruit ripening and leaf senescence in tomatoes expressing an antisense ethylene-forming enzyme transgene. Plant J. 3: 469–481.

Raven, P. H., Evert, R. F., and Eichhorn, S. E. 1986. Biology of Plants. New York: Worth Publishers.

Robichaud, C. S., Wong, J., and Sussex, I. M. 1980. Control of *in vitro* growth of viviparous embryo mutants of maize by abscisic acid. Develop. Genet. 1: 325–330.

Schmidt, R. J., Burr, F. A., Aukerman, M. J., and Burr, B. 1990. Maize regulatory gene *opaque-2* encodes a protein with a leucine-zipper motif that binds to zein DNA. Proc. Natl. Acad. Sci. USA 87: 46–50.

Schmidt, R. J., Ketudat, M., Aukerman, M. J., and Hoschek, G. 1992. Opaque-2 is a transcriptional activator that recognizes a specific target site in 22-kd zein genes. Plant Cell 4: 689–700.

Schnarf, K. 1929. Embryologie der Angiosprermen. In: Handbuch der Pflanzenanatomie II, part 2, pp. 321–372, Gebrüder Borntraeger, Berlin.

Sheehy, R. E., Kramer, M., and Hiatt, W. R. 1988. Reduction of polygalacturonase activity in tomato fruit by antisense RNA. Proc. Natl. Acad. Sci. USA 85: 8805–8809.

Smith, C.J.S., Watson, C. F., Ray, J., Bird, C. R., Morris, P. C., Schuch, W., and Grierson, D. 1988. Antisense RNA inhibition of polygalacturonase gene expression in transgenic tomatoes. Nature 334: 724–726.

Theologis, A., Oeller, P. W., Wong, L. M., Rottmann, W. H., and Gantz, D. M. 1993. Use of a tomato mutant constructed with reverse genetics to study fruit ripening a complex developmental process. Dev. Genet. 14: 282–295.

Thorne, J. H. 1985. Phloem unloading of C and N assimilates in developing seeds. Ann. Rev. Plant Physiol. 36: 317–343.

Unger, E., Parsons, R. L., Schmidt, R. J., Bowen, B., and Roth, B. A. 1993. Dominant negative mutants of *opaque2* suppress transactivation of a 22-kd zein promoter by *opaque2* in maize endosperm cells. Plant Cell 5: 831–841.

Vasil, V., Marcotte, W. R., Jr., Rosenkrans, L., Cocciolone, S. M., Vasil, I. K., Quatrano, R. S., and McCarty, D. R. 1995. Overlap of Viviparous1 (VP1) and abscisic acid response elements in the Em promoter: G-box elements are sufficient but not necessary for VP1 transactivation. Plant Cell 7: 1511–1518.

Walbot, V. 1996. Sources and consequences of phenotypic and genotypic plasticity in flowering plants. Trends Plant Sci. 1: 27–32.

West, M.A.L.W., Yee, K. M., Danao, J., Zimmerman, J. L., Fischer, R. L., Goldberg, R. B., and Harada, J. J. 1994. *LEAFY COTYLEDON1* is an essential regulator of late embryogenesis and cotyledon identity in Arabidopsis. Plant Cell 6: 1731–1745.

Wilkinson, J. Q., Lanahan, M. B., Yen, H. C., Giovannoni, J. J., and Klee, H. J. 1995. An ethylene-inducible component of signal transduction encoded by *Never-ripe*. Science 270: 1807–1809.

12

Root Development

Roots, like shoots, are tip-growing structures that grow in length through the action of tip meristems. The differences in form between roots and shoots are determined by their meristems. Unlike shoots, root apical meristems (RAMs) do not produce lateral organs (Fig. 12.1). Primary roots produce lateral roots, but lateral roots do not arise from the root tip; rather, lateral roots are outgrowths from more mature regions of the root. Unlike the shoot apical meristem (SAM), the RAM elaborates a root cap, which is a structure that covers the root tip and protects it as it pushes through the soil. The root cap is made up of cells that continually slough off as the root grows.

Like the SAM, the RAM is organized both radially and longitudinally (along the root axis). The radial organization of the root can be seen in either transverse or longitudinal sections (Fig. 12.2A and B). Roots are constructed from concentric layers of cells. In Arabidopsis, the layers are one cell thick near the root tip. In longitudinal projections, these layers appear as concentric cylinders composed of parallel rows of cell files that are oriented along the root axis. The root grows in length by transverse division and elongation of cells in different zones of the root tip. Behind the root cap, the root tip is organized along its length into successive zones of cell division, cell elongation, and cell specialization (Fig. 12.1). Although the zones are local regions of specialized cellular activity, the zones exist because cell division, elongation, and specialization occur as successive events in the life of a root cell. Newly divided cells in the "zone of cell division" do not elongate immediately; rather, they lengthen when the tip has grown a short distance away (a few hundred microns in Arabidopsis). These cells expand in what constitutes the "zone of cell elongation." After they have elongated, the cells begin to differentiate in the "zone of cell specialization," where epidermal cells produce root hairs.

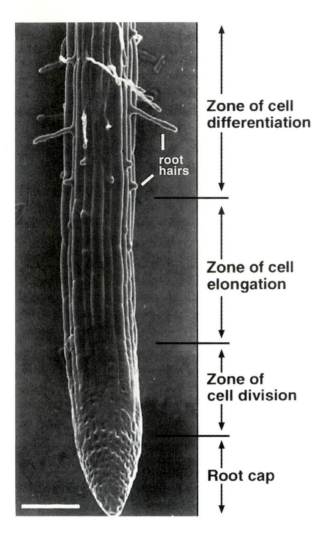

Zone of cell differentiation

root hairs

Zone of cell elongation

Zone of cell division

Root cap

FIGURE 12.1

Root tip of Arabidopsis. Specialized zones along the length of the root are indicated. Root hairs are formed in the zone of cell differentiation. Tip is covered by a cap; however, most of the lateral root cap has been sloughed off. Visualized by scanning electron microscopy. Bar = 100 µm. (From Dolan, L., et al. 1993. Cellular organisation of the *Arabidopsis thaliana* root. Development 119:74. Reprinted by permission of the Company of Biologists Ltd.)

THE ROOT APICAL MERISTEM AND THE PROMERISTEM

RAMs are quite simple in the aquatic fern (*Azolla*). A single large apical cell at the center of the root tip produces, by divisions at its various faces, new cells that make up both the root cap and root. RAMs are more complicated in higher plants with larger roots. For example, in radish cells of the RAM are organized in layers, and this pattern is referred to as a "closed"-type meristem (Clowes 1981). In pea (*Pisum sativa*), there are no obvious layers, and the RAM in pea is described as an "open" type.

In either case, RAMs are complicated because they contain both the cells that serve as initials for the various cell layers and the dividing cells involved in the elongation of the root. To simplify our understanding of the meristem, Clowes (1954) proposed the promeristem concept. The

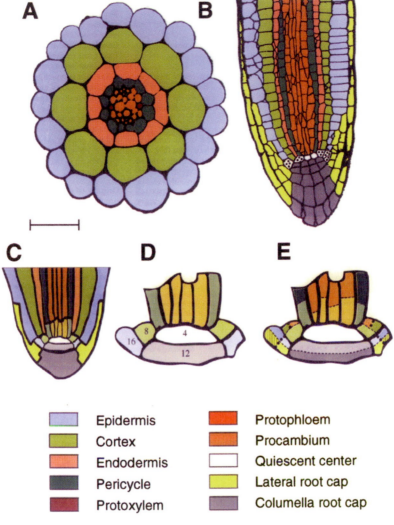

FIGURE 12.2

Arabidopsis root apical meristem (RAM) and promeristem region. Sections were stained with antipectin antibodies to visualize cell walls by fluorescent light microscopy. The image was reverse printed and colorized to highlight various tissue layers in the root. (A) Transverse section about 1 μm behind root tip and (B) median longitudinal section. Bar = 25 μm. (C,D) Schematic representation at higher magnification of median longitudinal section that shows detail of the promeristem region. (D) Individual cells are not shown; however, initials that give rise to different cell layers are represented as groups of cells, and the number of cells in each group of initials and the quiescent center (central cells) are indicated. (E) Pattern and sequence of cell divisions of cell layer initials. First division (dashed lines), second division (dotted lines), third or radial division (indicated by striped shading). Cell layers are color-coded according to legend. (From Dolan, L., et al., 1993. Cellular organisation of the *Arabidopsis thaliana* root. Development 119:74 and 81. Reprinted by permission of the Company of Biologists Ltd.)

promeristem is the small group of initial cells at the base of the vascular cylinder that gives rise to the radial pattern of the root and the root cap (Fig. 12.2C–E). The promeristem does not include cells in the zone of cell division. Cell divisions in this zone extend cell files but are not involved in forming the radial pattern of the root.

At the heart of the promeristem are cells much like the central mother cells in the SAM. These cells are thought to be the progenitors of all cells in the root, and they constitute a structure called the **quiescent center,** which is a group of slowly dividing cells flanked by more rapidly dividing initials. The doubling time for cells in the quiescent center in maize is less than 170 hours, whereas that for root cap initials is only about 12 hours (Clowes 1954). In Arabidopsis, Dolan et al. (1993) were unable to find any divisions at all in the quiescent center. They found instead high rates of division in the zone of cell division that lies behind the quiescent center and only modest rates of division in the initials adjacent to the central cells.

Thus, paradoxically, the cells that are thought to be the ultimate progenitors of cells in the root divide most slowly, whereas their immediate derivatives divide frequently. Given this situation, the question is whether the cells in the quiescent center are indeed true progenitors of cells in the root. Some have argued that the cells at the periphery of the quiescent center are true initials. Cells in the quiescent center, however, can be recruited to grow by damaging the root cap, or by nutrient refeeding of growth-arrested roots in culture. Feldman and Torrey (1976) excised quiescent centers from corn roots and found that they could generate organized roots in culture. The quiescent center, however, is very small and would be difficult to separate from the surrounding initials by microsurgery. For that reason, the matter is still an open question.

CELL FATES AND CELL LINEAGE ANALYSIS

A cell fate map was produced for the RAM of *Azolla*, the water fern, without a formal cell lineage analysis (Gunning et al. 1978). As stated earlier, *Azolla* roots are very small and uniform, and lineage relationships can be inferred from the simple structure of the root and the stereotypic patterns of cell division. *Azolla* roots have exactly fifty-six cells in cross-section, and these cells can be traced back through an orderly pattern of transverse and longitudinal divisions to a single apical cell.

The fate of RAM cells can also be predicted from the structure of the root in Arabidopsis, as was shown in an elegant study by Dolan et al.

(1993). In Arabidopsis, the number and arrangement of cells in the root tip is very uniform, and the ordered files of cells in the root can be easily traced back to a group of progenitor cells in the promeristem. Arabidopsis roots are composed of concentric layers of cells, which surround the central vascular cylinder in constant numbers and invariant pattern (Fig. 12.2A and B). The pericycle, for example, is composed of a ring of twelve cells and is surrounded by an endodermis made up of eight cells. The endodermis is in turn surrounded by a cortex of eight cells, then an epidermal layer of about sixteen cells. The central vascular cylinder or stele has two protophloem elements that are located perpendicular to two protoxylem elements (see Chap. 13 for more information on protoxylem and protophloem). The protoxylem abuts onto pericycle cells that each face two endoderm cells, while the protophloem abuts onto pericycle cells that face only one endoderm cell. The structure of the root cap is also regular. In cross-sections, the columella cells of the cap nearest the root tip are arranged in a pattern of four inner cells surrounded by eight cells, which are in turn surrounded by a single layer of lateral root cap cells composed of sixteen or thirty-two cells in circumference. Further from the tip, the lateral root cap forms one or two additional cell layers.

The highly organized structure of the Arabidopsis root arises from a stereotypic pattern of divisions of the promeristem that can be observed in anatomical studies of the root (Dolan et al. 1993). The Arabidopsis promeristem is composed of three tiers of cells (Fig. 12.2C and D). The root cap columella arises from periclinal divisions of the lower (distal) tier in the promeristem. The lateral root cap is derived from a ring of about sixteen initial cells that surround the columella initials. These initials undergo periclinal divisions to give rise to two different concentric layers of cells: the epidermal and lateral root cap cells (Fig. 12.2E). Divisions in the lateral root cap layer are coordinated with divisions of the root cap columella initials to maintain the cell files in register. Another ring of initials that surrounds the middle tier of cells in the promeristem also gives rise to two concentric layers: the cortical and endodermal cells (Fig. 12.2E). The cells in the center of the middle tier of promeristem cells are quiescent cells and are rarely seen in division (Dolan et al. 1993). The plate of cells in the upper (proximal) tier abut the central vascular cylinder and are initials for the stele (pericycle and vascular bundles).

Because the lateral root cap/epidermis layers and cortex/endodermis layers are derived from a single ring of initial cells, the periclinal cell divisions that produce two cell layers from one ring of initial cells are critical. Dolan et al. (1994) carried out a cell lineage analysis in Ara-

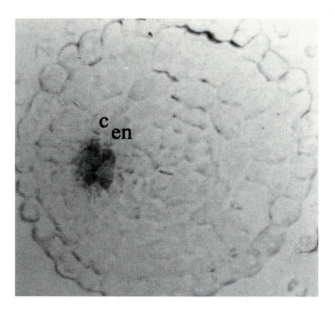

FIGURE 12.3

Sector in Arabidopsis root that results from the excision of an *Ac* transposable element from 35S:*Ac*:GUS construct. Transverse section in the meristematic region has been histologically stained for β-glucuronidase (GUS) activity. A sector of GUS expression spans both the cortical, c, and endothelial, en, cell layers, which indicates that Ac excision occurred in a precursor cell common to both cell layers. Interpretation has been verified by further serial sections toward the tip of the root. (From Dolan, L., et al. 1994. Clonal relationships and cell patterning in the root epidermis of Arabidopsis. Development 120:2468. Reprinted by permission of the Company of Biologists Ltd.)

bidopsis roots to confirm the apparent cell lineage pattern. They introduced a reporter gene marker into transgenic Arabidopsis plants. In the past, efforts to carry out cell lineage analysis in roots had been hampered by the absence of useful cell lineage markers. The transgenic marker in this study was the 35S:*Ac*:GUS construct described in Chapter 2 (Fig. 2.1B). The CaMV 35S promoter is normally expressed in roots, and the excision of the *Ac* transposable element from the construct activates GUS expression in sectors of the root tip.

In general, Dolan et al. (1994) confirmed the cell fates predicted from the anatomical analysis described above. For example, rare sectors spanned both epidermal and lateral root cap cells, which would have been expected if these two cell types were derived from the same initials (Fig. 12.3). (Most sectors, however, involved epidermal and lateral root cap cells exclusively because the probability was very low that an excision event would occur in the few initial cells that are common to both cell lineages.) In addition, rare sectors were found that included both cortical and endodermal cells, which is consistent with the prediction that cortical and endodermal cells are derived from common initials. No sectors were found that included both root cap columella and root cap lateral cells. From the histological analysis described earlier, these cells appeared to be derived from separate initials, and the absence of shared sectors in the cell lineage analysis bears that out.

THE ROLE OF POSITIONAL INFORMATION IN ROOT DEVELOPMENT

From the analysis described earlier, root development appears to be a highly determined process in which cells in the promeristem undergo stereotypic divisions to give rise to well-ordered arrays of cell files in the root. In order to investigate whether the fate of cells is irreversibly determined in the promeristem, microsurgery was performed on root tips to assess their ability to regenerate. Pellegrini (1957) found that if common bean (*Phaseolus*) root tips were surgically split, each side reorganized and produced a complete root. If the split was unequal, then two roots of different sizes were formed. Clowes (1953, 1954) made different incisions in the meristematic regions of a number of plants (broad bean, maize, and rye), and generally found that any part of the quiescent center could regenerate a normal apex.

The problem has been addressed again more recently, with higher spatial resolution. Van den Berg et al. (1995) used laser microbeam irradiation to ablate quiescent center cells in the Arabidopsis root tip (as described in Chap. 2). They found that the dead quiescent center cells were flattened and displaced toward the root tip by adjacent cells from the vascular cylinder. The vascular cells that replaced the quiescent center cells lost their vascular identity and took on a new one characteristic of the quiescent center and root cap cells. (The different cells were recognized by cell-type markers. The cell-type marker for vascular cells was an enhancer-trap construct, whereas the root-cap marker was a construct driven by a promoter containing a root-specific element from the CaMV 35S promoter.) Thus, the root tip was able to reorganize after the destruction of the quiescent center and recast adjacent vascular cells into a new role as quiescent center cells.

Van den Berg et al. (1995) also used laser cell ablation to determine whether contacts with adjacent cells control the critical periclinal divi-

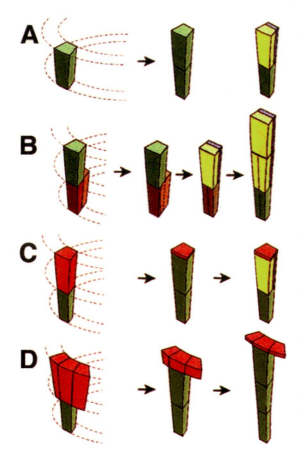

FIGURE 12.4

Schematic drawings of laser cell ablation experiments performed on Arabidopsis roots. (A) Cortical/endodermal initial, in green, in one of eight cell files in the Arabidopsis root. Initial undergoes a transverse division generating a daughter cell. Daughter cell then undergoes a periclinal division, which gives rise to cells in two cell layers: an inner endodermal cell, in green, and an outer cortical cell, in yellow. (B) Laser ablation of cortical/endodermal initial, red. Dead initial cell, red, is crushed and pushed to the outside and replaced by periclinal division of neighboring pericycle cell, light red. Replacement cell acts like cortical/endodermal initial. (C) Laser ablation of cortical/endodermal daughter cell, in red. Ablated cell is crushed. Ablation has no effect on divisions of cortical/endodermal initial, green. (D) Laser ablation of three cortical/endodermal daughter cells, red, in contact with cortical/endodermal initial, green. Initial undergoes normal transverse division, but does not carry out periclinal division to produce two cell files. (From van den Berg, C., et al. 1995. Cell fate in the Arabidopsis root meristem determined by directional signalling. Nature 378:65. Reprinted by permission of the author and Nature.)

sion in the cortical/endodermal cell initial that gives rise to the two cell files. They found that when three cortical daughter cells that abut onto the initial were ablated, the initial did not divide periclinally to produce both the cortical and endodermal files; rather, the initial generated a single cell file instead (Fig. 12.4D). They found that ablation of a single cortical daughter cell (Fig. 12.4C) or a different group of three cells (composed of endodermal, cortical, and pericycle daughter cells) adjacent to the initial had no such effect. It was concluded that information for the critical periclinal division in the cortical/endodermal cell initial was derived from cortical daughters cells in the same radial layer as the initial.

Thus, the cell ablation experiments tell us that despite its rigid organization and stereotypic patterns of division, the root is capable of considerable regulative development. The findings also raise a classic chicken-and-egg problem – does the promeristem dictate the structure of the root or does the root impose specific patterns of division on the promeristem?

THE ORIGIN OF THE PRIMARY
ROOT DURING EMBRYOGENESIS

Scheres et al. (1994) examined the anatomy of the embryo in an effort to determine the origin of the promeristem of the primary root. During the early globular stage, the embryo is divided into an upper tier and lower tier of cells, which are derivatives from the first transverse division of the initial proembryo cell (Fig. 12.5A). The lower tier is further partitioned by transverse divisions into an upper lower tier (ult) and a low lower tier (llt). The llt tier gives rise to the embryonic axis (i.e., the hypocotyl and root). At the late heart stage, the llt is further subdivided, and the most basal protoderm cells become RAM initials that can be recognized by periclinal divisions that give rise to the lateral root cap.

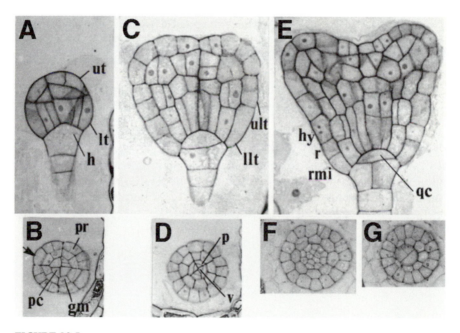

FIGURE 12.5

Radial organization of the root-forming region of the Arabidopsis embryo. Median longitudinal sections through the Arabidopsis embryo at the (A) globular stage, (C) early heart stage, (E) mid–early heart stage are shown. Transverse section through lower tier, lt, of cells in the (B) globular embryo and at the (D) early heart stage embryo. Transverse section through (F) root, r, and (G) root meristem initials in early–mid heart stage embryo. Ground meristem, gm; hypophysis, h; pericycle, p; procambium, pc; protoderm, pr; quiescent center, qc; root, r; root meristem initials, rmi; vascular primordium, v. Sections were stained with Astra blue and viewed in the light microscope. Bar = 50 μm. (From Scheres, et al. 1995. Mutations affecting the radial organisation of the Arabidopsis root display specific defects throughout the embryonic axis. Development 121:55. Reprinted by permission of the Company of Biologists Ltd.)

FIGURE 12.6

Schematic representation of alternating anticlinal and periclinal divisions in the embryonic axis that gives rise to the promeristem in the root region of Arabidopsis. Periclinal divisions produce concentric rings of cells, and anticlinal divisions subdivide the rings, increasing the numbers of cells. Thick lines represent most recent divisions. Ordered divisions account for the uniform pattern of cells and the constant number of cells in each tissue layer. (From Scheres, B., et al. 1995. Mutations affecting the radial organisation of the Arabidopsis root display specific defects throughout the embryonic axis. Development 121:56. Reprinted by permission of the Company of Biologists Ltd.)

Formation of the lateral root cap layer, which is the hallmark of RAM activity in the embryo, is first seen at this stage. The early lateral root cap is composed of a single layer, but periclinal divisions at later stages give rise to additional layers.

Arabidopsis roots have a well-ordered structure, and, as described earlier, the orderliness can be attributed largely to regular divisions in the promeristem. The promeristem, which forms during embryonic development, is itself produced from a series of regular periclinal and anticlinal divisions in the embryo (Scheres et al. 1995). Periclinal divisions in the promeristem progenitor cells are responsible for the formation of radial rings of initials in the promeristem, and anticlinal divisions determine the number of cells in each ring (Fig. 12.6). The orientation and number of these divisions must be highly regulated to generate constant numbers of layers and numbers of cells in each ring. At the globular stage, transverse sections of the lower tier of cells in the embryo show three concentric rings of cells that each consist of four procambium cells surrounded by eight ground meristem cells, which are in turn surrounded by sixteen protoderm cells (Fig. 12.5B). By the early heart stage, periclinal divisions in the procambium create separate initials for the future pericycle and vascular primordium (Fig. 12.5D). Similar divisions in the ground meristem give rise to the rings of cortical/epidermal initials, and divisions in the protoderm produce lateral root cap/epidermal initials with characteristic numbers of cells (Fig. 12.5F and G).

The root promeristem is derived from cells of the embryo proper and

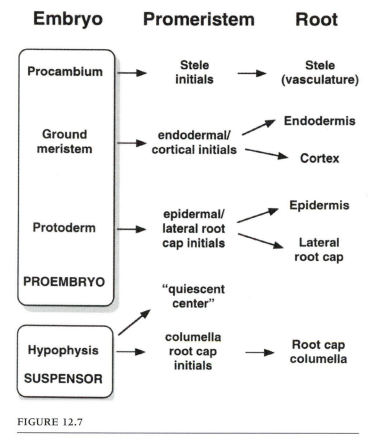

Embryo   **Promeristem**   **Root**

FIGURE 12.7

Origin of cells in various cell layers of the root. Cell lines are traced back through the promeristem to the embryo. Cells are derived in the embryo from both the proembryo and the suspensor.

the suspensor (Figs. 12.5 and 12.7). The central cells of the middle layer and columella initials are derived from the hypophysis, which is the upper cell of the suspensor. The hypophysis divides, giving rise to a distinct lenticular cell from which the quiescent center (central cells) arises, as well as a lower cell that divides to form columellar initials. The radial organization of initials in the hypophyseal region is produced through divisions much like those described for root initials in the proembryo. Is the dual origin of cells derived from the proembryo and suspensor required for the formation of a normal root? In one sense, it does not seem so. Adventitious roots with normal promeristems form from vegetative tissue, and lateral RAMs organize from the pericycle. Thus, the dual origin of the promeristem may have more to do with the organization of the embryo than with the requirements for the formation of the root promeristem.

LATERAL ROOTS

Plants produce different classes of roots that are distinguished by their site of origin in the plant and/or by the time when they are produced in development. Primary roots are formed in the embryo at the base of the hypocotyl. Lateral roots are formed from the primary root but not at the root tip by the RAM. In what seems like an afterthought, lateral roots emerge instead from the pericycle at some distance from the root tip (Fig. 12.8A and B). The lateral root primordia bulge out at predicted circumferential positions, which are usually adjacent to xylem poles. The emerging primordia burrow their way through the endodermis and cortex, and burst through the epidermis (Fig. 12.8C).

During this outgrowth the emerging lateral root primordium organizes a RAM. Lateral roots have meristems that are indistinguishable morphologically and functionally from primary RAMs. Lateral RAMs, however, are not formed in the embryo, like the primary RAM. Lateral roots arise from the pericycle, and a lateral root is derived from approximately ten or eleven pericycle cells (Laskowski et al. 1995). At some point in the outgrowth of the lateral root primordium, therefore, the new promeristem must form. Malamy and Benfey (1997) have chronicled the events that occur in that transition in an attempt to follow the

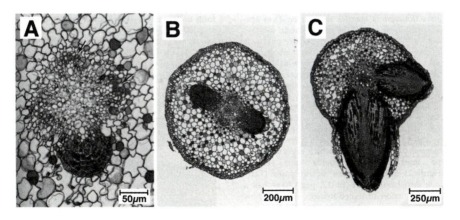

FIGURE 12.8

Lateral root formation in willow (*Salix*). (A) One lateral root primordium is well formed and two are just being initiated from the pericycle (see arrows). (B) Lateral root primordia grow through the cortex. (C) One lateral root primordium has burst through root epidermis and another one is about to emerge. Transverse sections of the primary root viewed by light microscopy. (From Raven, P. H., Evert, R. F., and Eichhorn, S. E. 1986. Biology of Plants, 4th ed. Worth Publishers, New York, p. 409. Reprinted by permission of the author and Worth Publishers, Inc.)

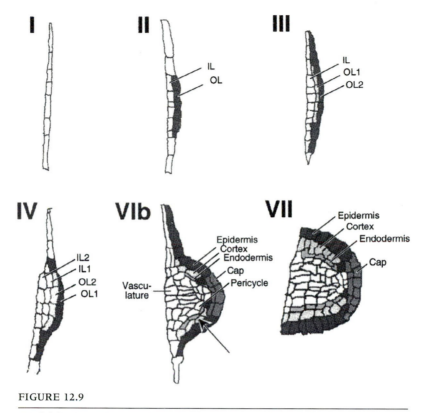

FIGURE 12.9

Model for lateral root primordia development in Arabidopsis. Figures are traced from cell outlines of primordia in actual roots, and the shading indicates cell layers that are defined by cell layer markers. All of the cell layers are present in the primordia by stage VIb. The cluster of unshaded cells toward the tip of the primordium at stages VIb and VII could not be identified by cell layer markers. The position of these cells, however, suggests that they will develop into the quiescent center and initials. Unshaded cells at the base of the primordia could not be identified either. (From Malamy, J. E., and Benfey, P. N. 1997. Organization and cell differentiation in lateral roots of *Arabidopsis thaliana*. Development 124:43. Reprinted by permission of the Company of Biologists Ltd.)

reformation of the various cell layers. To do so they have used cell-layer markers, several of which were derived from "enhancer traps," which are a type of gene trap (see Fig. 5.12). The cell layer markers allow one to trace the development of the cell layers.

Malamy and Benfey (1997) described various stages in the formation of the lateral roots in Arabidopsis and concluded that the process is highly ordered (Fig. 12.9). Stage I is characterized by anticlinal divisions that occur in the pericycle wall. At stage II, periclinal divisions generate two cell layers: an outer layer (OL) and inner layer (IL). In the next two stages, cells in the OL and IL divide periclinally, which again gives rise to four cell layers: OL1, OL2, IL1, and IL2. A number of stereotyped anticlinal and periclinal divisions follow, which results in a stage VIb primordium in which the three outer cell layers can be identified by markers, as well as by the core tissue of the stele and root cap. The root emerges at about stage VII, mostly due to the expansion of cells in the primordium. Most of the cell divisions, thereafter, take place at the apex, which indicates that the tip has acquired meristem function. The cell layer markers indicate that during this process epidermis is derived from OL1, such as is seen in stage III or IV, and the cortex and endodermis from OL2, as in stage III. Cortex and endodermis become separate cell layers through periclinal divisions at stage VIb.

An important transitional stage in the development of the lateral root primordium was identified by Laskowski et al. (1995). They found a stage in which further development of the lateral root primordium in tissue culture became independent of hormones. The development of lateral roots in culture can be induced by the hormone, auxin. Once induced, root primordium grow in culture without the addition of hormones. They found that at early stages of lateral root emergence, when the primordium was only about two cell layers thick, the explanted primordium would not produce roots in tissue culture in the absence of hormones. When the primordium was about three to five cell layers thick, however, it was able to grow in culture without hormones. Laskowski et al. argued that with respect to hormone dependence, formation of the lateral RAM was a two-step process. Formation of the primordium required hormones, but when the meristem organized, root growth was supported in the absence of hormones.

The application of auxin induces lateral root formation and synchronizes the time course of root development. Efforts have been undertaken by a number of investigators to study cell division activities in the early organization of the meristem and to identify genes expressed in actively dividing tissue. One such gene is an A-type cyclin (*cyc1At*), which is expressed at the site of presumptive lateral root formation (Ferreira et al. 1994). Cyclin gene expression was localized by detecting the expression of GUS histochemically in roots of transgenic Arabidopsis plants that bear a *cyc1At* promoter GUS fusion (*cyc1At*:GUS). As discussed in Chapter 5, cyclins are proteins that form complexes with and activate cell division protein kinases (CDPKs). A-type cyclins are S-phase markers, and *cyc1At* expression precedes the periclinal cell divisions that characterize the outgrowths of lateral roots. *Cyc1At* expression occurs at presumptive lateral root formation sites, even if the divisions are blocked by the mitotic poison, oryzalin.

GENETICS OF ROOT DEVELOPMENT

A number of interesting root mutants have been described in Arabidopsis. Root phenotypes can easily be identified because seedlings grow on agar medium, where their roots are visible. RAM mutants were sought by looking for seedlings with roots that undergo little or no cell division in the roots following germination (Cheng et al. 1995). Mutants were found that were called *root meristemless1* and *2* (*rml1* and *rml2*) and had very short primary roots (formed mostly by the expansion of cells in the embryonic root). These mutants were derived from embryos with normal embryonic roots and, therefore, differed from the embryonic mutants such as *monopteros* (Chap. 3), in which the basal part of the embryo and primary root were missing. The mutants had normal cell layer patterns in the root, and *rml1* initiated lateral roots that failed to elongate, much like the primary root. (Both mutants also produced unexplained nodulelike structures on their roots.) The *rml* mutants appear to be defective in functions required for cell division in the postembryonic root and not in other parts of the plant because embryos, shoots, and callus appeared to grow normally in culture.

Other interesting mutants that affect the cell layer pattern of the root have been selected simply on the basis of slow root growth (Benfey et al. 1993; Scheres et al. 1995). For example, roots on *short root* (*shr*) fail to maintain their ability to elongate because the zone of cell division degenerates or is somehow lost. In addition, the endodermal cell layer is completely missing (Fig. 12.10B), as is the Casparian strip, which is a suberized ring that delineates the boundary between the endodermis and the inner vascular cylinder. The endodermis is missing because the cortical/endodermal initials do not undergo the critical periclinal cell divisions that give rise to two cell layers (Scheres et al. 1995). The resulting single cell layer differentiates into cortical, not endodermal, cells. The lesion in *shr* incapacitates the RAM in some way so that it loses its generative ability.

To date, mutants in five complementation groups (including *shr*) have been identified that affect various root cell layers (Scheres et al. 1995). *Scarecrow* (*scr*) and *pinocchio* (*pic*) mutants are like *shr* in that the cortical/endodermal layers are only a single cell layer. In *pic*, it appears that the cortical layer is missing. In *gollum* (*glm*) and *wooden leg* (*wol*) (Fig. 12.10C), vascular tissue is affected. *Fass* (*fs*) mutants have enlarged roots with supernumerary cell numbers seen in cross-section (Fig. 12.10D). The vascular cylinder is enlarged in *fs* roots, and there are multiple cortical cell layers.

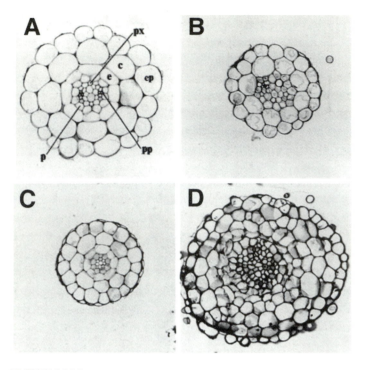

FIGURE 12.10

Mutants that affect the development of cell layers in Arabidopsis roots. Transverse sections through roots: (A) wild type; (B) *short root* (*shr*), which lacks an endodermis layer; (C) *wooden leg* (*wol*) has reduced vascular development; and (D) *fass* (*fs*) has supernumerary cell layers. Epidermis, ep; cortex, c; endodermis, e; pericycle, p; protophloem, pp; protoxylem, px. Sections stained with toluidine blue and viewed by light microscopy. (From Scheres, B., et al. 1995. Mutations affecting the radial organisation of the Arabidopsis root display specific defects throughout the embryonic axis. Development 121:57. Reprinted by permission of the Company of Biologists Ltd.)

Because *shr, scr,* and *pic* are missing radial cell layers that are derived from the ground meristem (either endodermal or cortical cell layers), it was reasoned by Scheres et al. (1995) that these mutants might have defects in embryonic development associated with the formation of the promeristem. In early heart stage embryos, it was found that the cells in the ground meristem of *shr, scr,* and *pic* mutants did not form two cell layers by periclinal divisions, as they did in wild type. In *glm* and *wol* mutants, defects were also observed in procambium formation in embryonic development. Thus, it appears that these root mutants with defects in formation of various cell layers have comparable problems in embryo development. It was quite surprising, therefore, that Scheres et al. (1995) found the same defects in lateral roots produced by the mutants. Lateral roots are derived from the pericycle, not from various embryonic tissues, as is the primary root. Lateral roots in the mutants nevertheless showed the same defects.

An interesting question about the cell layer mutants is whether they are defective in the production of a cell layer (or cell type) or in the identification of cells in the missing layer. The problem was addressed by generating double mutants made by crossing *fs* with either *scr* or *shr* (Scheres et al. 1995). Supernumerary cells and extra cell layers are produced in *fs* (12.10D); therefore, it might be predicted that *fs* would restore the normal phenotype of a mutant, as long as the defect was in the formation of the cell layer and not in the identification of cells in the layer. In *scr fs* double mutants, *fs* restored both cell layers missing in the *scr* mutant. *Scr* therefore appears to be defective in the production of the cortical cell layer, but not in the development of cortical cells, as long as the cell layer is formed. On the other hand, multiple cell layers were produced in *shr fs* double mutants by the action of *fs*, but the endodermal cell layer was not restored as indicated by the absence of the Casparian strip in presumptive double mutants. Hence, the defect in *shr* affects cell layer specification, not cell layer production.

In *scr*, the single layer between the epidermal and pericycle cell layers has attributes of both cortical/endodermal layers; therefore, it is not clear which layer is missing in the mutant. The Casparian strip characteristic of endodermal cells was present in the single cell layer. The cell layers, however, also stained with labeled monoclonal antibodies that were specific for either the cortex or the endodermis. Di Laurenzio et al. (1996) cloned the *SCR* gene and used it as a probe to detect its expression by *in situ* hybridization. It was found that *SCR* is expressed in the cortical/endodermal initials in endodermal, but not cortical cell files. *SCR*, therefore, is required for the critical periclinal divisions that separate the endodermal/cortical initials into two cell files, and the ability to express *SCR* partitions into the endodermal cell file. *SCR* appears to encode a transcription factor which suggests that it may have interesting regulatory functions. The issue remains unresolved about how *SCR* controls the critical periclinal division that partitions its own ability for expression in the endodermal cell line.

Another interesting pattern formation root mutant in Arabidopsis is *hobbit* (*hbt*). Hobbit has defects in determining the identity of root cap initials, which are cells derived from the hypophysis (Scheres et al. 1996). The mutants do not establish an active RAM, which indicates the importance of the root cap and root cap initials in RAM function. Because root cap initials are derived from the hypophysis and because *hbt* is defective in a hypophyseal function, does this mean that the dual origin of the promeristem from proembryo and hypophyseal cells is essential for primary root formation? Probably not. *Hbt* mutants have the same problem in the development of secondary roots that are not derived from hypophyseal progenitors; therefore, *hbt* is unable to stamp an identity on its root cap initials whether they are derived from hypophyseal cells or not.

Root growth mutants (not meristem or pattern mutants) have also been described in Arabidopsis, and one, *superroot* (*sur1*), produces an overabundance of lateral and adventitious roots (Boerjan et al. 1995). *Sur1* overproduces both free and conjugated auxins, and the effects of the mutant could be phenocopied (mimicked) by the application of auxin to wild type plants.

ROOT HAIRS

Root hairs are produced on epidermal cells in the zone of cell differentiation. In Arabidopsis, a root hair forms from a bulge at the distal end of an epidermal cell shortly before the cell reaches the zone of cell differentiation. The root hair grows in length when the epidermal cell stops elongating. Epidermal cells are arranged in cell files in roots. In Arabidopsis, root hairs appear on cells in alternating cell files (Dolan et al. 1994) (Fig. 12.1 and 12.12A). Root hair–forming cells are called *trichoblasts,* and files of trichoblasts are separated from each other by one or two files of non-root hair–forming, or *atrichoblast,* cells. Trichoblasts and atrichoblasts can be distinguished even before root hairs are formed. Trichoblasts are more cytoplasmically dense and do not elongate as much as atrichoblasts. The differentiation of epidermal cells into trichoblasts appears to depend on the position of trichoblasts with respect to other cells. Trichoblasts are found in cell files that lie over the anticlinal cell walls of the underlying cortical cell files (see Fig. 12.2A). From this it has been argued that the alternating pattern of trichoblast cell files is dictated by positional information. Tanimoto et al. (1995) proposed that the pattern may be due to the localized production of ethylene at the anticlinal wall boundaries in cortical cells. The argument was made because Arabidopsis *ctr1* mutants, which respond constitutively to ethylene (see Chap. 4), produce root hairs in all epidermal cell files. In wild type roots, the normal formation of root hairs in presumptive trichoblasts is prevented by ethylene inhibitors, which suggests that the local production of ethylene may be the positional information required for root hair formation.

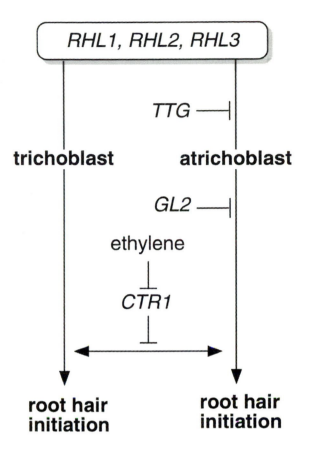

FIGURE 12.11

Scheme for the control of root hair initiation in Arabidopsis. *ROOT HAIRLESS (RHL1-RHL3)* genes promote root hair initiation in all epidermal cell files of the root. Their action, however, is suppressed in atrichoblast cells by the action of *TRANSPARENT TESTA GLABRA (TTG)* and *GLABROUS2 (GL2)*. *TTG* appears to suppress all aspects of atrichoblast differentiation, and *GL2* affects only root hair initiation. Ethylene stimulates root hair initiation as do mutations in *CONSTITUTIVE ETHYLENE RESPONSE1 (CTR1)*. (Activation indicated by an arrow, ↓, while inactivation is represented by ⊥.)(Based on Schneider, K., Wells, B., Dolan, L., and Robert, K. 1997. Structural and genetic analysis of epidermal cell differentiation in Arabidopsis primary roots. Development 124: 1789–1798.)

Arabidopsis mutants have been identified with defects in atrichoblast differentiation and/or root hair initiation. The effects of these mutants, along with that of ethylene, have been described in a model of root hair initiation (Fig. 12.11) (Schneider et al. 1997). Mutants, such as *glabrous2 (gl2)* and *transparent testa glabra (ttg)*, have exceptionally hairy roots because nearly all the epidermal cells in these roots develop root hairs (Galway et al. 1994; Masucci et al. 1996). These two loss-of-function mutations interfere with the differentiation of hairless cells, which indicates that the normal genes suppress root hair initiation in atrichoblasts. (The mutants *gl2* and *ttg* also affect trichome formation on leaves, but in an opposite way. Both eliminate or diminish trichome formation.) The two mutations appear to act at different stages in atrichoblast development (Fig. 12.11). Atrichoblast cells in *gl2* retain most other characteristics of the hairless cell type. On the other hand, cells in files that would normally differentiate in atrichoblasts are completely converted to trichoblasts in *ttg* mutants. From the characteristics of the mutants, it was suggested that *TTG* may be a general regulator of atri-

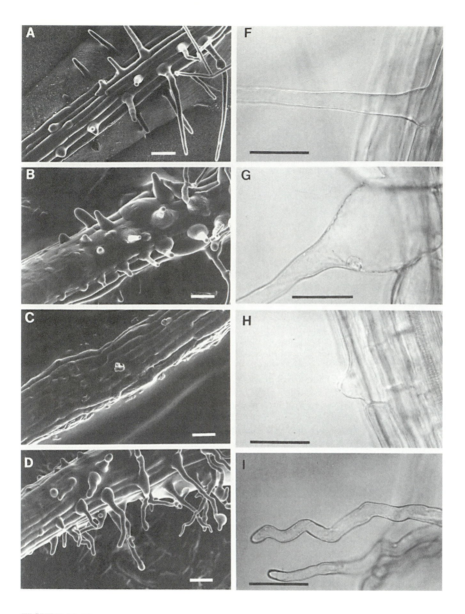

FIGURE 12.12

Root hair mutants in Arabidopsis. (A, F) Wild type. (B, G) *Root hair development1* (*rhd1*) mutant. Root hairs bulge at the base, and the mutation appears to affect the amount of epidermal cell expansion during root hair initiation. (C,H) *Rhd2* mutant. Mutation affects root hair elongation. (D,I) *Rhd3* mutant. Mutant root hairs are wavy, and the mutation appears to affect symmetrical growth at the root hair tip. Root hairs visualized by scanning electron microscopy (A–D) or by light microscopy (F–I). Bar = 50 μm. (From Schiefelbein, J. W., and Sommerville, C. 1990. Genetic control of root hair development in *Arabidopsis thaliana*. Plant Cell 2:235–243 and reprinted in Bowman, J. 1994. Arabidopsis: An Atlas of Morphology and Development. Springer-Verlag, New York, p. 117. Reprinted by permission of Springer-Verlag.)

choblast cell development, whereas *GL2* may only regulate root hair initiation.

GL2 has been cloned and the sequence predicts a homeodomain-containing transcription factor (Rerie et al. 1994). In wild type plants, *GL2* is preferentially expressed in atrichoblasts (the cell files in which root hair formation is suppressed), as demonstrated by *in situ* hybridization and by expression of the *GL2* promoter linked to the GUS reporter gene (*GL2*:GUS). The highest level of *GL2* expression was found in atrichoblasts in the zone of cell elongation. *TTG* has not been cloned, but *ttg* mutations in Arabidopsis can be functionally complemented by the maize *R* gene driven by the CaMV 35S promoter (35S:*R*) (Lloyd et al. 1992), which suggests that *TTG*, like *R*, encodes a transcription factor.

Another group of mutants that block root hair initiation are root hairless mutants (*rhl1–rhl3*) (Schneider et al. 1997). These mutants are loss-of-function mutations in genes that are required for root hair formation (Fig. 12.11). It is argued that these genes operate in both trichoblasts and atrichoblasts, and that their root hair–promoting effects in atrichoblasts are normally offset by the action of the downstream suppressors. Ethylene only weakly rescues the defects in the *rhl* mutants, which suggests that ethylene acts downstream from the *RHL* genes.

Other mutants largely affecting root hair elongation or root hair structure have been identified by Schiefelbein and Somerville (1990) (Fig. 12.12 B–D and G–I). The mutants belong to twelve complementation groups, and the analysis of epistatic relationships in double mutants has been used to construct a pathway of gene action in root hair elongation.

REFERENCES

Benfey, P. N., Linstead, P. J., Roberts, K., Schiefelbein, J. W., Hauser, M.-T., and Aeschbacher, R. A. 1993. Root development in Arabidopsis: Four mutants with dramatically altered root morphogenesis. Development 119: 57–70.

Boerjan, W., Cervera, M. T., Delarue, M., Beeckman, T., Dewitte, W., Bellini, C., Caboche, M., Van Onckelen, H., Van Montagu, M., and Inze, D. 1995. *Superroot*, a recessive mutation in Arabidopsis, confers auxin overproduction. Plant Cell 7: 1405–1419.

Bowman, J. 1994. Arabidopsis: An Atlas of Morphology and Development. New York: Springer-Verlag.

Cheng, J. C., Seeley, K. A., and Sung, Z. R. 1995. *RML1* and *RML2*, Arabidopsis genes required for cell proliferation at the root tip. Plant Physiol. 107: 365–376.

Clowes, F.A.L. 1953. The cytogenerative centre in roots with broad columellas. New Phytol. 52: 48–57.

Clowes, F.A.L. 1954. The promeristem and the minimal constructional centre in grass root apices. New Phytol. 53: 108–116.

Clowes, F.A.L. 1981. The difference between open and closed meristems. Ann. Bot. 48: 761–767.

Di Laurenzio, L., Wysock-Diller, J., Malamy, J. E., Pysh, L., Helariutta, Y., Freshour, G., Hahn, M. G., Feldmann, K. A., and Benfey, P. N. 1996. The *SCARECROW* gene regulates an asymmetric cell division that is essential for generating the radial organization of the Arabidopsis root. Cell 86: 423–433.

Dolan, L., Duckett, C. M., Grierson, C., Linstead, P., Schneider, K., Lawson, E., Dean, C., Poethig, S., and Roberts, K. 1994. Clonal relationships and cell patterning in the root epidermis of Arabidopsis. Development 120: 2465–2474.

Dolan, L., Janmaat, K., Willemsen, V., Linstead, P., Poethig, S., Roberts, K., and Scheres, B. 1993. Cellular organisation of the *Arabidopsis thaliana* root. Development 119: 71–84.

Feldman, L. J. and Torrey, J. G. 1976. The isolation and culture *in vitro* of the quiescent center of *Zea mays*. Am J. Bot. 63: 345–355.

Ferreira, P., Hemerly, A. S., de Almeida-Engler, J., Van Montagu, M., Engler, G., and Inze, D. 1994. Developmental expression of the Arabidopsis cyclin gene *cyc1At*. Plant Cell 6: 1763–1774.

Galway, M. E., Masucci, J. D., Lloyd, A. M., Walbot, V., Davis, R. W., and Schiefelbein, J. W. 1994. The *TTG* gene is required to specify epidermal cell fate and cell patterning in the Arabidopsis root. Devel. Biol. 166: 740–754.

Gunning, B.E.S., Hughes, J. E., and Hardham, A. R. 1978. Formation and proliferative cell divisions, cell differentiation and developmental changes in the meristem of *Azolla* roots. Planta 143: 125–144.

Laskowski, M. J., Williams, M. E., Nusbaum, H. C., and Sussex, I. M. 1995. Formation of lateral root meristems is a two-stage process. Development 121: 3303–3310.

Lloyd, A. M., Walbot, V., and Davis, R. W. 1992. Arabidopsis and Nicotiana anthocyanin production activated by maize regulators *R* and *C1*. Science 258: 1773–1775.

Malamy, J. E. and Benfey, P. N. 1997. Organization and cell differentiation in lateral roots of *Arabidopsis thaliana*. Development 124: 33–44.

Masucci, J. D., Rerie, W. G., Foreman, D. R., Zhang, M., Galway, M. E., Marks, M. D., and Schiefelbein, J. W. 1996. The homeobox gene *GLABRA2* is required for position-dependent cell differentiation in the root epidermis of *Arabidopsis thaliana*. Development 122: 1253–1260.

Pellegrini, O. 1957. Esperimenti chirugici sul comportamento del meristema radicle di Phaseolus vulgaris L. Delpinoa 10: 187–199.

Raven, P. H., Evert, R. F., and Eichhorn, S. E. 1986. Biology of plants. New York: Worth Publishers.

Rerie, W. G., Feldmann, K. A., and Marks, M. D. 1994. The *GLABRA2* gene encodes a homeodomain protein required for normal trichome development in Arabidopsis. Genes Develop. 8: 1388–1399.

Scheres, B., Di Laurenzio, L., Willemsen, V., Hauser, M.-T., Janmaat, K., Weisbeek, P., and Benfey, P. N. 1995. Mutations affecting the radial organisation of the Arabidopsis root display specific defects throughout the embryonic axis. Development 121: 53–62.

Scheres, B., McKhann, H. I., and van den Berg, C. 1996. Roots redefined: Anatomical and genetic analysis of root development. Plant Physiol. 111: 959–964.

Scheres, B., Wolkenfelt, H., Willemsen, V., Terlouw, M., Lawson, E., Dean, C., and Weisbeek, P. 1994. Embryonic origin of the Arabidopsis primary root and root meristem initials. Development 120: 2475–2487.

Schiefelbein, J. W. and Sommerville, C. 1990. Genetic control of root hair development in *Arabidopsis thaliana*. Plant Cell 2: 235–243.

Schneider, K., Wells, B., Dolan, L., and Roberts, K. 1997. Structural and genetic analysis of epidermal cell differentiation in Arabidopsis primary roots. Development 124: 1789–1798.

Tanimoto, M., Roberts, K., and Dolan, L. 1995. Ethylene is a positive regulator of root hair development in *Arabidopsis thaliana*. Plant J. 8: 943–948.

van den Berg, C., Willemsen, V., Hage, W., Weisbeek, P., and Scheres, B. 1995. Cell fate in the Arabidopsis root meristem determined by directional signalling. Nature 378: 62–65.

13

Vascular Development

CELLS OF THE VASCULATURE

The vasculature is an arterial system that transports water, minerals, photoassimilates, hormones, wounding signals, and so on, from one part of the plant to another. The vascular system moves substances from their sites of synthesis or uptake (sources) to the sites where they are utilized, stored, or released (sinks). The primary vasculature links shoots to roots, and serves as a major channel for the upward and downward movement of substances in a plant. The primary vasculature connects all organs and organ systems of the plant together – leaves, branches, flowers, and so on.

The vascular system is largely made up of vascular bundles or strands that contain two kinds of conducting tissue: phloem that moves photoassimilates (nutrients from photosynthesis) from sources to sinks, and xylem that conducts water and minerals from roots to other parts of the plant. The conducting cells of xylem are nonliving, **tracheary elements,** whereas the conducting elements of phloem are living, enucleate cells called **sieve elements** (Esau 1953). In addition, vascular bundles contain other nonconducting cells, and even minor veins in the leaf are complex, consisting of different cell types (Fig. 13.1). Not all conducting vessels are located in vascular bundles or strands. For example, several plant families have extrafasicular phloem that lies in the cortex beside vascular bundles.

Conducting or tracheary elements of the xylem are tracheids and vessel members (Fig. 13.2A) (Esau 1953). They differ from each other in that vessel members are perforated cells fused into long continuous tubes. Vessel members have single or multiple perforations in their end walls that form a perforation plate. Tracheids, on the other hand, usually do not have perforations in their end walls; rather, they have pits

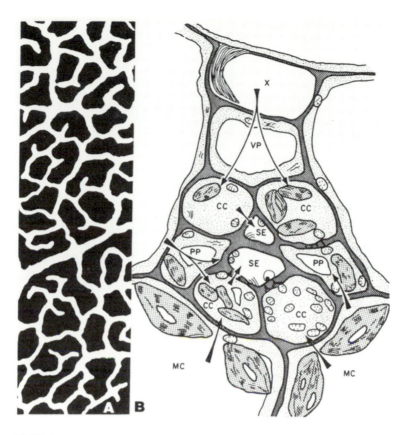

FIGURE 13.1

Minor vein in leaf. (A) Autoradiograph of leaf of common bean (*Phaseolus vulgaris*) that shows accumulation of (^{14}C)-sucrose in minor veins. Autoradiograph is reverse printed so that veins appear white. (B) Tracing of an electron microphotograph of a cross-section of minor vein in a tobacco leaf. Arrows indicate possible routes of entry of photoassimilates (sugars and other nutrients derived from photosynthesis) into the sieve element–companion cell complex. Sieve element, SE; companion cell, CC; phloem parenchyma, PP; xylem, X; vascular parenchyma, VP; mesophyll cell, MC. (From Giaquinta, R. T. 1983. Phloem loading of sucrose. Ann. Rev. Plant Physiol. 34:360. With permission from the Annual Review of Plant Physiology, Vol. 34, © 1983 by Annual Reviews Inc.)

that occur in pit-pairs on the common lateral walls with other xylem elements. Other cells in xylem include fibers and parenchyma cells.

At maturity, the tracheary elements of the xylem are large, nonliving cells (Fig. 13.2A). The differentiation of tracheary elements involves the formation of secondary cell walls with helical, annular, reticulate, or pitted wall thickenings. During development, tracheary elements undergo autolysis, losing their nuclei and cytoplasm, and leaving only

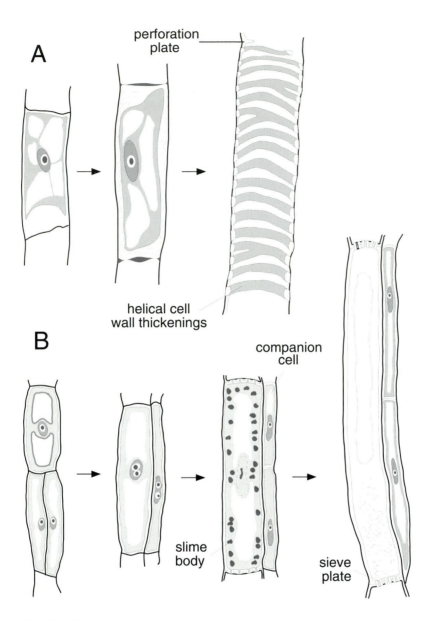

FIGURE 13.2

Development of conducting cells of the xylem and phloem. (A) Development of xylem vessel member in celery (*Cicuta*). Development of tracheary
element is shown from left to right, emphasizing the formation and thickening of the secondary wall. Left, young tracheary element without thickened end or secondary walls. Middle, secondary and end walls thicken
with the deposition of cell wall materials. Cell undergoes autolysis, and the
nucleus and all cytoplasmic contents are degraded. Right, functional vessel
member is devoid of cell contents. Cell wall with helical thickening
remains. End wall is perforated, leaving rim of perforation plate and allow-

(continued)

cell walls. The dead cells serve as conduits for the unimpeded movement of water and solutes. They also provide mechanical support for stems and other structures. (The differentiation of tracheary elements in culture will be discussed later in this chapter.)

Phloem consists of many different cell types; however, unlike xylem, conducting cells of the phloem are living. The conducting cells, which are sieve elements, are interconnected at their end walls by sieve areas or plates (Fig. 13.2B). The sieve areas or plates are specialized cell walls formed by the deposition of callose around pores that permit the flow of materials from cell to cell. Because functioning sieve elements are living cells with intact plasma membranes, the cells are interconnected through the sieve plate by cytoplasmic bridges or strands. Sieve areas or plates develop as an elaboration of plasmodesmata in the primary pit fields that interconnect sieve elements. The pores in the sieve plates are much larger than plasmodesmata, often 1–3 μm in diameter, but they may reach 15 μm in some species. Sieve elements are a heterogeneous collection of cells, ranging from less-developed sieve cells to specialized sieve cells that form long sieve tubes.

Because sieve elements are living cells, the phloem is actually a continuous stream of cytoplasm that courses through the plant. Although sieve elements retain a cytoplasm at maturity, their nuclei and other cellular components, including ribosomes, degenerate during development (Fig. 13.2B). Sieve elements are associated with nucleated companion cells, but are cordoned off during sieve element development. The link between companion cells and sieve elements, nonetheless, appears to be vital because companion cells are thought to produce proteins found in sieve elements. For example, the major protein found in sieve elements is a filamentous P protein, which is a constituent of the slime found in phloem, particularly in dicots. The function of the slime is not known; however, it may be involved in the immobilization of materials in these

FIGURE 13.2 *(continued)*

ing for the unimpeded movement of water and solutes. (B) Different stages of phloem sieve element differentiation in melon (*Curburbita*). Left, phloem cells in periclinal division. Second from left, developing sieve element and companion cell precursor. Second from right, young sieve element forms slime bodies. Nucleus degenerates and wall thickens. Mature sieve element retains cytoplasm, and has large vacuole and dispersed slime. Sieve plates are fully formed at ends of sieve element, allowing for the passage of materials from cell to cell through a continous cytoplasm. (Redrawn from Esau, K. 1953. Plant Anatomy. New York, John Wiley & Sons, Inc., pp. 226 and 278. Copyright © 1953 John Wiley and Sons. By permission of Wiley-Liss, a subsidiary of John Wiley and Sons, Inc.)

vessels or as clotting agents when the vascular system is severed. Using *in situ* hybridization and immunolocalization techniques in *Cucurbita maxima* (pumpkin), Bostwick et al. (1992) found P-proteins located in both companion cells and sieve elements; however, P-protein mRNA was found only in companion cells. They speculated that P-protein mRNA was transcribed and translated in companion cells, and the protein transported to sieve element cells. Monoclonal antibodies have been developed against P-proteins in *Streptanthus tortuosus,* and these antibodies are useful reagents to study the appearance of phloem elements in cell cultures (Toth et al. 1994).

In the development of phloem, sieve elements and companion cells are closely related in their ontogeny (Fig. 13.2B). The cell that gives rise to the sieve element divides one or more times longitudinally, and the larger cell that results from the divisions usually becomes a sieve element whereas the others become companion cells (Esau 1953). The number and size of companion cells can vary among species.

DEVELOPMENT OF THE STEM AND LEAF VASCULATURE

In higher plants, the primary stem vasculature is usually organized in separate vascular bundles or in a cylinder that forms a ring around the central pith. The vascular system in vegetative shoots is leaf-oriented (i.e., a vascular bundle at each node turns outward to serve the leaves and/or axillary buds) (Fig. 13.3). The vascular bundle from the base of the leaf to where it joins the stem vasculature is called a leaf trace. Other vascular bundles in the stem that continue their course to the next internode are called sympodial bundles. In many dicots, leaves are connected to the stem vasculature by three or more leaf traces, one central trace and two lateral traces that connect to sympodial bundles. Because the leaf traces turn outward to serve the leaf, the area of the stem above the leaf trace insertion points, which are devoid of vasculature, are called leaf gaps.

The vasculature of the stem is either primary or secondary, depending on its developmental origin. **Primary vasculature** arises from the procambium in the embryo or the shoot apical meristem (SAM). The primary vasculature of the root arises from stele initials that are located in the promeristem (described in Chap. 12). **Secondary vasculature,** which is found extensively in woody dicots (or herbaceous dicots with secondary growth), is derived from the vascular cambium, a lateral meristem generally sandwiched between the xylem and phloem. The

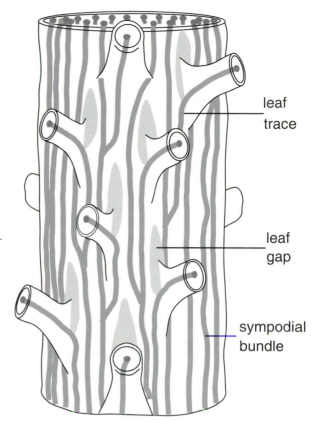

FIGURE 13.3

Dicot stem that has been cleared to show how vascular bundles in the stem connect to leaves. Leaf traces connect leaf vasculature from the base of the leaf to the stem vasculature below the point of leaf insertion. Leaf gap is the parenchymous area above the point of leaf insertion. (Based on flax [*Linum*] stem from Steeves, T. A., and Sussex, I. M. 1989. Patterns in Plant Development. Cambridge, Cambridge University Press.)

discussion in this chapter will focus on the development of the primary vasculature in the shoot.

Given the continuity between the stem and leaf vasculature, there is some debate about the origin of the primary vasculature in dicot shoots. The first event in the vacularization of the leaf primordium is the development of the midvein (in pinnately veined leaves). During midvein development, procambial cells (i.e., cells that give rise to the primary vasculature) differentiate acropetally (base to tip) along the midrib into the leaf primordium. The question is whether the primary vasculature arises in leaf primordia or in the shoot apex (see Steeves and Sussex 1989). The distinction is important because whether the leaf primordia initiate the development of the vasculature or whether the apex does so is at issue. The matter is not easy to resolve on morphological grounds because procambial cells can be difficult to identify and trace without proper markers. In general, procambial cells are elongated and dark staining; however, they may not be so in early stages. Warlaw (1946) addressed this issue by puncturing successive leaf primordia as they appeared in the shoot apex in *Dryopteris* and other ferns. He found that a complete ring of stem vasculature, which was uninterrupted by leaf gaps, was produced in new growth. The finding supported the argument that the apex initiates the development of the stem vasculature, and not the leaf primordia. It can be argued that early leaf primordia initiate vascular development before they emerge; however, the argument creates problems because the development of the leaf trace must be anticipated ahead of time since a leaf trace often diverges from the stem vasculature at a lower internode (see Fig. 13.3.) The alternative view is that the leaf trace arises from basipetal development of the leaf midvein vasculature, which then joins the stem vasculature at a node below. These alternatives probably cannot be sorted out without better markers for early vascular development.

The development of leaf venation in dicot leaves brings into focus interesting questions about patterning. Venation patterns in dicot leaves are often described as netlike, forming a reticulum of branching veins of discrete size orders (Fig. 13.4). Dicot leaves with pinnate venation are composed of a midvein that is continuous with the stem vasculature and secondary veins that branch from the midvein. Tertiary veins and veins of branching higher order form a reticulum in which some of the highest order veinlets end freely. In dicot leaf development, vascularization occurs in three phases (Nelson and Dengler 1997). In the first phase, as described earlier, provascular strands extend acropetally into the leaf primordium from the stem (Fig. 13.4). The midvein provasculature extends to the tip of the leaf primordium, then continues to grow by intercalary growth as the leaf primordium enlarges. During further development of the leaf, the central axis of the leaf thickens, which results in a midvein embedded in a midrib and petiole. In the second phase of leaf vascularization, secondary veins are formed. In leaves with pinnate venation, provascular strands of the secondary vasculature extend to the margins of the lamina as the lamina expands. In the third phase, higher orders of veins appear that create a reticulum of veins. An interesting feature about the vascularization process is that veins in each order appear sequentially, which indicates that whole leaf processes control the initiation of venation in each order.

Monocots have a different vein pattern from dicots, and the monocot pattern is usually described as parallel or striate (Nelson and Dengler 1997). The pattern in grasses is composed of longitudinal veins of different orders that include a large median vein embedded in a thickened midrib (see Figs. 6.8 and 13.7). The longitudinal veins are parallel throughout most of the leaf blade, but the smaller veins tend to anastomose toward the tip and base of the leaves. In addition, transverse and commisural veins interconnect adjacent longitudinal veins throughout the leaf blade, which creates a reticulum of smaller veins like dicots. As in dicots, vascularization in monocots is a process in which veins of dif-

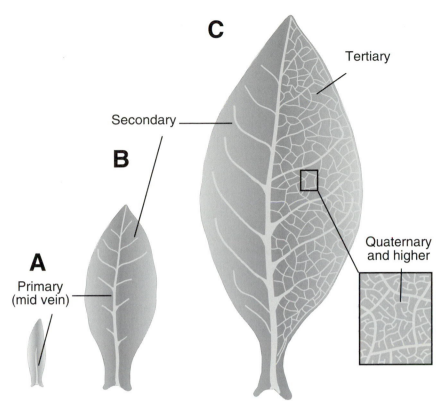

FIGURE 13.4

Development of leaf vasculature in tobacco (*Nicotiana tabacum*). Veins of higher branching orders are formed sequentially during leaf development. (A) Mid-vein (primary vein) develops acropetally (toward the leaf tip) of the leaf primordium. (B) Secondary veins appear (veins develop first at the tip and then toward the base) and extend toward leaf margins. (Right side of C) Tertiary veins produce a reticulum of interconnecting veins. (Inset) Minor veins (quaternary and higher) are spatially distributed and some veinlets end freely. (Based on Nelson, T., and Dengler, N. 1997. Leaf vascular pattern formation. Plant Cell 9:1126.)

ferent orders are produced sequentially (Nelson and Dengler 1997). First, the midvein and then the major lateral provascular strands extend acropetally into the leaf primordium. Next, provascular strands of prospective intermediate veins extend basetally from the primordium. Provascular strands of small interconnecting longitudinal and transverse veins then develop, starting from the tip of the primordium and proceeding basipetally down the blade. In grasses, the leaf vasculature is thought to begin its development isolated from the stem vasculature. Provascular strands from the midvein and large longitudinal veins begin to develop basipetally after initial acropetal development into the primordium. As with dicots, however, there is some uncertainty whether provascular connections to the stem vasculature might exist earlier but cannot be recognized.

Venation patterns in leaves are often species-specific and are used as taxonomic characters. No clear theories have emerged about how venation patterns are produced; however, in both dicots and monocots it can be inferred from the sequential appearance of veins in different orders that existing veins serve as "positional landmarks" for the differentiation of later veins (Nelson and Dengler 1997). To serve in the collection and dispersal of materials, the vasculature needs to be finely and evenly distributed. Spatial regulation is most evident in grasses, where there is regularity in the spacing of veins. There is spatial order even in dicots with a netlike venation pattern. For example, the distance measured between branch points is fairly constant within the fine reticulum of veins, regardless of the size order of vein in which the branch occurs (Russin and Evert 1984). It is clear that to achieve such spatial regularity within a reticulum requires a fractal-like patterning process with constraints on the minimal spacing between veins.

REGENERATION OF VASCULAR STRANDS

Many of the experimental studies on vascular development *in planta* have focused on the process of vascular regeneration rather than on the development of the primary vasculature. Because the shoot apex is small, the differentiation and joining of vascular strands has been studied more frequently during vascular regeneration in accessible parts of the stem. The vascular system has a remarkable capacity for regeneration, and revascularization can occur when a stem is wounded or when a graft is formed. When a vascular bundle is severed, new vascular tissue regenerates and bypasses the wound (Jacobs 1979). Vascular regeneration in mature tissue may be a reasonable model for vascular differentiation in developing tissue because vascular development, particularly the development of minor veins, often occurs later in the development of an organ.

In regeneration experiments, vascular development depends on the presence of an adjacent bud or leaf. To demonstrate that the bud or leaf is the source of factors promoting vascular development, Wetmore and Rier (1963) grafted a lilac bud onto a block of undifferentiated callus tissue in culture (Fig. 13.5). They observed that vascular tissue developed below the bud. The critical factor released by the bud was the plant hormone, auxin. Auxin (and sucrose) could be substituted for the bud, and when it was supplied continuously, a ring of vascular tissue was formed beneath the application site.

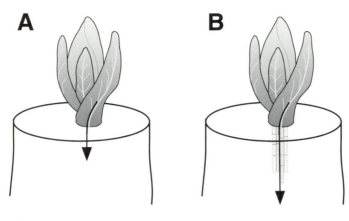

FIGURE 13.5

Bud induces the vascular development in a block of lilac (*Syringa*) callus tissue. (A) The basipetal movement out of the bud of auxin and possibly other nutrients induces vascular development. (B) The formation of vascular tissue canalizes the further movement of auxin and nutrients extending the channel. (Based on Wetmore, R. H., and Rier, J. P. 1963. Experimental induction of vascular tissues in callus of angiosperms. Amer. J. Bot. 50:418–430.)

The stem vasculature differentiates basipetally (from tip to base), and it has been shown in regeneration studies that the stem is polarized with respect to development of the vasculature (Sachs 1981). The direction is not determined by gravity because vascular differentiation can proceed upward locally in stem sections that have been reoriented by experimental manipulations. The direction of vascular strand differentiation results from basipetal transport of auxin. Sachs proposed that the formation of the vasculature is a canalization process that involves positive feedback between the signal flow and transport capacity (Fig. 13.5B) (Sachs 1991). He argued that local vascular tissue development forms a specialized path for auxin transport and produces more vascular tissue in advance of the site of auxin flow. As a result, continuous files of vascular cells are formed. The concept of canalization is important to understand because it explains how long and continuous channels of conducting cells develop through existing tissue.

A common misconception about the elongation of vasculature strands is that they grow like pollen tubes or nerve axons. Vascular strands do not actually grow; rather, they recruit existing cells at the growing front to differentiate into vascular elements. Differentiation may involve cell division and shape change, but the growing strand is not a moving tip. Thus, an important issue in vascular development is:

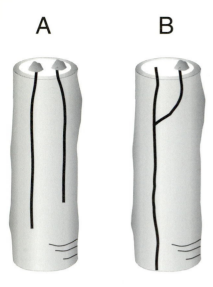

A **B**

FIGURE 13.6

Experimental demonstration of the joining of vascular strands in pea epicotyl tissue. Lanolin paste containing auxin (indoleacetic acid) was dotted on the top of cut epicotyl stems. Lanolin paste provides a continuous, localized source of auxin. Auxin promotes the basipetal development of vascular tissue, but prevents the strands from joining. Newly formed strand, however, will join vascular strand without a source of auxin. (Based on Sachs, T. 1981. The control of patterned differentiation of vascular tissue. Adv. Bot. Res. 9:151–262.)

How do elongating vascular strands induce the differentiation of cells at their growing front?

Vascular strand joining, which may be important in linking leaf traces to the stem vasculature, can also be studied in regenerating systems. Sachs (1981) decapitated pea (*Pisum sativa*) seedlings, and put dots of lanolin paste containing auxin onto the cut epicotyl stems (Fig. 13.6). When equal amounts of auxin were placed at two spots, two parallel strands were formed that did not join each other. When auxin was added at one spot, a new strand formed and joined an existing strand without an auxin source. The joining of strands was dependent on the difference in concentration of auxin supplied to the joining strand versus the main strand. Joining occurred at high frequency when the concentration of auxin supplied to the joining strand was much higher than the concentration of auxin supplied to the main strand. It was concluded from these observations that a new strand supplied with auxin appears to differentiate toward a sink for auxin, thereby avoiding another auxin source.

LINEAGE OF VASCULAR CELLS IN THE LEAF

What is the lineage relationship of vascular cells to each other and to the cells in the tissue through which the vascular bundles pass? Langdale et al. (1989) studied the lineage of vascular cells in the maize leaf and focused on the relationship of **bundle sheath** (BS) and **mesophyll** (M) **cells** (Fig. 13.7). Maize carries out a form of photosynthesis called C4 photosynthesis. To do so a special arrangement of BS and M cells in the leaves permits the intimate exchange of metabolites neces-

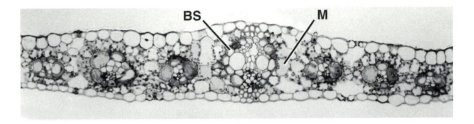

FIGURE 13.7

Transverse section through a young leaf of maize (*Zea mays*), a C_4 plant. Central vein and several lateral veins are shown in cross-section. Vascular bundles are surrounded by an inner layer of bundle sheath cells and an outer layer of mesophyll cells. Visualized by light microscopy. (Kindly provided by Laurie G. Smith.)

sary for this form of photosynthesis. The special arrangement of BS and M cells around the vasculature is called Kranz anatomy in which BS cells form an inner ring around the vascular bundle and M cells form an outer ring (Fig. 13.7). In C4 photosynthesis, CO_2 is fixed in M cells by phosphoenol pyruvate carboxylase (PEPCase), which is an enzyme with a high affinity for CO_2. Fixed CO_2 is converted to malate or pyruvate and shipped to adjacent BS cells. In BS cells, malate is decarboxylated, and the released CO_2, which is present in higher levels in BS cells, is fixed again by the standard, low-affinity CO_2 fixing enzyme, ribulose bisphosphate carboxylase (RuBPCase). This peculiar arrangement improves the CO_2 fixation capacity of RuBPCase by reducing the enzyme's predisposition for photorespiration rather than photosynthesis (burning up carbon compounds with O_2 rather than fixing CO_2).

The analysis of leaf development in maize is aided by the fact that the leaf largely grows at its base (as described in Chap. 6) rather than by intercalary growth. Thus, the developmental history of a maize leaf can be traced from the tip of the leaf to the base. Langdale et al. (1989) followed the lineage of BS and M cells in maize leaves. They did so by examining leaf sectors that arose spontaneously from the breakage of chromosomes caused by the action of a ring chromosome. One of the questions they asked was whether BS and M cells have separate cell lineages. The answer was no. They found that BS and M cells are usually derived from common precursors because sectors were shared more frequently between adjacent BS and M cells than from the subepidermal cells from which they are derived (Fig. 13.8). Equally surprising was the finding that BS cells in a cross section of a single vein were not derived from the same cell lineage. Instead, each half vein had a different origin.

It was an intriguing finding that M cells and cells of the vasculature are so closely related in their cell lineage histories because these cells

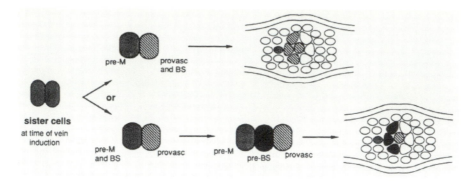

FIGURE 13.8

Interpretation of leaf mesophyll and bundle sheath cell lineages from leaf sector patterns in maize (*Zea mays*). A frequent sectoring pattern in transverse sections was observed in which one mesophyll cell, all the bundle sheath cells around half a vein, and a number of vascular elements comprised a sector. Pattern could arise by two routes in which one of two sister cells differentiated into a premesophyll cell, pre-M, and the other into a provascular and prebundle sheath cell, pre-BS (upper lineage), or a premesophyll and prebundle sheath cell and the other into a provascular cell (lower lineage). In either case, various cell types, including BS and M cells, do not have separate cell lineages. (Redrawn from Langdale, J. A., et al. 1989. Cell lineage analysis of maize bundle sheath and mesophyll cells. Develop. Biol. 133:138. By permission of the author and Academic Press, Inc.)

carry out such different biochemistries. One might think that BS cells are of one origin and M cells of another. Langdale and Nelson (1991) showed that these cells differentiate from each other by one of two routes (Fig. 13.8), and that positional information, particularly the proximity of the cells to the lumen of the vein, may be an important determinant in the differentiation of the two cell types. The findings echo the common theme discussed in Chapters 1 and 2 that positional information is a critical determinant of plant cell development.

Because BS and M cells are closely related in lineage and function, is the differentiation of one cell type dependent on the other? A number of mutations have been isolated in maize that specifically affect BS cell development without affecting M cells. In one such mutant, *bundle sheath defective2* (*bsd2*), BS cells are morphologically normal, but their chloroplasts swell and become distorted in plants grown in the light (Roth et al. 1996). On the other hand, chloroplast development is unaffected in M cells of the mutant. RuBPCase, which is a characteristic marker of BS cell development, does not accumulate in mutant BS cells; however, the RNA encoding the two RuBPCase subunits (derived from the transcription of the nuclear genes, *RBCS*, and chloroplast genes, *rbcL*) does accumulate, which suggests that there may be a defect in posttranscriptional or protein turnover mechanisms in the mutant. It is interesting that *rbcL* RNA transcripts also accumulate in M cells in the mutant, a cell type where the *rbcL* gene (a chloroplast gene) is not expressed in wild type. Thus, while M cell development is not disrupted in *bsd2*, the mutation does affect gene expression patterns in M cells. It is not clear, however, whether the altered gene expression pattern in M cells is a primary effect of the mutation or a secondary effect from the failure of BS cells to develop normally. Nonetheless, the *bsd2* mutation appears to be reasonably cell-specific and suggests that the biochemical differences between BS and M cells can arise fairly independently.

DIFFERENTIATION OF XYLEM AND PHLOEM AND PROGRAMMED CELL DEATH

Phloem and xylem differentiate at a late stage in vascular bundle development. In general, phloem differentiation precedes xylem. Furthermore, phloem is often found in the absence of xylem, but xylem is usually not found in the absence of phloem (except in some interesting mutants that will be described later). Does this mean that xylem requires phloem for differentiation? That does not appear to be the case because xylem differentiation, as described later, can be induced in tissue culture in the absence of phloem. Functional phloem that carries a stream of sugars, other nutrients, and growth regulators, however, may be required for xylem differentiation *in planta*. Xylem and phloem differentiate in response to similar hormonal signals, but the difference seems to be quantitative. In tissue culture, higher levels of plant hormones appear to be needed for xylem differentiation. For example, soybean (*Glycine max*) tissues in culture will form phloem in response to low levels of auxin, and xylem in response to higher auxin levels (Aloni 1980).

The development of the vasculature and differentiation of xylem and phloem involve issues in programmed cell death (Pennel and Lamb 1997). The conducting or tracheary elements in xylem are dead cells, and the death of these cells is a normal part of their development. There are aspects of death in these cells that are similar to apoptosis (a form of programmed cell death) in animal cell systems; however, there are other aspects that are not. Unlike cell death in animal cells, the process in tracheary cell maturation in Zinnia (*Zinnia elegans*) appears at a morphological level to be initiated by the rupture of the vacuole membrane (tonoplast) (Fukuda 1997). Following vacuole rupture other organelles quickly follow suit, in which organelles with single membranes such as

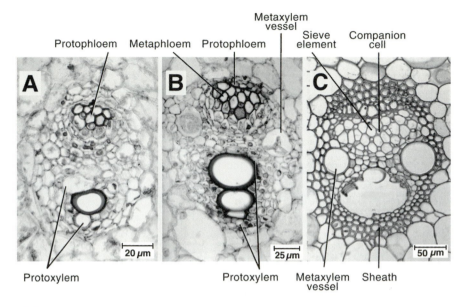

FIGURE 13.9

Differentiation of vascular bundles in maize (*Zea mays*). (A) Transverse sections of the stem show mature protophloem and protoxylem elements. (B) Protophloem elements are crushed and mature metaphloem is formed. Three protoxylem elements are mature and two metaxylem elements are almost fully expanded. (C) Mature vascular bundle sheathed by thick walled sclerenchyma cells. Metaphloem is composed entirely of mature sieve elements and companion cells. Metaxylem elements are mature and expanded. Part of protoxylem has been destroyed by an expanding air space. (From Raven, P. H., Evert, R. F., and Eichhorn, S. E. 1986. Biology of Plants. Worth Publishers, New York, 4th ed., pp. 420–421. Reprinted by permission of the author and Worth Publishers, Inc.)

endoplasmic reticulum and Golgi compartments lyse followed by organelles with double membranes, including the nucleus. Like cell death in animal cells, fragmentation of nuclear DNA (based on an *in situ* TUNEL assay, terminal deoxynucleotide transferase-mediated dUTP-biotin nick end-labeling assay) in developing vessel elements in pea (*Pisum sativa*) root tips is observed before the disruption of nuclei (Mittler and Lam 1995). Sieve elements, the conducting cells in phloem, are living cells, but they lose their nuclei and most organelles during development. As a result, they rely on companion cells for many vital functions.

Entire vascular bundles with conducting elements and all are routinely destroyed during plant growth. Vascular development is a successional form of growth. The first phloem formed in the developing vasculature of a newly forming organ, such as a leaf, is called protophloem (Esau 1953) (Fig. 13.9). Protophloem is a transient form that usually develops before an organ completes its elongation. Phloem formed after organ elongation is called metaphloem. In herbaceous dicots the metaphloem may constitute the phloem vasculature of the mature plant. In woody or herbaceous dicots with secondary growth, the metaphloem is usually destroyed by secondary growth. Protophloem is composed of less-specialized vascular elements than metaphloem. Sieve elements in protophloem are narrower with thinner walls, and they have less developed sieve plates. Companion and parenchyma cells that are present in dicot metaphloem are less frequently found in protophloem. Phloem fibers found in protophloem are likewise often missing in metaphloem.

The first xylem laid down in the vasculature of a newly forming organ is called protoxylem (Fig. 13.9). Protoxylem has few tracheary elements and more xylem parenchyma. The cell walls of tracheary elements are thinner with only annular or helical thickenings. During elongation of the organ, protoxylem is stretched, distorted, and often destroyed. Metaxylem is laid down when the organ is fully expanded. Tracheary elements are more abundant in metaxylem and their cell walls are reinforced with helical, scalariform, and reticulated thickenings.

GENE EXPRESSION DURING VASCULAR DIFFERENTIATION

During vascular differentiation genes involved in the differentiation of the vasculature are expressed, such as those genes required for the formation and lignification of the cell walls of conducting elements. A technique that has been very valuable in studying gene expression during vascular development in stems is a procedure called tissue printing (Ye and Varner 1991). In tissue printing stems are cut in cross-section, and the exposed ends are stamped onto pieces of filter paper (Fig. 13.10). Because stems are rigid, they leave imprints on the paper in which one can see morphological detail. From the cut end, residues containing RNA and protein are deposited on the paper. These macromolecules can be fixed to the paper and their whereabouts with respect to the structure of the stem can be visualized with very specific hybridization probes or antibodies.

The expression of genes that encode cell wall proteins was localized using this technique (Ye and Varner 1991). Two major groups of structural proteins found in cell walls are hydroxyproline-rich glycoproteins (HRGPs) and glycine-rich glycoproteins (GRPs). They found that HRGP genes in new stems are most highly expressed in the cambium, in cortex cells surrounding primary phloem, and in parenchyma cells about

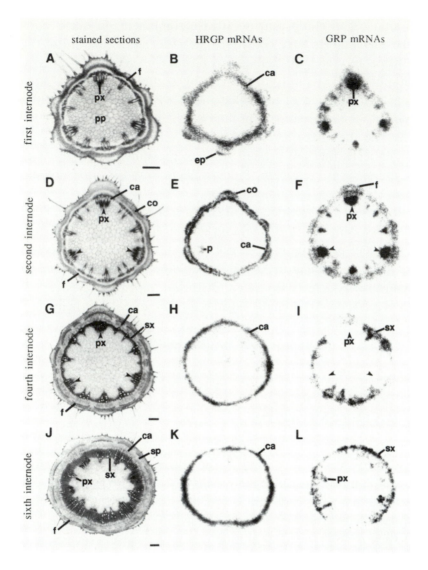

FIGURE 13.10

Tissue print localization of RNA encoding cell wall glycoproteins in developing soybean (*Glycine max*) stems. Cut ends of stems were printed (pressed) onto nylon membranes and hybridized to probes for RNA encoding hydroxyproline-rich glycoproteins (HRGP) and glycine-rich glycoproteins (GRP). Stem vasculature is more mature in lower (higher numbered) internodes. Note that HRGP mRNA is found mostly in association with the cambium and primary phloem, and GRP mRNA is highly expressed in protoxylem in younger internodes and in secondary xylem in older internodes. Cambium, ca; cortex, co; epidermis, ep; primary phloem, f; parenchyma, p; pith parenchyma, pp; primary xylem, px; secondary phloem, sp; secondary xylem, sx. Bars = 300 μm. (From Ye, Z.-H., and Varner, J. E. 1991. Tissue-specific expression of cell wall proteins in developing soybean tissues. Plant Cell 3:25. Reprinted by permission of the Society for Plant Physiologists.)

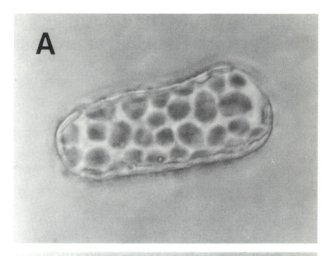

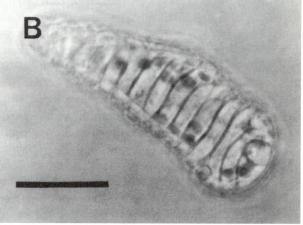

FIGURE 13.11

Leaf mesophyll protoplasts from zinnia (*Zinnia elegans*) can be induced to form tracheary elements in culture. Addition of hormones (auxin and cytokinin) to the culture induces cells to differentiate into tracheary elements without undergoing cell division. Single mesophyll cell (A) and tracheary element (B) formed about 2.5 days after hormone addition. Note that annular or helical cell wall thickenings that are characteristic of tracheids are formed. Bar = 25 μm. (From Fukuda, H. 1992. Tracheary element formation as a model system of cell differentiation. Int. Rev. Cytol. 136:290. Reprinted by permission of the author and Academic Press, Inc.)

the primary xylem (Fig. 13.10). In older stems, HRGP genes are expressed exclusively in the cambium. GRP genes are expressed in new stems in primary phloem and xylem, and they are highly expressed in tracheid cells of protoxylem. In older stems, GRPs are expressed in secondary xylem.

Genes expressed during xylem development have also been identified in a cell culture system by Fukuda and coworkers (1992). They found that single leaf mesophyll cells from Zinnia protoplasts will differentiate in culture to form xylem tracheary elements. Thirty to sixty percent of the cells synchronously differentiate, without dividing, into tracheary elements after cytokinin and auxin are added to the culture. The formation of tracheary elements is marked by the development of thickenings in the secondary cell walls, which produces spiral or annular patterns (Fig. 13.11). Fukuda (1997) divided the differentiation of tracheary elements into three stages. In stage I, the mesophyll cells dedifferentiate and lose their ability to carry out photosynthesis. Stage II occurs before cell wall thickening, but it involves the activation of expression of a number of genes concerned with cell wall biosynthesis. The transition from stage II to III appears to be a regulated stage in that brassinosteroid synthesis inhibitors (uniconazol) and calmodulin antagonists will block this transition. In stage III, secondary cell walls are formed, and the cell contents undergo autolysis. This stage involves intense biosynthesis of components of secondary cell walls, cellulose and other polysaccharides, cell proteins, and lignin compounds. No means of separating secondary cell wall formation from cell death has been discovered to date, which suggests that the processes may have a common mechanism or be causal with respect to each other (Fukuda 1997).

A special advantage of the Zinnia system is that one can use cDNAs from differentiating cell cultures as probes to locate the sites of gene expression in developing vascular tissue of seedlings by *in situ* hybridization (Demura and Fukuda 1994). The vascular system in the root matures progressively from the root tip to the junction between the root and the hypocotyl. By performing *in situ* hybridization on sections at different points along the root, one can therefore observe the timing of gene expression with respect to the development of the root vasculature. For example, a cDNA probe called TED3, which represents a gene expressed in stage II and one that encodes a GRP, hybridizes to sections a short distance up the root (1 mm from the root cap) in a region where the first protoxylem elements will differentiate. Thus, the gene is a useful marker for tracheary element precursor cells.

Other gene expression markers have been sought to detect early vascular development because vascular cells are largely formed by the recruitment and differentiation of nonvascular cells into the vasculature. Baima et al. (1995) found an Arabidopsis gene (*Athb-8*) that encodes a homeodomain-containing transcription factor was a good marker for early vascular development. By *in situ* hybridization, it was found that *Athb*-8 was expressed in procambial cells of the developing embryo and in regions of active vascularization in the developing plant. The gene was not expressed in mature vascular cells.

The *Athb*-8 promoter was linked to the GUS reporter gene in order to examine the expression of the transgene during revascularization in tobacco (Baima et al. 1995). In these experiments, regions of the stem were cut, and the reformation of vascular tissue was followed. The stem was cut in such a way that the path of vascularization was redirected (Fig. 13.12). It was found that the gene was expressed at the wound site soon after cutting. After a few days, however, the gene was expressed along the new path of revascularization. In the early stages of recovery,

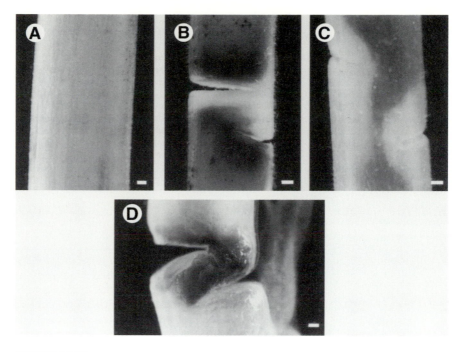

FIGURE 13.12

Expression pattern of *Athb-8* gene in transgenic tobacco (*Nicotiana tabacum*) plant bearing an *Athb-8* promoter linked to the GUS reporter gene (*Ath-8*:GUS). Incisions are made in the stem to stimulate regeneration of the vasculature. Reporter gene expression is localized by histological staining for β-glucuronidase (GUS) activity. (A) Uncut control stem, (B) 1 hour after cutting, (C) 8 hour after cutting, and (D) 1 week after cutting. Note that GUS staining follows path of regenerating vasculature. Bar = 500 μm. (From Baima, S., et al. 1995. The expression of the *Athb-8* homeobox gene is restricted to provascular cells in *Arabidopsis thaliana*. Development 121:4179. Reprinted by permission of the Company of Biologists Ltd.)

Athb-8 expression was observed in parenchyma cells in the pith along the new path. The authors suggested that the cells expressing the *Athb*-8 construct were most likely the cells that had been recruited for vascular differentiation. *Athb*-8 was also found to be activated by auxin, which suggests that auxin might be the stimulus for *Athb*-8 expression in the revascularization experiments.

MUTANTS THAT AFFECT VASCULAR DEVELOPMENT

There are very few reports of mutants that have been selected on the basis of defects in the vasculature. Turner and Somerville (1997) screened populations of mutagenized Arabidopsis for mutants with

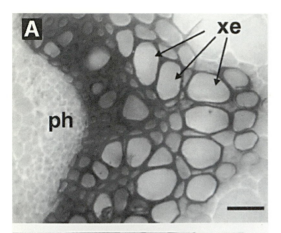

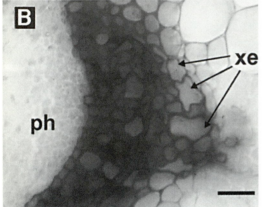

FIGURE 13.13

Collapsed xylem in *irregular xylem1* (*irx1*) mutant of Arabidopsis. Transverse sections through inflorescence meristem in (A) wild type and (B) *irx1* mutant. Xylem elements, xe, with collapsed cell walls are noted in the *irx1* mutant. Phloem, ph. Hand-cut inflorescence stem sections stained with phloroglucinol. Bar = 25 μm. (From Turner, S. R., and Somerville, C. R. 1997. Collapsed xylem phenotype of Arabidopsis identified mutants deficient in cellulose deposition in the secondary cell wall. Plant Cell 9:692. Reprinted by permission of Plant Cell.)

altered inflorescence stem vascular morphology. Mutants called *irregular xylem* (*irx*) were isolated that showed evidence of collapsed xylem (Fig. 13.13). Transporting xylem cells are under negative pressure, and collapsed xylem was taken to indicate a defect in xylem cell wall development. The mutants representing three different loci (*irx1–irx3*) all had reduced cellulose content in inflorescence stems compared with wild type. The xylem cell walls of the mutants were generally thinner or were uneven in thickness compared with wild type. In general, the mutants had less stiff stems, and *irx3* had a reclining growth habit. Collapsed xylem is also a characteristic of plants in which lignin biosynthesis has been blocked; however, lignin content in the mutant stems was nearly normal. The mutants should be very useful to study the role of cellulose in mechanical stiffening of the stem, secondary wall formation, and cellulose biosynthesis.

Turner and Somerville (1995) also identified two other groups of Arabidopsis mutants that affect vascular development: one that acts on the patterning of vascular bundles and another that affects the differentiation of vascular cells. The patterning mutant called *continuous vascular ring* (*cov*) does not form discrete vascular bundles as in wild type; rather, it forms a continuous vascular ring. Other interesting mutants that affect vascular development were identified on the basis of other phenotypes. For example, *wooden leg* (*wol*) mutants in Arabidopsis were picked up as slow root growth mutants. There are fewer cells in the vascular cylinders of these mutants, and all of the cells differentiate into xylem, but none differentiates into phloem (Scheres et al. 1995).

Given the importance of auxin in vascular development, auxin mutants should have profound effects on the development of the vasculature. Indeed, the vasculature in auxin-resistant mutants, such as *axr1*, is less well developed (Hobbie and Estelle 1994). More striking effects are observed in transgenic plants with altered auxin levels. Romano et al. (1991) reported that vascular development was inhibited in transgenic tobacco plants with reduced levels of free auxin due to the expression of an *iaaL* transgene. (The *iaaL* gene encodes indoleacetic acid–lysine synthetase, which is an activity that converts IAA, an auxin, into an inactive conjugate.) In *iaaL*-expressing plants, fewer but larger xylem elements were formed. The phenotype is consistent with the tissue culture experiments described earlier in which it was found that higher threshold levels of auxin were required for xylem formation. Just the opposite was observed in transgenic plants with elevated auxin levels brought about by the action of the *iaaM* gene (i.e., vascularization was enhanced) (Hobbie and Estelle 1994).

Arabidopsis mutants have been found that appear to be defective in the basipetal transport of auxin (Carland 1996). The mutants are called *lopped* (*lop*) mutants because occasional leaves showed bladeless regions along the midrib. *Lop* mutants were selected on the basis of abnormal morphology of leaves that were deformed and dwarfed. The midrib in leaves was often disoriented and split in two veins. The mutants also had abnormal development of lateral roots and disorganized patterns of epidermal cell expansion in the root elongation zone. *Lop* mutants had normal auxin levels, but they also had a substantially reduced capacity to transport indoleacetic acid (an auxin) in a basipetal direction. It is not known whether the defect occurs directly in the mechanism of auxin transport or whether some cellular disorganization interferes indirectly with polar auxin transport. Directed transport of auxin is one of the fundamental requirements for the canalization of vascular development. Although *lop* mutants show many abnormalities, they still form interconnected, although disoriented, vascular elements. The fact that mutants with reduced auxin response or transport capacity still produce functional vasculature indicates that either the mutants are "leaky" or there is redundancy in the control of vascular development.

REFERENCES

Aloni, R. 1980. Role of auxin and sucrose in the differentiation of sieve and tracheary elements in plant tissue cultures. Planta 150: 255–263.

Baima, S., Nobili, F., Sessa, G., Lucchetti, S., Ruberti, I., and Morelli, G. 1995. The expression of the *Athb-8* homeobox gene is restricted to provascular cells in *Arabidopsis thaliana.* Development 121: 4171–4182.

Bostwick, D. E., Dannenhoffer, J. M., Skaggs, M. I., Lister, R. M., Larkins, B. A., and Thompson, G. A. 1992. Pumpkin phloem lectin genes are specifically expressed in companion cells. Plant Cell 4: 1539–1548.

Carland, F. C. 1996. LOP1: A gene involved in auxin transport and vascular patterning in Arabidopsis. Development 122: 1811–1819.

Demura, T. and Fukuda, H. 1994. Novel vascular cell-specific genes whose expression is regulated temporally and spatially during vascular system development. Plant Cell 6: 967–981.

Esau, K. 1953. Plant anatomy. New York: John Wiley & Sons, Inc.

Fukuda, H. 1997. Tracheary element differentiation. Plant Cell 9: 1147–1156.

Fukuda, H. 1992. Tracheary element formation as a model system of cell differentiation. Int. Rev. Cytol. 136: 289–332.

Giaquinta, R. T. 1983. Phloem loading of sucrose. Ann. Rev. Plant Physiol. 34: 347–387.

Hobbie, L. and Estelle, M. 1994. Genetic approaches to auxin action. Plant Cell Environ. 17: 525–540.

Jacobs, W. P. 1979. Plant hormones and development. Cambridge: Cambridge University Press.

Langdale, J. A., Lane, B., Freeling, M., and Nelson, T. 1989. Cell lineage analysis of maize bundle sheath and mesophyll cells. Devel. Biol. 133: 128–139.

Langdale, J. A. and Nelson, T. 1991. Spatial regulation of photosynthetic development in C4 plants. Trends Genet 7: 191–196.

Mittler, R. and Lam, E. 1995. *In situ* detection of nDNA fragmentation during the differentiation of tracheary elements in higher plants. Plant Physiol. 108: 489–493.

Nelson, T. and Dengler, N. 1997. Leaf vascular pattern formation. Plant Cell 9: 1121–1135.

Pennel, R. and Lamb, C. 1997. Programmed cell death in plants. Plant Cell 9: 1157–1168.

Raven, P. H., Evert, R. F., and Eichhorn, S. E. 1986. Biology of Plants. New York: Worth Publishers.

Romano, C. P., Hein, M. B., and Klee, H. J. 1991. Inactivation of auxin in tobacco transformed with the indoleacetic acid-lysine synthetase gene of *Pseudomonas savastanoi.* Gene Develop 5: 438–446.

Roth, R., Hall, L. N., Brutnell, T. P., and Langdale, J. A. 1996. *Bundle sheath defective2,* a mutation that disrupts the coordinated development of bundle sheath and mesophyll cells in the maize leaf. Plant Cell 8: 915–927.

Russin, W. A. and Evert, R. F. 1984. Studies on the leaf of *Populus deltoides* (Saliceae): Morphology and anatomy. Am. J. Bot. 71: 1398–1415.

Sachs, T. 1981. The control of patterned differentiation of vascular tissue. Adv. Bot. Res. 9: 151–262.

Sachs, T. 1991. Pattern formation and plant tissues. Cambridge: Cambridge University Press.

Scheres, B., Di Laurenzio, L., Willemsen, V., Hauser, M.-T., Janmaat, K., Weisbeek, P., and Benfey, P. N. 1995. Mutations affecting the radial organisation of the Arabidopsis root display specific defects throughout the embryonic axis. Development 121: 53–62.

Steeves, T. A. and Sussex, I. M. 1989. Patterns in plant development. Cambridge: Cambridge University Press.

Toth, K. F., Wang, Q., and Sjolund, R. D. 1994. Monoclonal antibodies against phloem P-protein from plant tissue cultures: I. Microscopy and biochemical analysis. Am. J. Bot. 81: 1370–1377.

Turner, S. and Somerville, C. 1995. Analysis of vascular tissue differentiation. 6th International Conference on Arabidopsis Research. 93.

Turner, S. R. and Somerville, C. R. 1997. Collapsed xylem phenotype of Arabidopsis identified mutants deficient in cellulose deposition in the secondary cell wall. Plant Cell 9: 689–701.

Wardlaw, C. W. 1946. Experimental and analytical studies of pteridophytes. VII. Stelar morphology: The effect of defoliation on the stele of *Osmunda* and *Todea*. Ann. Bot. 10: 97–107.

Wetmore, R. H. and Rier, J. P. 1963. Experimental induction of vascular tissues in callus of angiosperms. Am. J. Bot. 50: 418–430.

Ye, Z.-H. and Varner, J. E. 1991. Tissue-specific expression of cell wall proteins in developing soybean tissues. Plant Cell 3: 23–37.

Index

1-aminocyclopropane-1-carboxylate (ACC)
    synthase, role in ethylene biosynthesis,
    283
a1-m locus, as maize pigmentation gene, 224
amphibian embryos, "primary organizer" of,
    39
amylose, *waxy (Wx)* locus control of, 266
angiosperms
    cell layers in, 44–45, 58, 104
    seed development in, 263
ANGUSTIFOLIA (AN) gene, in Arabidopsis,
    161–162
angustifolia (an) mutant, of Arabidopsis, 131,
    149–150, 161–162
animals
    gene expression patterns in, 59
    germ line development in, 9
animal development
    pattern formation in, 4–7, 8
    plant development compared to, 9–18
    positional information in, 38–40
    postembryonic, 11
"anlage," of leaf primordium, 141
antennapedia mutants, in Drosophila, 7
ANT gene, expression of, 239–240
anther development, 224–228
    genetics of, 228–232
    in maize
        cell layer invasion in, 46, 224–225
        sector boundary analysis of, 32–34
    nuclear genes acting on, 229
antherless (at) mutant, in maize, 220
anthesis, pistil parts at, 232, 233
anthocyanin, 34
    accumulation in *fus* mutants, 72, 92–93,
        279
    C1 gene in regulation of, 275
    chromosome for synthesis of, 28
    marker for, in cotton, 32
    R gene role in biosynthesis of, 159
    role in vegetative growth, 130
    VP1 gene role in biosynthesis of, 275
anthocyanin biosynthetic pathway, gene
    expression in, 273, 276
anthocyanin gainer mutant, of tomato, 46
anticlinal divisions
    in embryo, 58
    in lateral root development, 302
    in plants, 11–12
    in promeristem development, 298
antipodal cells, in plants, 15
Antirrhinum. See snapdragon *(Antirrhinum
    majus)*

antisense mRNA, in studies of S-locus glyco-
    proteins, 256–257
antisense 35S:*ACS2* constructs, effect on ethyl-
    ene synthesis, 283, 284–285
ant mutants, of Arabidopsis, 239, 240
Antp gene, in Drosophila, 5
AP3:DT-A, as "suicide gene," 214
APETALA1 (AP1) gene
    of Arabidopsis, 178, 179, 180, 181, 183,
        184, 201, 205, 211
    as floral homeotic gene, 197, 211
APETALA2 (AP2) gene family, *ANT* gene
    related to, 239
apetala (ap) mutants, in Arabidopsis, 197–198,
    199
AP3 gene
    activation of, 211, 212
    SUP gene antagonism of, 215
apical-basal axis
    leaf growth along, 143
    mutations affecting, 56, 64, 66–69
apical cells
    formation of, 56, 57, 58
    in root tip, 290
apical growing zone, in pollen tip, 247
apical hook, exaggerated coiling of, 97
apical initial cells
    in shoot apical meristem, 107
    STM stimulation of, 115–116
apical meristem, mutations affecting, 71
ap1 mutant, of Arabidopsis, 178, 183, 200,
    203
ap2 mutant, of Arabidopsis, 204
ap3 mutant, of Arabidopsis, 260
apomixis, 258–260, 263
    adventitious embryony, 258, 259
    aposporic, 258
    control in plant breeding programs, 260
    diplosporic, 258, 260
    facultative, 259, 260
    gametophytic, 258
    sporophytic, 258, 259
apoproteins, of phytochromes, 86, 87
apospory, as type of apomixis, 258
apparent cell number (ACN)
    at different developmental stages, 141
    estimation of, 30–31
    in maize embryos, 36, 37, 119
    in shoot apical meristem, 119
AP1 protein, interaction with UFO protein,
    213
AP3 protein, expression of, 206–207
AP proteins, as MADS box factors, 209